MW00760065

Sensors, Signal and Image Processing in Biomedicine and Assisted Living

Sensors, Signal and Image Processing in Biomedicine and Assisted Living

Editor

Dimitris K. Iakovidis

MDPI • Basel • Beijing • Wuhan • Barcelona • Belgrade • Manchester • Tokyo • Cluj • Tianjin

Editor
Dimitris K. Iakovidis
University of Thessaly
Greece

Editorial Office
MDPI
St. Alban-Anlage 66
4052 Basel, Switzerland

This is a reprint of articles from the Special Issue published online in the open access journal *Sensors* (ISSN 1424-8220) (available at: https://www.mdpi.com/journal/sensors/special_issues/sensors_biomedicine).

For citation purposes, cite each article independently as indicated on the article page online and as indicated below:

LastName, A.A.; LastName, B.B.; LastName, C.C. Article Title. *Journal Name* **Year**, *Article Number*, Page Range.

ISBN 978-3-03943-418-3 (Hbk)
ISBN 978-3-03943-419-0 (PDF)

Contents

About the Editor

Dimitris K. Iakovidis, Ph.D., received his BSc in Physics in 1997, his MSc in Cybernetics in 2001, and his Ph.D. in Informatics in 2004, all from the University of Athens in Greece. In 2015, Dr. Iakovidis was appointed as Associate Professor at the Department of Computer Science and Biomedical Informatics of the University of Thessaly in Greece. His research interests include signal and image processing, decision support systems, and intelligent systems and applications. In this context, he has co-authored over 150 papers in international journals, conferences, and books. Dr. Iakovidis is the Associate Editor of IEEE Transactions on Fuzzy Systems, IET Signal Processing, and is an Editorial Board Member of Measurement Science and Technology and Sensors journals. He has also acted as a reviewer of several international scientific journals and as an evaluator for National and European research projects.

Editorial

Sensors, Signal and Image Processing in Biomedicine and Assisted Living

Dimitris K. Iakovidis

Department of Computer Science and Biomedical Informatics, University of Thessaly, 35131 Lamia, Greece; diakovidis@uth.gr

Received: 2 September 2020; Accepted: 2 September 2020; Published: 7 September 2020

Sensor technologies are crucial in biomedicine, as the biomedical systems and devices used for screening and diagnosis rely on their efficiency and effectiveness. In this context digital signal and image processing methods play an important role in feature or quality enhancement, and compression of the acquired, transmitted, or received biomedical signals and images. Today, smart sensor systems, incorporating such methods, are entering our life through smartphones and other wearable devices to monitor our health status and help us maintain a healthy lifestyle. The impact of such technologies can be even more significant for the elderly, or for people with disabilities, such as the visually impaired.

This special issue gathers a broad range of novel contributions on sensors, systems, and signal/image processing methods for biomedicine and assisted living. These include methods for heart, sleep and vital sign measurement [1–5]; human motion-related signal analysis in the context of rehabilitation and tremor assessment [6–8]; assistive systems for color deficient and visually challenged individuals, as well as for wheelchair control by people with motor disabilities [9–12]; and, image and video-based diagnostic systems [13–16].

1. Heart, Sleep, and Vital Sign Measurement

Contactless measurement of biomedical signals, *i.e.*, measurement performed from a distance from the subjects, are more hygienic, and usually, they can be performed faster on larger populations. A methodology for distant heart rate measurement is presented in [1]. It aims to continuous monitoring of the user's heart rate during typical human-computer interaction (HCI) scenarios, *e.g.*, during reading texts and playing games. Monitoring is based on the principles of video plethysmography (VPG), a non-invasive, low-cost technique used to detect volumetric changes in the peripheral blood circulation, using a high frame rate RGB camera. Novel contributions of that study include experimental assessment of the impact of human activities during various HCI scenarios, and a novel image representation enabling more accurate and faster measurements.

Another novel contactless heart rate measurement approach is presented in [2], where heart rate is detected with a continuous-wave Doppler radar and an artificial neural network (ANN). Its experimental evaluation on healthy volunteers indicates that this approach is viable for the fast detection of individual heartbeats without heavy signal preprocessing.

To overcome the drawbacks of infrared thermography, which is characterized from low sensitivity for fever-based screening, a novel measurement system combining RGB with thermal image sensors is presented and evaluated in [3]. This system measures multiple vital signs, including body temperature, heart rate and respiration rate, contactlessly. Another contribution of that study is the analysis of the acquired signals using a machine learning methodology to discriminate patients with seasonal influenza from healthy individuals.

A novel methodology for adaptive sampling of electrocardiographic (ECG) signals is proposed in [4]. In that study generalized perceptual features, extracted from the statistics of experts' scanpaths obtained by gaze tracking, are used to control the ECG sampling process. The main advantage of the proposed method is that the temporal distribution of the distortions, caused by local data loss of the ECG trace, is based on medical features of the signals. Its experimental evaluation indicates a compression efficiency suitable for clinical applications.

Sleep-related metrics from different sleep smart monitoring devices, including a smartwatch, a mattress sensor pad, and a smart ring, are comparatively assessed in [5]. The metrics address sleep staging and total sleep duration, and they are derived via proprietary algorithms that utilize various physiological recordings. The results of that study indicate moderate correlations between the different devices, which indicates that there are still open issues with respect to standardization of such metrics across different devices.

2. Human Motion-Related Signal Analysis

Human motion analysis can be useful for the diagnosis and management of various medical conditions, such as the Parkinson's disease (PD). In [6], a machine learning -based method for automatic assessment of human motor behavior from video, is proposed. Novel contributions of that work include the development of a robust hierarchical multiple-target pose tracking method in uncontrolled environments in the presence of multiple human actors; the introduction of an explicit body movement representation, called pose evolution, that can be used to complement appearance and motion cues for action recognition; and, a target-specific action classification architecture applied on video recordings of patients with PD. The experimental evaluation of the proposed method indicates that it can provide accurate target-specific classification of activities in the presence of other human actors, robust to changing environments.

In a relevant context, [7] proposes a novel device, called the Rehapiano, for the fast and quantitative assessment of action tremor, such as PD tremor. The device uses strain gauges to measure force exerted by individual fingers, and it is applied and assessed for the measurement and monitoring of the development of upper limb tremor.

Signals related to human motor activities can also be obtained by electroencephalography (EEG). In fact, EEG can provide information not only about motor activities that actually happen, but also about motor activities that are imagined by a human subject. The work presented in [8] presents a methodology for improved classification of such motor imagery signals. It is based on a combination of a blind source separation to obtain estimated independent components, a 2D representation of these component signals using the continuous wavelet transform, and a Classification Stage Using A Convolutional Neural Network (CNN).

3. Assistive Systems

Assistive systems can have a significant impact on the quality of life of individuals with disabilities or deficiencies. A novel application for improving color discrimination for individuals with color vision deficiency is presented in [9]. In that study, the proposed approach is to automatically recolor the images by a color warping technique, so that the different colors of an image become perceivable by individuals with color vision deficiency.

To assist outdoor navigation of visually challenged individuals, a novel wearable visual perception system (VPS) is presented in [10]. By wearing that system, which is equipped with a stereoscopic RGB camera that can assess depth, the users receive information enabling them to avoid obstacles and safely navigate in outdoor environments. The proposed system goes beyond the state-of-the-art by following a novel uncertainty-aware approach to obstacle detection, incorporating salient regions generated using a Generative Adversarial Network (GAN) trained to estimate saliency maps based on human eye-fixations. The estimated eye-fixation maps, expressing the human perception of saliency in the scene, adds to the intuition of the obstacle detection methodology.

Video coding and transmission methods can be useful in the context of such assistive systems, *e.g.*, for transmitting the images acquired by the system as a video stream to remote users that could provide navigational instructions to the visually challenged. These methods can be inspired from other application domains, *e.g.*, robotics or vehicular technologies. In this light, a low-complexity H.265/HEVC encoder for ad-hoc vehicular networks, is presented in [11].

In the landscape of assistive systems, challenging, still unresolved issues are associated with wheelchair control. In this context, [12] proposes an intelligent, low-cost eye-tracking system for motorized wheelchair control, *e.g.*, for cases with complete paralysis of the four limbs. The input of that system is images of the user's eye that are processed to estimate the gaze direction and move the wheelchair accordingly to different directions.

4. Image and Video-Based Diagnostic Systems

Medical imaging and medical image interpretation, constitute essential tools for healthcare, contributing to patient safety, usually in a cost-efficient way. Among the recently upcoming medical imaging techniques, hyperspectral imaging is among the most promising for the discrimination of malignant tissues. In this context, [13] investigates the most relevant spectral bands for identification of brain cancer from in vivo brain images, by applying heuristic optimization algorithms. An important outcome of that study is the identification of specific spectral ranges for brain cancer detection.

A CNN-based approach is proposed for hyperspectral analysis of histopathological images, aiming to the detection of glioblastoma tumor cells [14]. The main goal of that work is to differentiate between high-grade gliomas (glioblastoma) and non-tumor tissue. Its results indicate a slight advantage in the use of hyperspectral, instead of RGB images.

In the context of endoscopy, narrow band imaging (NBI) constitutes a promising technique for the discrimination of different kinds of lesions as it can provide intraoperative real-time visualization of the vascular changes in the mucosa. A relevant work [15], presents an automatic approach for classification of laryngeal lesions, based on vascular patterns in Contact Endoscopy NBI images. The proposed approach demonstrated its capacity to act as an aid when there are disagreements among otolaryngologists, or when they all misclassify the patients.

In [16], the use of RGB endoscopic video sequences of the gastrointestinal tract is investigated to improve the detection of early malignant lesions appearing in patients with Barrett's esophagus. Such lesions are very difficult to distinguish, and they are often missed due to subtle visual features. Unlike previous methodologies that use still endoscopic images for lesion detection, the methodology presented in that study focuses on temporal feature extraction from endoscopic video for enhanced robustness of tissue classification and lesion detection. This is achieved by a two-stage ANN architecture performing feature extraction and classification, based on CNNs and recurrent neural networks. The results validate that the proposed approach improves temporal stability and accuracy of tissue classification.

Funding: The work performed by the Guest Editor for this special issue has been co-financed by the European Union and Greek national funds through the Operational Program Competitiveness, Entrepreneurship and Innovation, under the call RESEARCH—CREATE—INNOVATE (project code: T1EDK-02070).

Acknowledgments: The Guest Editor would like to thank all the authors for their contributions to this special issue, the referees, who have timely and professionally performed the peer-reviews of all articles, and the editorial staff of Sensors journal for their continuous support during the issue development.

Conflicts of Interest: The author declares no conflict of interest.

References

1. Przybyło, J. Continuous Distant Measurement of the User's Heart Rate in Human-Computer Interaction Applications. *Sensors* **2019**, *19*, 4205. [CrossRef] [PubMed]
2. Malešević, N.; Petrović, V.; Belić, M.; Antfolk, C.; Mihajlović, V.; Janković, M. Contactless Real-Time Heartbeat Detection via 24 GHz Continuous-Wave Doppler Radar Using Artificial Neural Networks. *Sensors* **2020**, *20*, 2351. [CrossRef] [PubMed]

3. Negishi, T.; Abe, S.; Matsui, T.; Liu, H.; Kurosawa, M.; Kirimoto, T.; Sun, G. Contactless Vital Signs Measurement System Using RGB-Thermal Image Sensors and Its Clinical Screening Test on Patients with Seasonal Influenza. *Sensors* **2020**, *20*, 2171. [CrossRef] [PubMed]
4. Augustyniak, P. Adaptive Sampling of the Electrocardiogram Based on Generalized Perceptual Features. *Sensors* **2020**, *20*, 373. [CrossRef] [PubMed]
5. Chaudhry, F.; Danieletto, M.; Golden, E.; Scelza, J.; Botwin, G.; Shervey, M.; De Freitas, J.; Paranjpe, I.; Nadkarni, G.; Miotto, R.; et al. Sleep in the Natural Environment: A Pilot Study. *Sensors* **2020**, *20*, 1378. [CrossRef] [PubMed]
6. Rezaei, B.; Christakis, Y.; Ho, B.; Thomas, K.; Erb, K.; Ostadabbas, S.; Patel, S. Target-Specific Action Classification for Automated Assessment of Human Motor Behavior from Video. *Sensors* **2019**, *19*, 4266. [CrossRef] [PubMed]
7. Ferenčík, N.; Jaščur, M.; Bundzel, M.; Cavallo, F. The Rehapiano—Detecting, Measuring, and Analyzing Action Tremor Using Strain Gauges. *Sensors* **2020**, *20*, 663. [CrossRef] [PubMed]
8. Ortiz-Echeverri, C.; Salazar-Colores, S.; Rodríguez-Reséndiz, J.; Gómez-Loenzo, R. A New Approach for Motor Imagery Classification Based on Sorted Blind Source Separation, Continuous Wavelet Transform, and Convolutional Neural Network. *Sensors* **2019**, *19*, 4541. [CrossRef] [PubMed]
9. Lin, H.; Chen, L.; Wang, M. Improving Discrimination in Color Vision Deficiency by Image Re-Coloring. *Sensors* **2019**, *19*, 2250. [CrossRef] [PubMed]
10. Dimas, G.; Diamantis, D.; Kalozoumis, P.; Iakovidis, D. Uncertainty-Aware Visual Perception System for Outdoor Navigation of the Visually Challenged. *Sensors* **2020**, *20*, 2385. [CrossRef] [PubMed]
11. Jiang, X.; Feng, J.; Song, T.; Katayama, T. Low-Complexity and Hardware-Friendly H.265/HEVC Encoder for Vehicular Ad-Hoc Networks. *Sensors* **2019**, *19*, 1927. [CrossRef] [PubMed]
12. Dahmani, M.; Chowdhury, M.; Khandakar, A.; Rahman, T.; Al-Jayyousi, K.; Hefny, A.; Kiranyaz, S. An Intelligent and Low-Cost Eye-Tracking System for Motorized Wheelchair Control. *Sensors* **2020**, *20*, 3936. [CrossRef] [PubMed]
13. Martinez, B.; Leon, R.; Fabelo, H.; Ortega, S.; Piñeiro, J.; Szolna, A.; Hernandez, M.; Espino, C.; O'Shanahan, A.J.; Carrera, D.; et al. Most Relevant Spectral Bands Identification for Brain Cancer Detection Using Hyperspectral Imaging. *Sensors* **2019**, *19*, 5481. [CrossRef] [PubMed]
14. Ortega, S.; Halicek, M.; Fabelo, H.; Camacho, R.; Plaza, M.; Godtliebsen, F.; Callicó, G.M.; Fei, B. Hyperspectral Imaging for the Detection of Glioblastoma Tumor Cells in H&E Slides Using Convolutional Neural Networks. *Sensors* **2020**, *20*, 1911. [CrossRef]
15. Esmaeili, N.; Illanes, A.; Boese, A.; Davaris, N.; Arens, C.; Navab, N.; Friebe, M. Laryngeal Lesion Classification Based on Vascular Patterns in Contact Endoscopy and Narrow Band Imaging: Manual Versus Automatic Approach. *Sensors* **2020**, *20*, 4018. [CrossRef] [PubMed]
16. Boers, T.; van der Putten, J.; Struyvenberg, M.; Fockens, K.; Jukema, J.; Schoon, E.; van der Sommen, F.; Bergman, J.; de With, P. Improving Temporal Stability and Accuracy for Endoscopic Video Tissue Classification Using Recurrent Neural Networks. *Sensors* **2020**, *20*, 4133. [CrossRef] [PubMed]

Article

Low-Complexity and Hardware-Friendly H.265/HEVC Encoder for Vehicular Ad-Hoc Networks

Xiantao Jiang [1,*], Jie Feng [2], Tian Song [3] and Takafumi Katayama [3]

1 Department of Information Engineering, Shanghai Maritime University, Shanghai 201306, China
2 State Key Laboratory of Integrated Services Networks, Department of Telecommunications Engineering, Xidian University, Xi'an 710071, China; jiefengcl@163.com
3 Department of Electrical and Electronics Engineering, Tokushima University, 2-24, Shinkura-cho, Tokushima 770-8501, Japan; tiansong@ee.tokushima-u.ac.jp (T.S.); katayama@ee.tokushima-u.ac.jp (T.K.)
* Correspondence: xtjiang@shmtu.edu.cn

Received: 19 March 2019; Accepted: 22 April 2019; Published: 24 April 2019

Abstract: Real-time video streaming over vehicular ad-hoc networks (VANETs) has been considered as a critical challenge for road safety applications. The purpose of this paper is to reduce the computation complexity of high efficiency video coding (HEVC) encoder for VANETs. Based on a novel spatiotemporal neighborhood set, firstly the coding tree unit depth decision algorithm is presented by controlling the depth search range. Secondly, a Bayesian classifier is used for the prediction unit decision for inter-prediction, and prior probability value is calculated by Gibbs Random Field model. Simulation results show that the overall algorithm can significantly reduce encoding time with a reasonably low loss in encoding efficiency. Compared to HEVC reference software HM16.0, the encoding time is reduced by up to 63.96%, while the Bjontegaard delta bit-rate is increased by only 0.76–0.80% on average. Moreover, the proposed HEVC encoder is low-complexity and hardware-friendly for video codecs that reside on mobile vehicles for VANETs.

Keywords: high efficiency video coding; low complexity; hardware friendly; vehicular ad-hoc networks

---

1. Introduction

Vehicular ad-hoc networks (VANETs) can provide multimedia communication between vehicles with the aim of providing efficient and safe transportation [1]. Vehicles with different sensors can exchange and share information for safely breaking, localization and obstacle avoiding. Moreover, the sharing of traffic accident's live video can improve the rescue efficiency and alleviate traffic jams. However, video transmission has been considered as a challenging task for VANETs, because video transmission over VANETs can significantly increase bandwidth [2]. This work focuses on the development of video codec that supports real-time video transmission over VANETs for road safety applications.

The demanding challenges of VANETs are bandwidth limitations and opportunities, connectivity, mobility, and high loss rates [3]. Because of the resource-demanding nature of video data in road safety applications, bandwidth limitations is the bottleneck for real-time video transmission over VANETs [4,5]. Moreover, due to the limited vehicle node's battery lifetime, video delivery over VANETs remains extremely challenging. Essentially, the low-complexity video encoder can accelerate the video transmission in real-time, and achieve low delay for video streaming. Hence, in order to transmit real-time video with bandwidth constraint, it is vital to develop an efficient video encoder. A video encoder with high encoding efficiency and low encoding complexity is the core requirement of VANETs [6].

Concurrently, the main video codecs are High Efficiency Video Coding (HEVC, or H.265) that are developed by the Joint Collaborative Team on Video Coding (JCT-VC) Group, and AV1 that is developed by Alliance for Open Media (AOMedia) [7]. Nevertheless, video codec is faced with several challenges. AV1 is a newer codec, royalty free and open sourced. However, the hardware implementation of AV1 encoder will take a long time. H.265/HEVC is the state-of-the-art standardized video codecs [8]. Compared with H.265/HEVC, AV1 increases the complexity significantly without achieving an increase in coding efficiency. When supporting the most available (resource-limited/mobile) devices or having a need for real-time, low latency encoding, it would be better to stick to H.265/HEVC. However, the encoding complexity of H.265/HEVC encoder increases dramatically due to its recursive quadtree representation [9]. Although previous excellent works have been proposed for reducing H.265/HEVC encoder complexity [10–14], most of them balance the encoding complexity and encoding efficiency unsuccessfully.

To address this issue, spatial and temporal information is widely used to reduce the computation redundancy of the H.265/HEVC encoder. However, the spatiotemporal correlation between the coding unit and the neighboring coding unit is not better used. To the author's best knowledge, complexity reduction in real-time coding with the available computational power at the VANET nodes has not been well studied, especially from the viewpoint of hardware implementation. The key contributions of this work are summarized as follows:

- We propose a low-complexity and hardware-friendly H.265/HEVC encoder. The proposed encoder allows the encoding complexity to be reduced significantly so that low delay requirements for video transmission in power-limited VANETs nodes are satisfied.
- A novel spatiotemporal neighboring set is used to predict the depth range of the current coding tree unit. The prior probability of coding unit splitting or non-splitting is calculated with the spatiotemporal neighboring set. Moreover, the Bayesian rule and Gibbs Random Field (GRF) are used to reduce the encoding complexity for H.265/HEVC encoder with the combination of the coding tree unit depth decision and prediction unit modes decision.
- The proposed algorithm can balance the encoding complexity and encoding efficiency successfully. The encoding time can be reduced by 50% with negligible encoding efficiency loss, and the proposed encoder is suitable for real-time video applications.

The rest of paper is organized as follows. Related works are reviewed in Section 2. Section 3 discusses background details. In Section 4, the fast CU decision algorithm is presented. Simulation results are discussed in Section 5. Section 6 concludes this work.

2. Related Work

2.1. Video Streaming in Vehicular Ad-Hoc Networks

Real-time Video transmission over VANETs can improve emergency responses' effectiveness for road safety applications. The high rates and low delay are the most challenging aspects of video transmission over VANETs [15]. Recently, some works have been proposed to solve this problem. Different video applications over VANETs need different resources. The collaborative vehicle to vehicle communication approach is presented to enhance the scalable video quality in intelligent transportation systems (ITS), and different methods are developed to enhance the quality of experience and quality of service during scalable video transmission over VANETs [16]. Meanwhile, the use of redundancy for video streaming over VANETs has been analyzed, and a selective additional redundancy approach is proposed to improve the video quality [17]. Moreover, in Ref. [18], a vehicle rewarding method for video transmission over VANETs using real neighborhood and relative velocity is presented to optimize video transmission. Although previous works have studied the issues of video streaming over VANETs, the performance of video codec to support real-time video transmission is missing.

2.2. Low Complexity Algorithm for H.265/HEVC Encoder

To reduce the encoding complexity of the H.265/HEVC encoder, some fast algorithm of coding unit (CU) size decision and prediction unit (PU) mode decision is presented. In previous works, the main spatiotemporal parameters that are used for fast CU size decisions include the neighboring CU depth, rate-distortion (RD) cost, motion vector (MV), coded block flag (cbf), and sample-adaptive-offset (SAO) information. Moreover, some other statistical learning based CU selection methods are proposed include Bayesian classifier, support vector machine (SVM), decision tree (DT), AdaBoost classifier and artificial neural network (ANN).

Jiang et al. presented a fast encoding complexity method based on the probabilistic graphical model [19,20]. These proposed algorithms consist of CU early termination and CU early skip methods to reduce the redundant computing of inter-prediction in H.265/HEVC. However, these methods cannot achieve better trade-off between the encoding efficiency and encoding complexity. Refs. [21,22] focused on decreasing the CU depth to reduce the encoding complexity of the H.265/HEVC encoder. In Ref. [23], the unimodal stopping model-based early skip mode decision was used to speed up the process of mode decision. This proposed early skip mode decision method can reduce encoding time significantly. In Ref. [24], a fast algorithm for the H.265/HEVC encoder was based on the Markov Chain Monte Carlo (MCMC) model and Bayesian classifier. Even though the above fast CU size decision methods utilized the spatiotemporal correlations, the fast PU mode decision methods are ignored.

Tai et al. introduced three novel methods including early CU split, early CU termination and search range adjustment to reduce the computation complexity for H.265/HEVC [25]. This proposed algorithm can outperform previous works with respect to both the speed and the RD performance. In Ref. [26], a fast inter CU decision was proposed based on the latent sum of absolute differences (SAD) estimation. This proposed algorithm achieved an average of 52% and 58.4% reductions of the encoding time. Refs. [27,28] focused on CU size decision and PU mode decision, the fast encoding algorithms based on statistical analysis were proposed to reduce the encoding complexity for the H.265/HEVC encoder. This method can reduce about 57% and 55% of the encoding time of the H.265/HEVC encoder. The above methods can significantly reduce the encoding complexity with the joining of the CU depth and PU modes prediction, however, these previous works cannot balance the encoding complexity and encoding efficiency successfully. Moreover, the cost of hardware implementation is higher for previous works.

All in all, this paper focus on the development of video encoder that supports real-time video streaming over VANETs. We design a low-complexity and hardware-friendly encoder to allow video transmission to adapt to the VANETs environment. In addition, compared with the current literature, the proposed encoder can achieve better performance trade-off.

3. Technical Background

H.265/HEVC

H.265/HEVC standard was released in 2013 by JCT-VC, which can reduce bit-rates by about 50% over H.264. In addition, H.265/HEVC adopts hybrid video compression technology, and the typical structure of H.265/HEVC encoder is shown in Figure 1. The main modules of H.265/HEVC encoder include: (1) Intra-prediction and inter-prediction, (2) transform (T), (3) quantization (Q) and (4) context-adaptive binary arithmetic coding (CABAC) entropy coding. Moreover, inter-prediction and intra-prediction modules are used to decrease the spatial and temporal redundancy. Transform and quantization modules are used to decrease visual redundancy. The entropy coding module is used to decrease the information entropy redundancy. It is noted that the inter-prediction module is the most critical tool, which consumes about 50% computation complexity. Then, in order to achieve real-time coding, the computation complexity of H.265/HEVC encoder should be reduced by decreasing spatiotemporal redundancy.

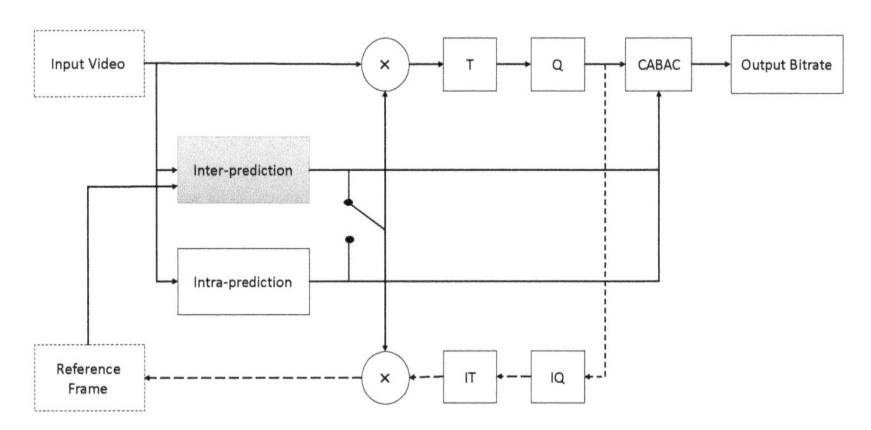

Figure 1. The structure of high efficiency video coding (HEVC, or H.265) encoder.

The video frame is divided into a lot of coding tree units (CTUs) in H.265/HEVC standard. A CTU includes a coding tree block (CTB) of the luma samples, two CTBs of the chroma samples, and associated syntax elements. The CTU size can be adjusted from 16 × 16 to 64 × 64. Each CTU can be divided into four square CUs, and a CU can be recursively divided into four smaller CUs. A CU consists of a coding block (CB) of the luma samples, two CBs of the choma samples, and the associated syntax elements. The CU size can be 8 × 8, 16 × 16, 32 × 32, or 64 × 64. Figure 2 shows an example of the CTB structure for a given CTU. The CTU in Figure 2a is divided into different sized CUs. Correspondingly, the CTB structure is shown in Figure 2b. In each depth of CTB, the rate-distortion (RD) cost of each node is checked until the RD cost is minimum.

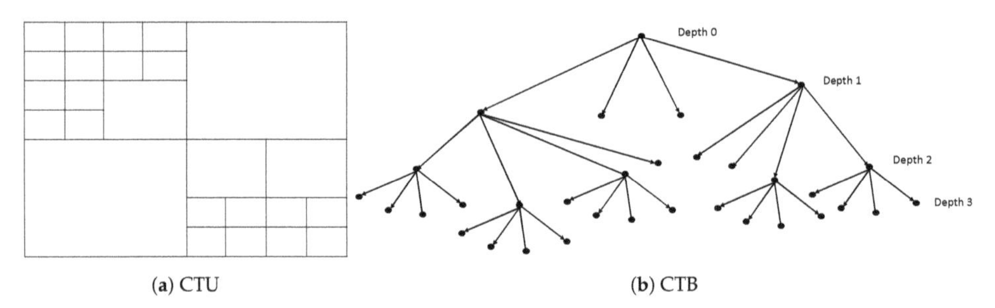

(**a**) CTU (**b**) CTB

Figure 2. coding tree unit (CTU) partitioning and coding tree block (CTB) structure.

The prediction unit (PU) can be transmitted in the bitstream, which identifies the prediction mode of CU. A PU consists of a prediction block (PB) of the luma, two PB of the chroma, and associated syntax elements. Figure 3 shows the eight partition modes that may be used to define the PUs for a CU in H.265/HEVC inter-prediction. For a CU configured to use inter-prediction, all eight partitions include four symmetry modes ($2N * 2N$, $2N * N$, $N * 2N$, $N * N$) and four asymmetric modes ($2N * nU$, $2N * nD$, $nL * 2N$, $nR * 2N$).

A CU can be recursively divided into transform units (TUs) according to the quadtree structure, and CU is the root of the quadtree. The TU is a basic representative block having residual or transform coefficients. In TU, one syntax element named coded block flag (cbf) indicates whether at least one non-zero transform coefficient is transmitted for the whole CU. When there is a non-zero coefficient, cbf is equal to 0. When there is no non-zero coefficients, cbf is equal to 1. Moreover, cbf is an important factor for the CU size decision [14].

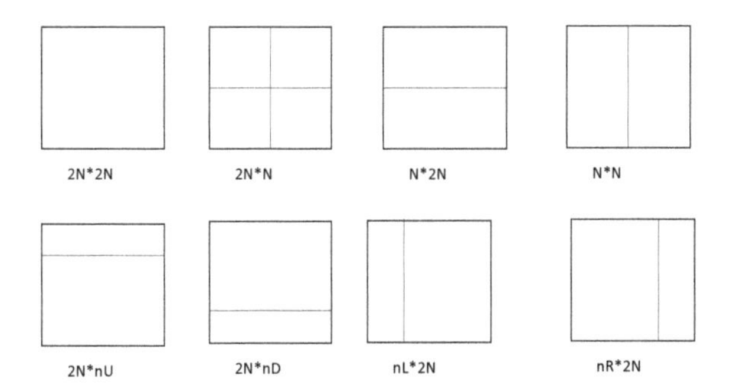

Figure 3. PU modes in H.265/HEVC inter-prediction.

The advantage of block partitioning structure is that the arbitrary size of CTU enables the codec to be readily optimized for various contents, applications, and devices. However, the recursive structure of coding block causes lots of redundant computing. In order to support the real-time video transmission over VANETs, the redundant computing of the H.265/HEVC encoder should be decreased significantly.

4. The Proposed Low-Complexity and Hardware-Friendly H.265/HEVC Encoder for VANETs

4.1. The Novel Spatiotemporal Neighborhood Set

The object motion is regular in video sequences and there is some continuity in the depth between adjacent CUs. If the depth range of the current CU can be inferred from the encoded neighboring CU, then some hierarchical partitioning is directly skipped or terminated. Therefore, the computational complexity has been reduced, significantly.

In order to utilize the spatiotemporal correlation, the four neighborhood set G is defined as

$$G = \{CU_L, CU_{TL}, CU_{TR}, CU_{CO}\}. \quad (1)$$

Set G is shown in Figure 4, where CU_L, CU_{TL}, CU_{TR}, and CU_{CO} denote the left, top-left CU, top-right, and collocated of the current CU, respectively.

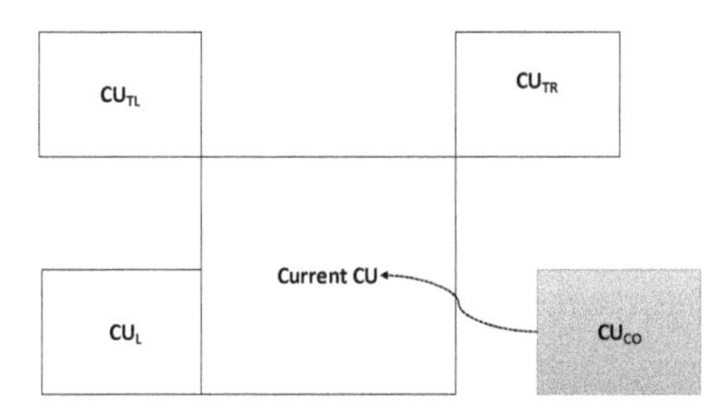

Figure 4. Spatiotemporal neighborhood set.

4.2. CTU Depth Decision

For video compression techniques, a smooth coding block popularly has the smaller CU depth. By contrast, the larger depth value is suitable for a complex area. Previous works show that the object motion in the same frame remains directional, and the motion and texture of the neighboring CUs are similar. In this work, the depths of neighboring CTU in the set G are used to predict the depth range of current CTU, and the predicted depth of current CTU is calculated as

$$\widehat{Dep}_{CTU} = \sum_{k=0}^{3} \theta_k \times Dep_k, \tag{2}$$

where k is the index of neighboring CTU in set G, Dep_k is the depth of neighboring CTU in the set G, and θ_k is a weight factor of neighboring CTU's depth, respectively, in the set G. In the H.265/HEVC standard, the range of CTU is depth 0, 1, 2, and 3. Hence, the calculated depth of the current CTU ($\widehat{Dep}_{CTU}$) satisfies

$$\widehat{Dep}_{CTU} \leq 3. \tag{3}$$

In Equation (3), $Dep_k \leq 3$. Therefore, weight factor θ_k satisfies

$$\sum_{i=0}^{3} \theta_k \leq 1. \tag{4}$$

If the range of current CTU is depth 0, 1, 2, and 3, then the sum of weight factor θ_k is 1 in this work. Moreover, Zhang's work confirms that, when the weight factor of the spatial neighboring CTU's depth is more than the weight factor of the temporal neighboring CTU's depth, the calculated CTU depth is closer to the actual depth of the current CTU [29]. In this work, each weight factor of the spatial neighboring CTU's depth is equal, and the weight factor of spatial neighboring CTU's depth is more than the weight factor of temporal CTU's depth. Then θ_k satisfies

$$\theta_k = \begin{cases} 0.3, & \text{if } k = 0, 1, 2 \\ 0.1, & \text{if } k = 3 \end{cases}. \tag{5}$$

However, the calculated value of $\widehat{Dep}_{CTU}$ is a non-integer most of the time. It is not suitable to directly predict the depth of current CTU by the value of $\widehat{Dep}_{CTU}$. Therefore, the rule of CTU depth range has been formulated as Table 1, and the depth range of current CTU can be generated with the value of $\widehat{Dep}_{CTU}$.

Table 1. The CTU depth range.

CTU Type	$\widehat{Dep}_{CTU}$ Range	The CTU Depth Range
$T1$	$\widehat{Dep}_{CTU} \leq 1.5$	[0, 1, 2]
$T2$	$1.5 < \widehat{Dep}_{CTU} \leq 2.5$	[1, 2, 3]
$T3$	$2.5 < \widehat{Dep}_{CTU} \leq 3$	[2, 3]

Due to the predicted depth of the current CTU, each CTU can be divided into three types: $T1, T2, T3$. The CTU depth range can be decided from Table 1. The expressions of the relation between CTU type, $\widehat{Dep}_{CTU}$, and CTU depth are as follows.

(1) when the predicted depth of current CTU $\widehat{Dep}_{CTU}$ satisfies $\widehat{Dep}_{CTU} \leq 1.5$, it means that the motions of neighboring CTUs are smooth and the depths of neighboring CTUs are small. The current CTU belongs to the still or homogeneous motion region and is classified as type $T1$. In this case, the minimum depth of current CTU Dep_{min} is equal to "0", and the maximum depth of current CTU Dep_{max} is equal to "2".

(2) when the predicted depth of current CTU $\widehat{Dep}_{CTU}$ satisfies $1.5 < \widehat{Dep}_{CTU} \leq 2.5$, it means that the depths of neighboring CTUs are middle. The current CTU belongs to the moderate motion region

and is classified as type $T2$. In this case, the minimum depth of current CTU Dep_{min} is equal to "1", and the maximum depth of current CTU Dep_{max} is equal to "3".

(3) when the predicted depth of current CTU $\widehat{Dep}_{CTU}$ satisfies 2.5 < $\widehat{Dep}_{CTU} \leq 3$, it means that the motions of neighboring CTUs are intense and the depths of neighboring CTUs are high. The current CTU belongs to the fast motion region and is classified as type $T3$.

In this case, the minimum depth of current CTU Dep_{min} is equal to "2", and the maximum depth of current CTU Dep_{max} is equal to "3".

4.3. PU Mode Decision

The CU splitting or non-splitting is formulated as a binary classification problem ω_i, where i = 0, 1. In this work, ω_0 and ω_1 respectively represent CU non-splitting and CU splitting, and the variable x represents the RD-cost of the PU. According to the Bayes' rule, the posterior probability $p(\omega_i|x)$ can be calculated as follows:

$$p(\omega_i|x) = \frac{p(x|\omega_i)p(\omega_i)}{p(x)}. \tag{6}$$

According to Bayesian decision theory, the prior probability $p(\omega_i)$ and the conditional probability $p(x|\omega_i)$ values must be known. Therefore, CU non-splitting (ω_0) will be chosen if the following condition holds true:

$$p(\omega_0|x) > p(\omega_1|x). \tag{7}$$

Otherwise, CU splitting (ω_1) will be chosen.

The conditional probability $p(x|\omega_0)$ and $p(x|\omega_1)$ are the probability density function of the RD cost, and they are approximated by normal distributions. Defining the mean values and covariance of RD cost of CU non-splitting and splitting as $N(\mu_0, \sigma_0)$ and $N(\mu_1, \sigma_1)$, the normal function can be given by

$$p(x|\omega_0) = \frac{1}{\sqrt{2\pi}\sigma_0} exp\{-\frac{(x-\mu_0)^2}{2\sigma_0{}^2}\}, \quad p(x|\omega_1) = \frac{1}{\sqrt{2\pi}\sigma_1} exp\{-\frac{(x-\mu_1)^2}{2\sigma_1{}^2}\}. \tag{8}$$

The prior probability $p(\omega_i)$ is modeled with Gibbs Random Fields (GRF) model in set G [30], and $p(\omega_i)$ will always have the Gibbsian form

$$p(\omega_i) = Z^{-1} exp(-E(\omega_i)), \quad E(\omega_i) = \sum_{k \in G} \varphi(\omega_i, \overline{\omega}_k). \tag{9}$$

where Z is a normalization constant, and $E(\omega_i)$ is cost function. k is the index of set G, and $\overline{\omega}_k$ denotes the non-splitting or splitting value of the neighborhood k-CU ($\overline{\omega}_k = -1, 1$). The CU size decision deals with the binary classification problem ($\omega_i = -1, 1$), and the clique potential $\varphi(\omega_i, \overline{\omega}_k)$ obeys the Ising model [31]:

$$\varphi(\omega_i, \overline{\omega}_k) = -\gamma \times (\omega_i \times \overline{\omega}_k), \tag{10}$$

where the parameter γ is the coupling factor, which denotes the strength of current CU correlation with neighborhood k-CU in set G. In this work, γ is set to "0.75". Then, the prior $p(\omega_i)$ can be written in the factorized form:

$$p(\omega_i) \propto exp(-E(\omega_i)) = exp(\sum_{k \in G} -\gamma \times (\omega_i \times \overline{\omega}_k)). \tag{11}$$

At last, the Equation (6) can be written as

$$p(\omega_i|x) \propto p(x|\omega_i)p(\omega_i) \propto exp(\sum_{k \in G} -\gamma \times (\omega_i \times \overline{\omega}_k)) \times \frac{1}{\sigma_i} exp\{-\frac{(x-\mu_i)^2}{2\sigma_i^2}\}. \tag{12}$$

Finally we can define the final CU decision function as $S(\omega_i)$, which can be written in the exponential form

$$S(\omega_i) = exp(\sum_{k \in G} -\gamma \times (\omega_i \times \overline{\omega}_k)) \times \frac{1}{\sigma_i} exp\{-\frac{(x-\mu_i)^2}{2\sigma_i^2}\}. \tag{13}$$

It should be noted that the statistical parameters $p(x|\omega_i)$ are estimated by using a non-parametric estimation with online learning, and are stored in a lookup table (LUT). The frames used for online updating of the values of (μ_0, σ_0) and (μ_1, σ_1) are shown as in Figure 5. In each group of pictures(GOP), the 1st frame that can be encoded by using the original H.265/HEVC coding will be used for the online update, while the successive frames are coded by using the proposed algorithm.

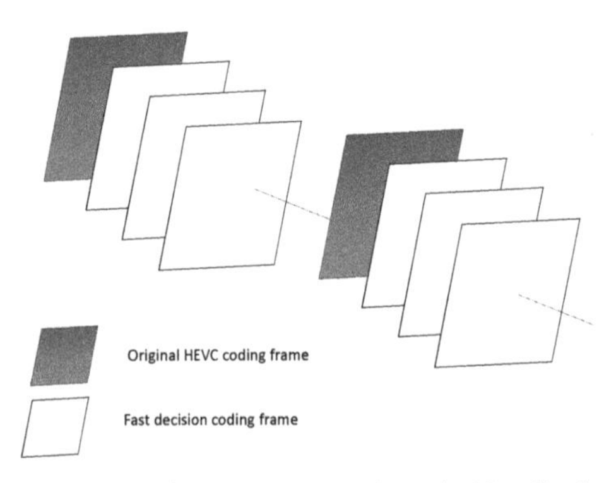

Figure 5. The statistical parameters are estimated with online learning.

Through the above analysis, the proposed PU decision based on Bayes' rule includes the CU termination decision (inter $2N * 2N$) and CU skip decision (inter $2N * 2N$, $N * 2N$, $2N * N$). In the case of the CU termination decision, the current CU is not divided into sub-CUs in the sub-depth. In the case of the CU skip decision, the current PU mode in current CU depth is determined at the earliest possible stage. Therefore, the flowchart of the proposed PU mode decision is described as follows.

(1) At the encoding time for inter prediction, first of all, look up the statistical parameters in LUT. Then, the RD cost of the inter $2N * 2N$ PU mode is checked. If the condition satisfies $S(\omega_0) > S(\omega_1)$ and cbf = 0, the CU termination decision is processed. Otherwise, if the condition is satisfying $S(\omega_1) > S(\omega_0)$ and cbf = 1, a CU skip decision is made.
(2) RD cost of the inter $2N * N$ PU mode is checked. If the condition is satisfying $S(\omega_1) > S(\omega_0)$ and cbf = 1, a CU skip decision is made.
(3) RD cost of the inter $N * 2N$ PU mode is checked. If the condition is satisfying $S(\omega_1) > S(\omega_0)$ and cbf = 1, a CU skip decision is made.
(4) Other PU modes are checked according to the H.265/HEVC reference model.

4.4. The Overall Framework

Based on the above analysis, the proposed overall algorithm incorporates the CTU depth decision and the PU mode decision algorithms to reduce the computation complexity of the H.265/HEVC encoder. The flowcharts are shown in Figures 6 and 7, respectively. The proposed CTU depth decision and PU mode decision algorithms have been discussed in Sections 4.2 and 4.3.

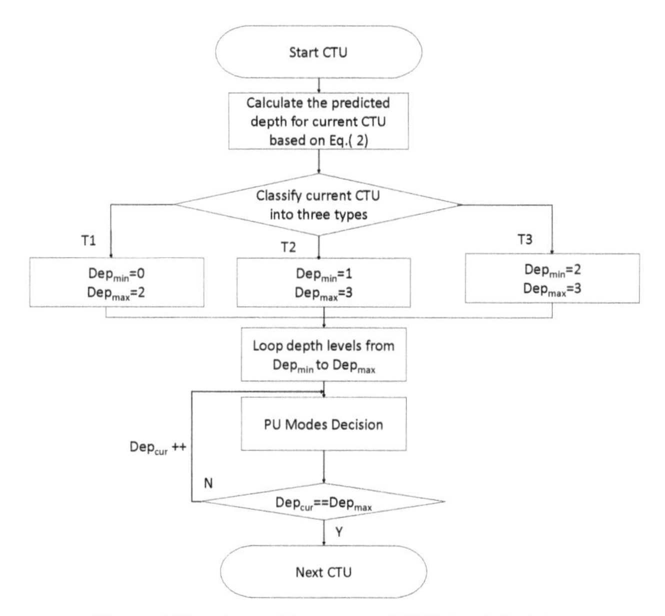

Figure 6. Flowchart of the proposed CTU depth decision.

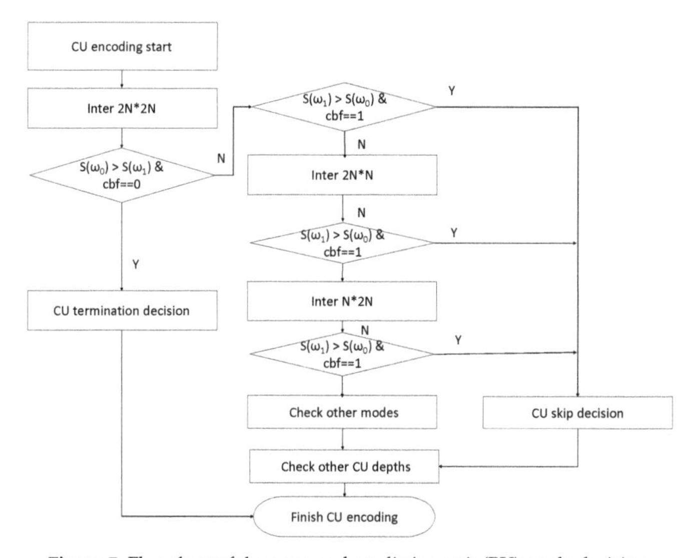

Figure 7. Flowchart of the proposed prediction unit (PU) mode decision.

It is noted that the maximum GOP size is equal to "8" in this work, and the value of (μ_0, σ_0) and (μ_1, σ_1) are updated every GOP for PU mode decision.

4.5. Encoder Hardware Architecture

Figure 8 shows the core architecture of the H.265/HEVC with mode decision. By using the architecture, inter-frame prediction is used to eliminate the spatiotemporal redundancy. The proposed CU decision method can accelerate the inter-prediction module before fast rate-distortion optimization (RDO). The novel spatiotemporal neighboring set is used to reduce the complexity of inter encoder which leads to a very low-power cost. Moreover, video codec on mobile vehicles for VANETs need to be more energy efficient and more reliable, so reducing the complexity of the video encoder is important. Then, the proposed low-complexity and hardware-friendly H.265/HEVC encoder can ensure the reliability of the video codec for VANETs significantly. Moreover, as a benefit of the high complexity reduction rate, the energy consumption can be reduced for hardware design, significantly.

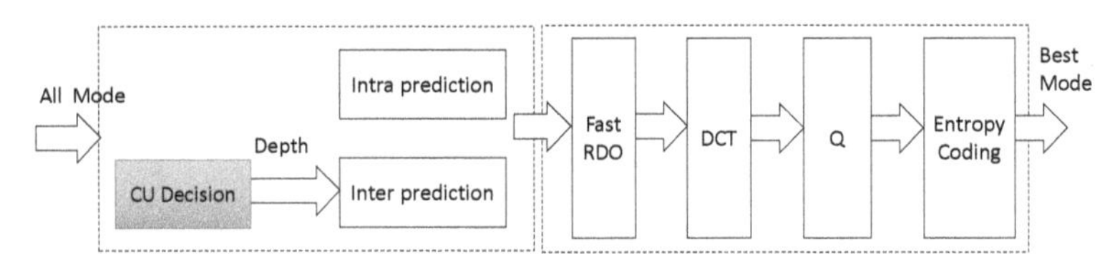

Figure 8. Mode decision process.

5. Experimental Results

To evaluate the performance of the proposed low-complexity and hardware-friendly H.265/HEVC encoder for VANETs, this section shows the experimental results by implementing the proposed algorithms with the H.265/HEVC reference software [32]. The simulation environments are shown in Table 2.

Table 2. The simulation environments.

Items	Descriptions
Software	HM16.0
Video Size	2560 × 1600, 1920 × 1080, 1280 × 720, 832 × 480, 416 × 240
Configurations	random access (RA), low delay (LD)
Quantization Parameter (QP)	22, 27, 32, 37
Maximum CTU size	64 × 64

The Bjontegaard delta bit-rate (BDBR) is used to represent the average bit-rate [33], and the average time saving (TS) is calculated as

$$TS = \frac{1}{4} \times \sum_{i=1}^{4} \frac{Time_{HM16.0}(QP_i) - Time_{proposed}(QP_i)}{Time_{HM16.0}(QP_i)} \times 100\% \quad (14)$$

where $Time_{HM16.0}(QP_i)$ and $Time_{proposed}(QP_i)$ denote the encoding time of using HM16.0 and the proposed algorithm with different QP.

In this work, the scenarios have been chosen carefully. This work focuses on the development of a video codec that supports real-time video transmission over VANETs for road safety applications. The common test conditions (CTC) are provided to conduct experiments [34]. The test sequences in CTC have different spatial and temporal characteristics and frame rates. Furthermore, the video sequences of traffic scenarios including 'Traffic' and 'BQTerrace' (as in Figure 9) are tested in this work. Moreover, we selected low delay (LD) configuration to reflect the real-time application scenario for all encoders.

(**a**) Traffic

(**b**) BQTerrace

Figure 9. Traffic scenario.

Tables 3 and 4 show the performance results of the CTU depth decision, PU mode decision and the overall (proposed) methods, compared to H.265/HEVC reference software in random access (RA) and low delay (LD) configurations. From the experimental results on Table 3, It can be seen that the encoding time can be reduced by 15.59%, 55.79%, and 50.96% on average for CTU depth decision, PU mode decision, and overall methods, while the BDBR can be incremented by only 0.11%, 0.96%, and 0.80%, respectively. From the experimental results in Table 4, the encoding time can be reduced by 14.05%, 50.28%, and 50.23% on average for CTU depth decision, PU mode decision, and overall methods, while the BDBR can be incremented by only 0.15%, 0.79%, and 0.76%, respectively. For high-resolution of sequences such as "BQTerrace", and "Vidyo4", the time saving is particularly high. Therefore, the overall (proposed) algorithm can significantly reduce the encoding complexity and rarely affects encoding efficiency. Moreover, the proposed method can achieve the trade-off between the encoding complexity and the encoding efficiency. In addition, the optimal tradeoff of encoding performance can be adjusted by the coupling factor γ. Therefore, the optimal tradeoff of encoding performance is that the encoding complexity can be reduced significantly with less than or equal to 0.8% encoding efficiency, and less low delay (LD) and random access (RA) configuration. In order to find the optimal tradeoff with coupling factor γ, the γ is set to "0.5", "0.75" and "0.85" under the same simulation environments. The compared results of the average efficiency and time saving are shown as in Table 5. From this table we can see that, in this case of $\gamma = 0.75$, the encoding performance is optimal in this work.

Table 3. Performance comparison of different parts of the proposed method (random access (RA)).

		CTU Depth Decision		**PU Mode Decision**		**Overall (Proposed)**	
Size	**Sequence**	**BDBR(%)**	**TS(%)**	**BDBR(%)**	**TS(%)**	**BDBR(%)**	**TS(%)**
2560 × 1600	Traffic	0.19	12.46	1.15	58.03	1.00	52.31
	SteamLocomotive	0.12	13.79	0.82	56.32	0.72	52.65
1920 × 1080	ParkScene	0.14	12.83	1.03	56.93	0.83	51.76
	Cactus	0.13	12.25	1.34	52.57	1.19	47.31
	BQTerrace	0.02	14.18	0.84	57.54	0.69	54.09
832 × 480	BasketballDrill	−0.13	14.11	0.73	51.55	0.50	46.10
	BQMall	0.18	15.97	0.92	56.76	0.73	51.37
	PartyScene	0.06	17.34	0.75	50.09	0.61	44.96
	RaceHorses	0.02	13.59	1.31	44.27	1.08	37.10
416 × 240	BasketballPass	0.26	7.09	0.90	54.73	0.60	46.40
	BQSquare	0.05	14.37	0.57	54.08	0.44	45.93
	BlowingBubbles	0.17	8.54	1.26	48.12	1.09	39.53
1280 × 720	Vidyo1	0.11	16.19	1.18	66.60	0.77	63.30
	Vidyo3	0.11	14.81	0.57	63.90	0.75	61.74
	Vidyo4	0.20	16.29	1.04	65.36	1.04	63.96
Average		0.11	13.59	0.96	55.79	0.80	50.96

Table 4. Performance comparison of different parts of the proposed method (low delay (LD)).

		CTU Depth Decision		PU Mode Decision		Overall (Proposed)	
Size	**Sequence**	**BDBR(%)**	**TS(%)**	**BDBR(%)**	**TS(%)**	**BDBR(%)**	**TS(%)**
2560 × 1600	Traffic	0.12	9.34	0.92	54.54	0.89	54.81
	SteamLocomotive	−0.19	11.65	0.33	51.62	0.29	51.84
1920 × 1080	ParkScene	0.11	10.15	1.07	52.66	1.08	53.02
	Cactus	0.06	10.19	1.03	47.47	0.85	47.82
	BQTerrace	0.01	12.83	0.58	54.33	0.62	54.46
832 × 480	BasketballDrill	0.18	10.75	0.71	43.95	0.79	44.22
	BQMall	0.42	8.09	0.93	51.06	0.86	49.39
	PartyScene	0.22	9.16	0.58	40.92	0.55	41.29
	RaceHorses	0.02	6.73	0.79	37.04	0.89	37.03
416 × 240	BasketballPass	0.94	6.99	0.91	51.21	0.91	51.80
	BQSquare	0.10	4.33	0.54	44.85	0.36	42.24
	BlowingBubbles	0.24	3.92	1.16	40.16	1.15	40.57
1280 × 720	Vidyo1	0.18	20.30	0.68	64.11	0.70	64.13
	Vidyo3	0.25	12.33	1.04	58.55	1.01	58.95
	Vidyo4	−0.37	14.03	0.55	61.67	0.48	61.95
Average		0.15	14.05	0.79	50.28	0.76	50.23

Table 5. Performance comparison of different γ.

	(BDBR, TS)		
	$\gamma = 0.5$	$\gamma = 0.75$ (Proposed)	$\gamma = 0.85$
Random Access	(1.01, 55.25)	(0.80, 50.96)	(0.98, 51.83)
Low Delay	(0.86, 50.63)	(0.76, 50.23)	(0.82, 50.55)

Video objective quality evaluation can be expressed by rate–distortion (R–D) curve. The R–D curve is fitted through four data points, and PSNR/bit-rate are assumed to be obtained for QP = 22, 27, 32, 37. In addition, when an error on the predicted depth of the current CTU occurs, the bit-rate will increase. In this paper, the video objective quality is evaluated by using bit-rate and PSNR. Then the lower the accuracies of the predicted depth of the current CTU algorithm, the more the bit-rate increases. Figure 10 shows the R–D curve of the proposed method, compared with the H.265/HEVC reference software. It can be noticed that the enlarged part of the figure shows the proposed algorithm is close to HM16.0 under the LD and RA configurations. In addition, Figure 11 shows the time saving of the sequences "Cactus" and "BlowingBubbles". It is noted that the encoding time can be reduced under different configurations.

The performance comparison of the proposed method is shown in Table 6, compared to previous works [12–14,24–26]. Goswami's work is based on Bayesian decision theory and Markov Chain Monte Carlo model (MCMC). Zhang's work is based on the Bayesian method and Conditional Random Fields (CRF). Tai's algorithm is based on depth information and RD cost. Zhu's algorithm is based on the machine learning method. Ahn's work is based on spatiotemporal encoding parameters. Xiong's work is based on the latent sum of absolute differences (SAD) estimation. However, the proposed approach is based on Bayesian rule and Gibbs Random Field. Although Zhu's method can achieve a 65.60% encoding time reduction, the BDBR is higher than the proposed method. Moreover, the increasing of the BDBR is smaller than state-of-the-art works, while the time saving is more than 50% on average. Compared with previous works [19,20], the proposed work can trade-off the encoding complexity and encoding efficiency successfully.

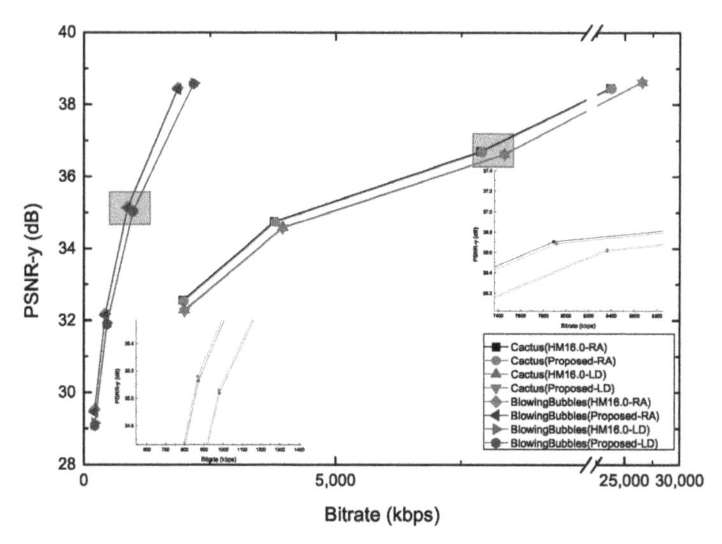

Figure 10. Rate–distortion (R–D) curve of the proposed method for "Cactus" and "BlowingBubbles".

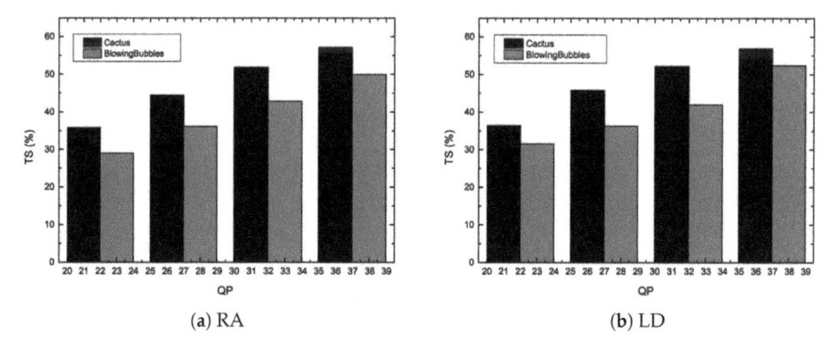

Figure 11. Time savings of the proposed method for "Cactus" and "BlowingBubbles".

Table 6. Performance comparison of the proposed method compared to previous works.

	Method	**(BDBR, TS)**
RA	Proposed	(0.80, 50.96)
	Zhang's [12]	(1.19, 54.93)
	Zhu's [13]	(3.67, 65.60)
	Ahn's [14]	(1.40, 49.60)
	Goswami's [24]	(1.11, 51.68)
	Tai's [25]	(1.41, 45.70)
	Xiong's [26]	(2.00, 58.40)
LD	Proposed	(0.76, 50.23)
	Zhu's [13]	(3.84, 67.30)
	Ahn's [14]	(1.00, 42.70)
	Tai's [25]	(0.75, 37.90)
	Xiong's [26]	(1.61, 52.00)

6. Conclusions

In order to develop the low-complexity and hardware-friendly H.265/HEVC encoder for VANETs, based on a novel spatiotemporal neighborhood set, the Bayesian rule and Gibbs Random Field are used to reduce the encoding complexity for the H.265/HEVC inter-prediction in this work. The proposed algorithm consists of CTU depth decision and PU mode decision methods. Experimental results demonstrate that the proposed approach can reduce the average encoding complexity of H.265/HEVC encoder by about 50% for VANETs, while the increasing of BDBR is less than or equal to 0.8% on average.

Author Contributions: X.J. designed the algorithm, conducted all experiments, analyzed the results, and wrote the manuscript. J.F. analyzed the results. T.S. conceived the algorithm. T.K. conducted all experiments.

Funding: This work was supported by in part of by the National Natural Science Foundation of China under grant 61701297, in part of by the China Postdoctoral Science Foundation under grant 2018M641982, in part of by the China Scholarship Council and Mitacs, in part of by JSPS KAKENHI under grant 17K00157, and in part of by the Shanghai Sailing Program under grant 1419100.

Conflicts of Interest: The authors declare no conflict of interest.

References

1. Bazzi, A.; Masini, B.M.; Zanella, A.; De Castro, C.; Raffaelli, C.; Andrisano, O. Cellular aided vehicular named data networking. In Proceedings of the 2014 IEEE International Conference on Connected Vehicles and Expo (ICCVE), Vienna, Austria, 3–7 November 2014; pp. 747–752.
2. Paredes, C.I.; Mezher, A.M.; Igartua, M.A. Performance Comparison of H. 265/HEVC, H. 264/AVC and VP9 Encoders in Video Dissemination over VANETs. In Proceedings of the International Conference on Smart Objects and Technologies for Social Good, Venice, Italy, 30 November–1 December 2016; pp. 51–60.
3. Shaibani, R.F.; Zahary, A.T. Survey of Context-Aware Video Transmission over Vehicular Ad-Hoc Networks (VANETs). *EAI Endorsed Trans. Mob. Commun. Appl.* **2018**, *4*, 1–11.[CrossRef]
4. Torres, A.; Piñol, P.; Calafate, C.T.; Cano, J.C.; Manzoni, P. Evaluating H.265 real-time video flooding quality in highway V2V environments. In Proceedings of the 2014 IEEE Wireless Communications and Networking Conference (WCNC), Istanbul, Turkey, 6–9 April 2014; pp. 2716-2721.
5. Mammeri, A.; Boukerche, A.; Fang, Z. Video streaming over vehicular ad hoc networks using erasure coding. *IEEE Syst. J.* **2016**, *10*, 785–796. [CrossRef]
6. Pan, Z.; Chen, L.; Sun, X. Low complexity HEVC encoder for visual sensor networks. *Sensors* **2015**, *15*, 30115–30125. [CrossRef] [PubMed]
7. Laude, T.; Adhisantoso, Y.G.; Voges, J.; Munderloh, M.; Ostermann, J. A Comparison of JEM and AV1 with HEVC: Coding Tools, Coding Efficiency and Complexity. In Proceedings of the 2018 Picture Coding Symposium (PCS), San Francisco, CA, USA, 24–27 June 2018; pp. 36–40.
8. Bossen, F.; Bross, B.; Suhring, K.; Flynn, D. HEVC complexity and implementation analysis. *IEEE Trans. Circuits Syst. Video Technol.* **2012**, *22*, 1685–1696. [CrossRef]
9. Jiang, X.; Song, T.; Zhu, D.; Katayama, T.; Wang, L. Quality-Oriented Perceptual HEVC Based on the Spatiotemporal Saliency Detection Model. *Entropy* **2019**, *21*, 165. [CrossRef]
10. Xu, Z.; Min, B.; Cheung, R.C. A fast inter CU decision algorithm for HEVC. *Signal Process. Image Commun.* **2018**, *60*, 211–223. [CrossRef]
11. Duan, K.; Liu, P.; Jia, K.; Feng, Z. An Adaptive Quad-Tree Depth Range Prediction Mechanism for HEVC. *IEEE Access* **2018**, *6*, 54195–54206. [CrossRef]
12. Zhang, J.; Kwong, S.; Wang, X. Two-stage fast inter CU decision for HEVC based on bayesian method and conditional random fields. *IEEE Trans. Circuits Syst. Video Technol.* **2018**, *28*, 3223–3235. [CrossRef]
13. Zhu, L.; Zhang, Y.; Pan, Z.; Wang, R.; Kwong, S.; Peng, Z. Binary and multi-class learning based low complexity optimization for HEVC encoding. *IEEE Trans. Broadcast.* **2017**, *63*, 547–561. [CrossRef]
14. Ahn, S.; Lee, B.; Kim, M. A novel fast CU encoding scheme based on spatiotemporal encoding parameters for HEVC inter coding. *IEEE Trans. Circuits Syst. Video Technol.* **2015** *25*, 422–435. [CrossRef]
15. Sharma, P.; Kaul, A.; Garg, M.L. Performance analysis of video streaming applications over VANETs. *Int. J. Comput. Appl.* **2015**, *112*, 13–18.

16. Yaacoub, E.; Filali, F.; Abu-Dayya, A. QoE enhancement of SVC video streaming over vehicular networks using cooperative LTE/802.11 p communications. *IEEE J. Sel. Top. Signal Process.* **2015**, *9*, 37–49. [CrossRef]
17. Rezende, C.; Boukerche, A.; Almulla, M.; Loureiro, A.A. The selective use of redundancy for video streaming over Vehicular Ad Hoc Networks. *Comput. Netw.* **2015**, *81*, 43–62. [CrossRef]
18. Yousef, W.S.M.; Arshad, M.R.H.; Zahary, A. Vehicle rewarding for video transmission over VANETs using real neighborhood and relative velocity (RNRV). *J. Theor. Appl. Inf. Technol.* **2017**, *95*, 242–258.
19. Jiang, X.; Wang, X.; Song, T.; Shi, W.; Katayama, T.; Shimamoto, T.; Leu, J.S. An efficient complexity reduction algorithm for CU size decision in HEVC. *Int. J. Innov. Comput. Inf. Control* **2018**, *14*, 309–322.
20. Jiang, X.; Song, T.; Shi, W.; Katayama, T.; Shimamoto, T.; Wang, L. Fast coding unit size decision based on probabilistic graphical model in high efficiency video coding inter prediction. *IEICE Trans. Inf. Syst.* **2016**, *99*, 2836–2839. [CrossRef]
21. Zhang, J.; Kwong, S.; Zhao, T.; Pan, Z. CTU-level complexity control for high efficiency video coding. *IEEE Trans. Multimed.* **2018**, *20*, 29–44. [CrossRef]
22. Jiang, X.; Song, T.; Katayama, T.; Leu, J.S. Spatial Correlation-Based Motion-Vector Prediction for Video-Coding Efficiency Improvement. *Symmetry* **2019**, *11*, 129. [CrossRef]
23. Li, Y.; Yang, G.; Zhu, Y.; Ding, X.; Sun, X. Unimodal stopping model-based early SKIP mode decision for high-efficiency video coding. *IEEE Trans. Multimed.* **2017**, *19*, 1431–1441. [CrossRef]
24. Goswami, K.; Kim, B.G. A Design of Fast High-Efficiency Video Coding Scheme Based on Markov Chain Monte Carlo Model and Bayesian Classifier. *IEEE Trans. Ind. Electron.* **2018**, *65*, 8861–8871. [CrossRef]
25. Tai, K.H.; Hsieh, M.Y.; Chen, M.J.; Chen, C.Y.; Yeh, C.H. A fast HEVC encoding method using depth information of collocated CUs and RD cost characteristics of PU modes. *IEEE Trans. Broadcast.* **2017**, *43*, 680–692. [CrossRef]
26. Xiong, J.; Li, H.; Meng, F.; Wu, Q.; Ngan, K.N. Fast HEVC inter CU decision based on latent SAD estimation. *IEEE Trans. Multimed.* **2015**, *17*, 2147–2159. [CrossRef]
27. Liu, Z.; Lin, T.L.; Chou, C.C. Efficient prediction of CU depth and PU mode for fast HEVC encoding using statistical analysis. *J. Vis. Commun. Image Represent.* **2016**, *38*, 474–486. [CrossRef]
28. Chen, M.J.; Wu, Y.D.; Yeh, C.H.; Lin, K.M.; Lin, S.D. Efficient CU and PU Decision Based on Motion Information for Interprediction of HEVC. *IEEE Trans. Ind. Inform.* **2018**, *14*, 4735–4745. [CrossRef]
29. Zhang, Y.; Wang, H.; Li, Z. Fast coding unit depth decision algorithm for interframe coding in HEVC. In Proceedings of the IEEE Data Compression Conference, Snowbird, UT, USA, 20–22 March 2013; pp. 53–62.
30. Clifford, P. Markov random fields in statistics. in *Disorder in Physical Systems: A Volume in Honour of John M. Hammersley*; Oxford University Press: Oxford, UK, 1990; p. 19.
31. Kruis, J.; Maris, G. Three representations of the Ising model. *Sci. Rep.* **2016**, *6*, 34175. [CrossRef]
32. Rosewarne, C. High Efficiency Video Coding (HEVC) Test Model 16 (HM 16); Document JCTVC-V1002, JCT-VC. October 2015. Available online: http://phenix.int-evry.fr/jct/ (accessed on 15 March 2019).
33. Bjontegaard, G. Calculation of Average PSNR Differences between RD-Curves. In Proceedings of the ITU-T Video Coding Experts Group (VCEG) Thirteenth Meeting, Austin, TX, USA, 2–4 April 2001. Available online: https://www.itu.int/wftp3/av-arch/video-site/0104_Aus/ (accessed on 15 March 2019).
34. Bossen, F. Common Test Conditions and Software Reference Configurations, Joint Collaborative Team on Video Coding (JCT-VC), Document JCTVC-L1110, Geneva, January 2014. Available online: https://www.itu.int/wftp3/av-arch/video-site/0104_Aus/ (accessed on 15 March 2019).

Article

Improving Discrimination in Color Vision Deficiency by Image Re-Coloring

Huei-Yung Lin [1,*], Li-Qi Chen [2] and Min-Liang Wang [3]

[1] Department of Electrical Engineering, Advanced Institute of Manufacturing with High-Tech Innovation, National Chung Cheng University, Chiayi 621, Taiwan
[2] Department of Electrical Engineering, National Chung Cheng University, Chiayi 621, Taiwan; kenbroms@gmail.com
[3] Asian Institute of TeleSurgery/IRCAD-Taiwan, Changhua 505, Taiwan; ccuvislab@hotmail.com
* Correspondence: lin@ee.ccu.edu.tw; Tel.: +886-5-272-0411

Received: 19 April 2019 ; Accepted: 13 May 2019; Published: 15 May 2019

Abstract: People with color vision deficiency (CVD) cannot observe the colorful world due to the damage of color reception nerves. In this work, we present an image enhancement approach to assist colorblind people to identify the colors they are not able to distinguish naturally. An image re-coloring algorithm based on eigenvector processing is proposed for robust color separation under color deficiency transformation. It is shown that the eigenvector of color vision deficiency is distorted by an angle in the λ, *Y-B*, *R-G* color space. The experimental results show that our approach is useful for the recognition and separation of the CVD confusing colors in natural scene images. Compared to the existing techniques, our results of natural images with CVD simulation work very well in terms of RMS, HDR-VDP-2 and an IRB-approved human test. Both the objective comparison with previous works and the subjective evaluation on human tests validate the effectiveness of the proposed method.

Keywords: color vision deficiency; image re-coloring; visual assistance

---

1. Introduction

Most human beings have the ability of color vision perception, which senses the frequency of the light reflected from object surfaces. However, color vision deficiency (CVD) is a common genetic condition [1]. It is in general not a fatal or serious disease, but still brings inconvenience to most patients. People with color vision deficiency (or so-called color blindness) cannot observe the colorful world due to the damage of color reception nerves. Whether caused by genetic problems or chemical injury, the damaged nerves are not able to distinguish certain colors. There are a few common types of color vision deficiency such as protanomaly (red weak), deuteranomaly (green weak) and tritanomaly (blue weak). They can be detected and verified easily by some special color patterns (e.g., Ishihara plates [2]), but, unfortunately, cannot be cured by medical surgery or other treatments. Compared to the human population, people with color vision deficiency are still a minority, and they are sometimes ignored and restricted by our society.

In many places, colorblind people are not allowed to have a driver's license. A number of careers in engineering, medicine and other related fields have set some restrictions on the ability of color perception. The display and presentation of most media on devices and in many forms do not specifically take color vision deficiency into consideration. Although the weakness in distinguishing different colors does not obviously affect people's learning and cognition, there is still a challenge in terms of color-related industries. In this work, we propose an approach to assist people with color vision deficiency to tell the difference among the confusing colors as much as possible. A simple yet reasonable technique, "color reprint", is developed and used to represent the CVD-proof colors.

The algorithm does not only preserve the naturalness and details of the scenes, but also possess real-time processing capability. It can, therefore, be implemented on low-cost or portable devices, and brought to everyday life.

Human color vision is based on three light-sensitive pigments [3,4]. It is trichromatic and presented in three dimensions. The color stimulus is specified by the power contained at each wavelength. Normal trichromacy is because that the retina contains three classes of cone photo-pigment neural cells, L-, M-, and S-cones. A range of wavelengths of the light stimulate each of these receptor types at various degrees. For example, yellowish green light stimulates both L- and M-cones equally strongly, but S-cones weakly. Red light stimulates more L-cones than M-cones, and S-cones hardly at all. Our brain combines the information from each type of cone cells, and responds to different wavelengths of the light as shown in Table 1. The color processing is carried out in two stages. First, the stimulus from the cones is recombined to form two color-opponents and luminance. Second, an adaptive signal regulation processes within the operating range and stabilizes the illumination changes of the object appearance. When any kind of sensitive pigments is broken or loses the functionality [1], people can only view a part of the visible spectrum compared to those with normal vision capability [5] (see Figure 1).

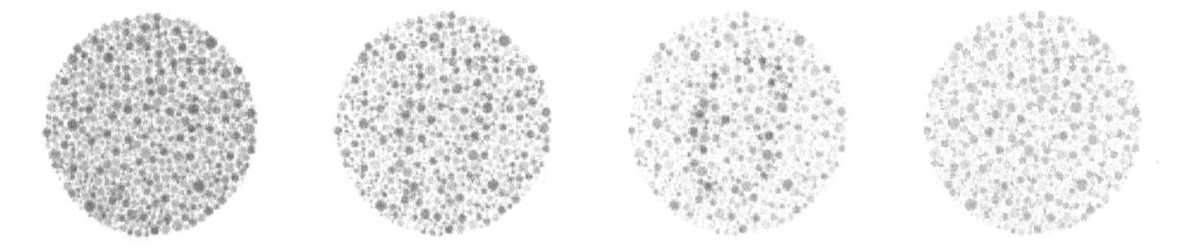

(a) One of the images in Ishihara plates (left), and the images enhanced by the proposed re-coloring algorithm for protanomaly, deuteranomaly and tritanomaly, respectively (the rest).

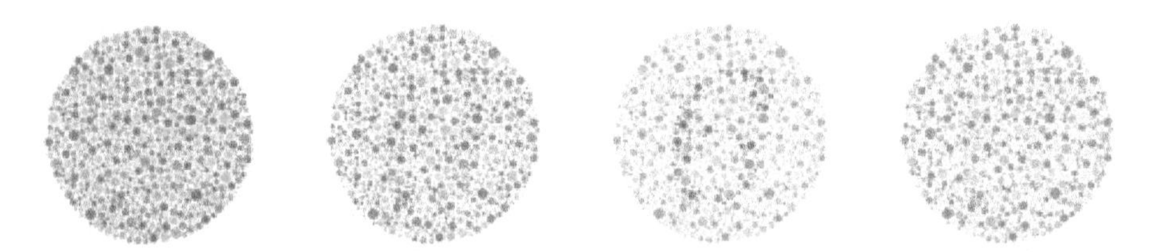

(b) The images in (a) generated by color vision deficiency simulation. The first image is deuteranomaly simulation of the original Ishihara plate. The rest images are the simulation results of protanomaly, deuteranomaly and tritanomaly on the re-colored images, respectively.

Figure 1. (**a**) An original image from Ishihara plates and the enhanced images using our re-coloring algorithms for protanomaly, deuteranomaly and tritanomaly. (**b**) The images generated from a color vision deficiency simulation tool [6]. The results show that our image enhancement technique is able to improve check pattern recognition under various types of color vision deficiency.

Table 1. Cone cells in the human eyes and the response to the light wavelength.

Type	Range	Peak Wavelength
S	400–500 nm	420–440 nm
M	450–630 nm	534–555 nm
L	500–700 nm	564–580 nm

There are studies about the molecular genetics of human color vision in the literature. Nathans et al. have described the isolation and sequencing of genomic and complementary DNA clones which encode the apoproteins of the red, green and blue pigments [4]. With newly refined methods, the number and ratio of genes are re-examined in men with normal color vision. A recent report reveals that many males have more pigment genes on the X chromosome than previously studied, and many have more than one long-wave pigment gene [7]. The loss of characteristic sensitivities of the red and green receptors introduced into the transformed sensitivity curves also indicates the appropriate degrees of luminosity deficit for deuteranopes and protanopes [8].

Color vision deficiency is mainly caused by two reasons: natural genetic factors and impaired nerves or brain. A protanope suffers from a lack of the L-cone photo-pigment, and is unable to discriminate reddish and greenish hues since the red–green opponent mechanism cannot be constructed. A deuteranope does not have sufficient M-cone photo-pigment, so the reddish and greenish hues are not distinguishable. People with tritanopia do not have the S-cone photo-pigment, and, therefore, cannot discriminate yellowish and bluish hues [9]. The literature shows that more than 8% of the world population suffer from color vision deficiency (see Table 2). For color vision correction, gene therapy which adds the missing genes is sufficient to restore full color vision without further rewiring of the brain. It has been tested on a monkey with colorblindness since birth [10]. Nevertheless, there are also non-invasive alternatives available by means of computer vision techniques.

Table 2. Approximate percentage occurrences of various types of color vision deficiency [11].

Type	Male (%)	Female (%)
Protanopia	1.0	0.02
Deuteranopia	1.1	0.01
Trianopia	0.002	0.001
Protanomaly	1.0	0.02
Deuteranomaly	4.9	0.38
Tritanomaly	∼0	∼0
Total	**8.002**	**0.44**

In [12], Huang et al. propose a fast re-coloring technique to improve the accessibility for the impaired color vision. They design a method to derive an optimal mapping to maintain the contrast between each pair of the representative colors [13]. In a subsequent work, an image re-coloring algorithm for dichromats using the concept of key color priority is presented [14]. A color blindness plate (CBP) is presented by Chen et al., which is a satisfactory way to test color vision in the computer vision community [15]. The approach is adopted to demonstrate normal color vision, as well as red–green color vision deficiency. Rasche et al. propose a method to preserve the image details while reducing the gamut dimension, and seek a color to gray mapping to maintain the contrast and luminance consistency [16]. They also describe a method which allows the re-colored images to deliver the content with increased information to color-deficient viewers [17]. In [18], Lau et al. present a cluster-based approach to optimize the transformation for individual images. The idea is to preserve the information from the source space as much as possible while maintaining the natural mapping as faithfully as possible. Lee et al. develop a technique based on fuzzy logic and correction of digital images to improve the visual quality for individuals with color vision disturbance [19]. Similarly, Poret et al. design a filter based on the Ishihara color test for color blindness correction [20].

Most algorithms for color transformation aim to preserve the color information in the original image while maintaining the re-colored image as naturally as possible. This might be different from some image processing and computer vision tasks; the images appearing natural after enhancement is an important issue for color vision deficiency correction. It is not only to keep the image details intact, but also to maintain the colors as smooth as those without the re-coloring process. These conditions re-range in the color distribution space to let the colorblind people to discriminate different colors [21,22]. Moreover, it is generally agreed that color perception is subjective and will not be exactly the same for different people. In this work, the proposed method is carried out on color vision deficiency simulation tools and adopts human tests for evaluation. We use RMS (root mean squares) to calculate the change after re-coloring, and HDR-VDP (visual difference predictor) [23] to compare the visibility and quality of subjective human feeling. Our algorithms not only present the naturalness and details of the images, but process almost in real-time.

2. Approach

In this paper, a technique called *color warping* (CW) is proposed for effective image re-coloring. It uses the orientation of the eigenvectors of the color vision deficiency simulation results to warp the color distribution. In general, the acquired images are presented in the RGB color space for display. This is, however, not suitable for color vision-related processing. For human color perception related tasks, the images are first transformed to the λ, *Y-B*, *R-G* color space based on the CIECAM02 model [24]. It consists of a transformation from RGB to LMS [25] using

$$\begin{bmatrix} L \\ M \\ S \end{bmatrix} = \begin{bmatrix} 0.7328 & 0.4296 & -0.1624 \\ -0.7036 & 1.6975 & 0.0061 \\ 0.0030 & 0.0136 & 0.9834 \end{bmatrix} \begin{bmatrix} R \\ G \\ B \end{bmatrix} \tag{1}$$

followed by a second transformation from LMS to λ, *Y-B*, *R-G* with

$$\begin{bmatrix} \lambda \\ Y-B \\ R-G \end{bmatrix} = \begin{bmatrix} 0.6 & 0.4 & 0.0 \\ 0.24 & 1.05 & -0.7 \\ 1.2 & -1.6 & 0.4 \end{bmatrix} \begin{bmatrix} L \\ M \\ S \end{bmatrix}. \tag{2}$$

Since the above transformations are linear, it is easily to verify the relationship between the RGB and λ, *Y-B*, *R-G* color spaces is given by

$$\begin{bmatrix} \lambda \\ Y-B \\ R-G \end{bmatrix} = \begin{bmatrix} 0.3479 & 0.5981 & -0.3657 \\ -0.0074 & -0.1130 & -1.1858 \\ 1.1851 & -1.5708 & 0.3838 \end{bmatrix} \begin{bmatrix} R \\ G \\ B \end{bmatrix} \tag{3}$$

and

$$\begin{bmatrix} R \\ G \\ B \end{bmatrix} = \begin{bmatrix} 1.2256 & -0.2217 & 0.4826 \\ 0.9018 & -0.3645 & -0.2670 \\ -0.0936 & -0.8072 & 0.0224 \end{bmatrix} \begin{bmatrix} \lambda \\ Y-B \\ R-G \end{bmatrix}. \tag{4}$$

A flowchart of the proposed method is illustrated in Figure 2. The "Eigen-Pro" stage represents the eigenvector processing. The *color warping* is the key idea of this work, and the color constraints are used to make the distortion decrease after the color space transformation.

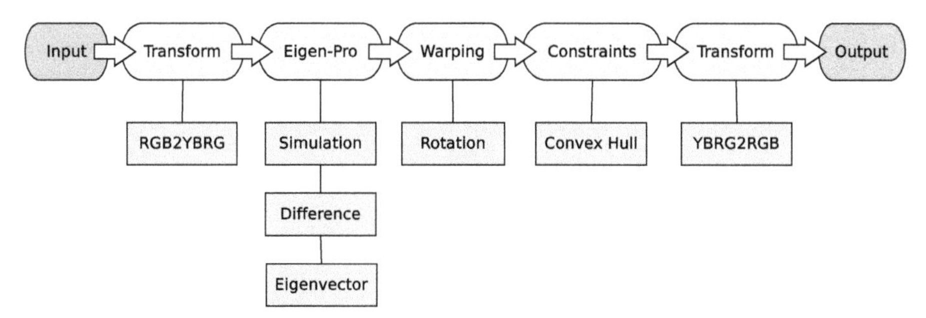

Figure 2. The flowchart of the proposed technique. In the pipeline, the images are first transformed to the λ, *Y-B*, *R-G* color space for the re-color processing, followed by a transformation back to the original RGB color space.

2.1. Color Transform

The physical property of the light used for color perception is the distribution of the spectral power [26]. In principle, there are many distinct spectral colors, and the set of all physical colors may be thought of as a large-dimensional vector space. A better alternative to the commonly adopted tristimulus coordinates for the spectral property of the light is to use L-, M-, and S-cone cells coordinates as a 3-space. To form a model for human perceptual color space, we can consider all the resulting combinations as a subset of the 3-space. The property of the cones covers the region away from the origin corresponding to the intensity of the S, M and L lights proportionately. A digital image acquisition device consists of different elements [27,28]. The characteristics of the light and the material of the observed object determine the physical properties of its color [27,29,30]. For color transformation, Huang et al. [31] present a method to warp images to the CIELab color space by rotating a matrix. Dana et al. [32] and Swain et al. [33] propose to use an antagonist space which does not take the non-linear human eye response into consideration. Instead, we transform the color space to (WS, RG, BY) based on the electro-physiological study [34].

2.2. Eigenvector Processing

Color vision deficiency cannot be understood easily by most people with normal vision. Thus, it is necessary to use simulation tools to create synthetic images for ordinary viewers to understand what are seen by the colorblind people [35]. Some well-known tools include Colblindor [6] and LMS [25], and there are also several websites to perform the simulation online (For example, Coblis Color Blindness Simulator (http://www.color-blindness.com/coblis-color-blindness-simulator), Color Blindness Simulator (http://www.etre.com/tools/colourblindsimulator), and Vision Simulator (http://www.webexhibits.org/causesofcolor/2.html)). In this work, Machado's approach is adopted for our color vision deficiency simulation [35]. It utilizes a physiology based simulation model to achieve the sensation of cones in human visual perception. The simulation is to shift the pigments of the responding curve of spectral sensitivity functions as shown in Figure 3. Anomalous trichromacy can be simulated by shifting the sensitivity of the L, M, and S cones in the following ways:

- Protanomaly: Shift L cone toward M cone, $L(\lambda)_a = L(\lambda + \Delta\lambda_L)$.
- Deuteranomaly: Shift M cone toward L cone, $M(\lambda)_a = M(\lambda + \Delta\lambda_M)$.
- Trianomaly: Shift S cone, $S(\lambda)_a = S(\lambda + \Delta\lambda_S)$.

The elements of the transformation matrix Γ can be derived by

$$f(\lambda, R, G, B)_{WS,YB,RG} = \rho_{R,G,B} \int \phi_{R,G,B}(\lambda) f(\lambda)_{WS,YB,RG} \, \mathrm{d}\lambda \tag{5}$$

where $\phi_{R,G,B}$ is the spectral power distribution function, and $\rho_{R,G,B}$ is a normalization factor. Thus, Γ is the projection of the spectral power distributions of RGB primaries onto a set of basic functions $f(\lambda, R, G, B)_{WS,YB,RG}$. That is,

$$\Gamma = \begin{bmatrix} f(R)_{WS} & f(G)_{WS} & f(B)_{WS} \\ f(R)_{YB} & f(G)_{YB} & f(B)_{YB} \\ f(R)_{RG} & f(G)_{RG} & f(B)_{RG} \end{bmatrix}. \quad (6)$$

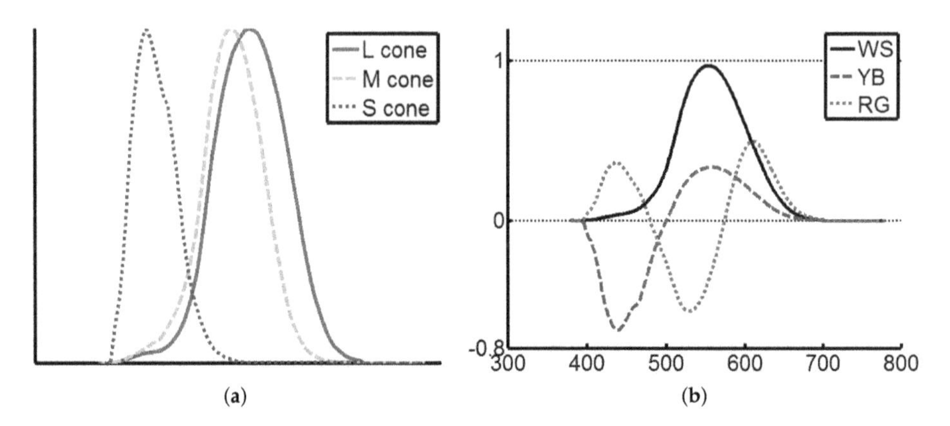

Figure 3. The cone spectral sensitivity functions at all wavelengths in the visible range. (**a**) Responding curve [36]. (**b**) Spectral response functions for the opponent channels [34].

This model is based on the stage theory of human color vision, and is derived from the data reported in electro-physiological study [34]. Let Φ_{CVD} be the matrix that maps RGB to the opponent-color space of normal trichromacy, then the simulation of dichromatic vision is obtained by the transformation

$$\begin{bmatrix} R_s \\ G_s \\ B_s \end{bmatrix} = \Phi_{CVD} \begin{bmatrix} R \\ G \\ B \end{bmatrix}. \quad (7)$$

By definition, an eigenvector is the non-zero vector mapped by a given linear transformation of a vector space onto a vector that is the product of the original vector multiplied by a scalar. Thus, the algorithm counts the eigenvectors of the covariance matrix from the images in *Y-B*, *R-G* of the λ, *Y-B*, *R-G* opponent color space, i.e.,

$$[v, d] = eig(cov(I_{Y-B}, I_{R-G})) \quad (8)$$

where *eig* is the function of eigenvalue and eigenvector, and I_{Y-B} and I_{R-G} are the *Y-B* and *R-G* images, respectively. On the left hand side of the equation, *d* is the generalized eigenvalue, and v is a 2×2 matrix since the covariance *cov* is a 2×2 matrix derived from a pair of $n \times 1$ images given by the covariance matrix

$$cov(X, Y) = \frac{\sum_{i=1}^{n}(X_i - \bar{X})(Y_i - \bar{Y})}{(n-1)} \quad (9)$$

For the original and CVD simulation images shown in Figure 4, the characteristics of the associated eigenvectors are illustrated in Figure 5. The black line (at about 91°) indicates the eigenvector of the original image. For protanopia (red line about 150°) and deuteranopia (green line about 140°), the eigenvectors lead the one associated with the original image. The eigenvector of tritanopia (blue line at about 80°) is behind the original image case. Our objective is to recover the angle difference between the normal and color vision deficiency images, and use it to re-color the image. The difference image when observed by normal viewers and the color vision deficiency simulation is defined by

$$I_{diff} = \sqrt{(I_n(YB) - I_c(YB))^2 + (I_n(RG) - I_c(RG))^2} \quad (10)$$

where I_n and I_c represent the intensity observed by a normal viewer and obtained from the color vision deficiency simulation, respectively.

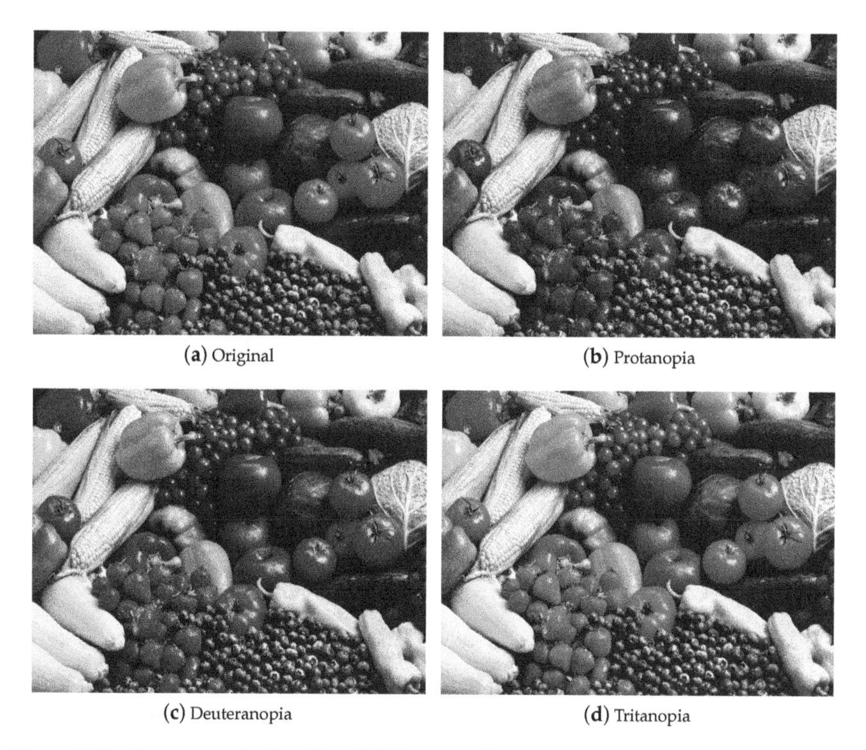

(**a**) Original (**b**) Protanopia

(**c**) Deuteranopia (**d**) Tritanopia

Figure 4. The three types of color vision deficiency simulation using Machado's approach [35] with sensitive 0.6 and the matrix Φ_{CVD} as shown in Table 3.

Table 3. Machado's Simulation Matrices Φ_{CVD}.

Sensitivity	**0.6**
Protanopia	$\begin{bmatrix} 0.385 & 0.769 & 0.154 \\ 0.101 & 0.830 & 0.070 \\ 0.007 & 0.022 & 1.030 \end{bmatrix}$
Deuteranopia	$\begin{bmatrix} 0.499 & 0.675 & 0.174 \\ 0.205 & 0.755 & 0.040 \\ 0.011 & 0.031 & 0.980 \end{bmatrix}$
Tritanopia	$\begin{bmatrix} 1.105 & 0.047 & 0.058 \\ 0.032 & 0.972 & 0.061 \\ 0.001 & 0.318 & 0.681 \end{bmatrix}$

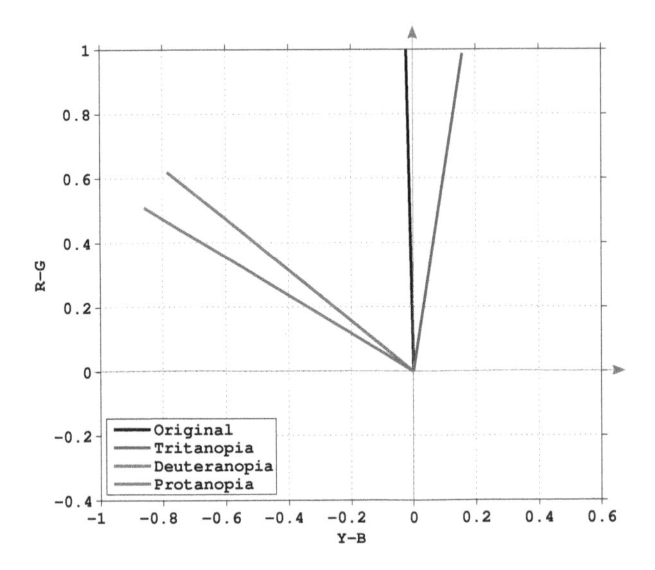

Figure 5. Eigenvectors of the covariance matrix of the images shown in Figure 4.

An example of protanopia simulation is illustrated in Figure 6. The difference image is shown in Figure 6c, and a binary image for better illustration is shown in Figure 6d. Our objective is to recover the angle and difference between the normal and CVD images and use the information to re-color the CVD simulation images.

(**a**) Original (**b**) Protanopia simulation (**c**) Difference image (**d**) Binary image

Figure 6. An example of Protanopia simulation. (**a**) is the original image and (**b**) is the Protanopia simulation result. (**c**) is the difference of (**a**,**b**) computed in the λ, *Y-B*, *R-G* color space. (**d**) is the binarized version of (**c**) for better illustration.

2.3. Color Warping

The color values of the images are transformed from RGB to the opponent color space λ, *Y-B*, *R-G* using Equations (3) and (4). It is assumed that color vision deficiency does not affected by the brightness, so the value λ corresponding to the luminance is keep intact. To define the warping range with the angle of an eigenvector, we construct twelve pure colors in RGB using the values of 0, 150 and 255. Figure 7 shows the range of missing chroma of color vision deficiency. The area within the two green lines is the red chroma missing for protanopia and deuteranopia, and the area within th two purple lines is the blue chroma missing for tritanopia. The color points represented in the Cartesian coordinates are then transformed to the polar coordinates by

$$\theta = \tan^{-1} \frac{y}{x} \tag{11}$$

for processing. The angle θ associated with the eigenvector in the λ, Y-B, R-G color space is used to derive the range to be processed. Since the image is now in the opponent color space, the range is defined by the angle of the simulation vector to the opposite angle of the simulation vector. Finally, the warping range is defined by the vertical angle of the original vector to the opposite angle of the simulation vector. An example is illustrated in Figure 8, the green area is warped to the red area for image re-coloring.

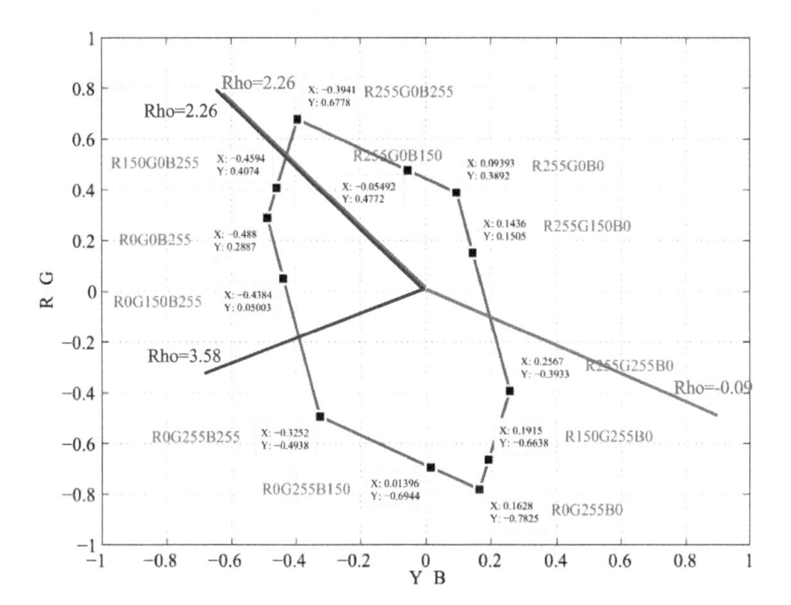

Figure 7. The pure colors RGB (0, 150, and 255).

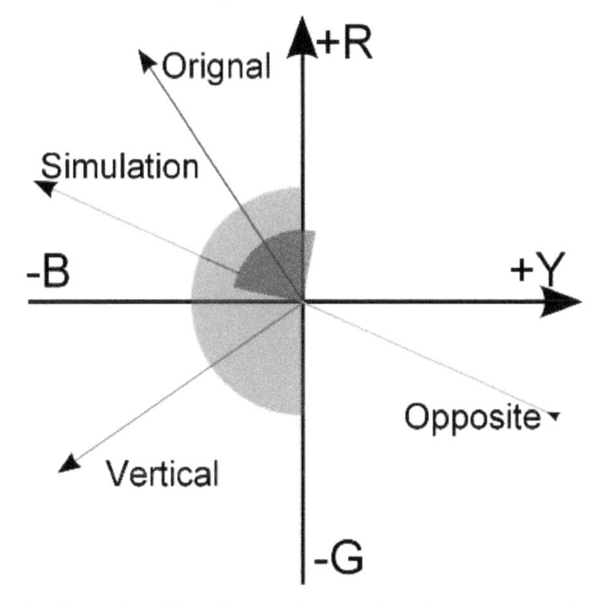

Figure 8. An illustration of the color warping range from the green area to the red area.

The new color angle is derived from the original color angle by

$$\theta_{new} = \frac{\theta_{\perp} - \theta_{op}}{\pi} \cdot (\theta - \theta_{op}) \tag{12}$$

where the angles of color points are defined in the range of $[-\pi, \pi]$, $\theta_{\perp}$ is the angle of vector orthogonal to the original vector, and θ_{op} is the angle of vector opposite to the color vision deficiency simulation vector.

When the image is converted from RGB to the λ, *Y-B*, *R-G* color space, it is in a limited range of color space representation. We need a constraint to avoid the luminance overflow problem, which will make colors not smooth after converted back to the RGB color representation. In our approach, a convex hull is adopted for the color constraint due to its simplicity for boundary derivation. Figure 9a–d illustrate the full-color images constructed using 256^3 pixels, i.e., the resolution of 4096×4096, and the corresponding convex hull is shown in Figure 9e (the red lines). The formula used for conversion is given by

$$\rho_{new} = \rho \times \frac{\rho(\theta_{new})}{\rho(\theta)} \tag{13}$$

where ρ is the original value, $\rho(\theta_{new})$ is the value of the convex hull at θ_{new}, and $\rho(\theta)$ is the value of the convex hull at θ. The resulting image in the λ, *Y-B*, *R-G* color space is then transformed back to the RGB color space for display.

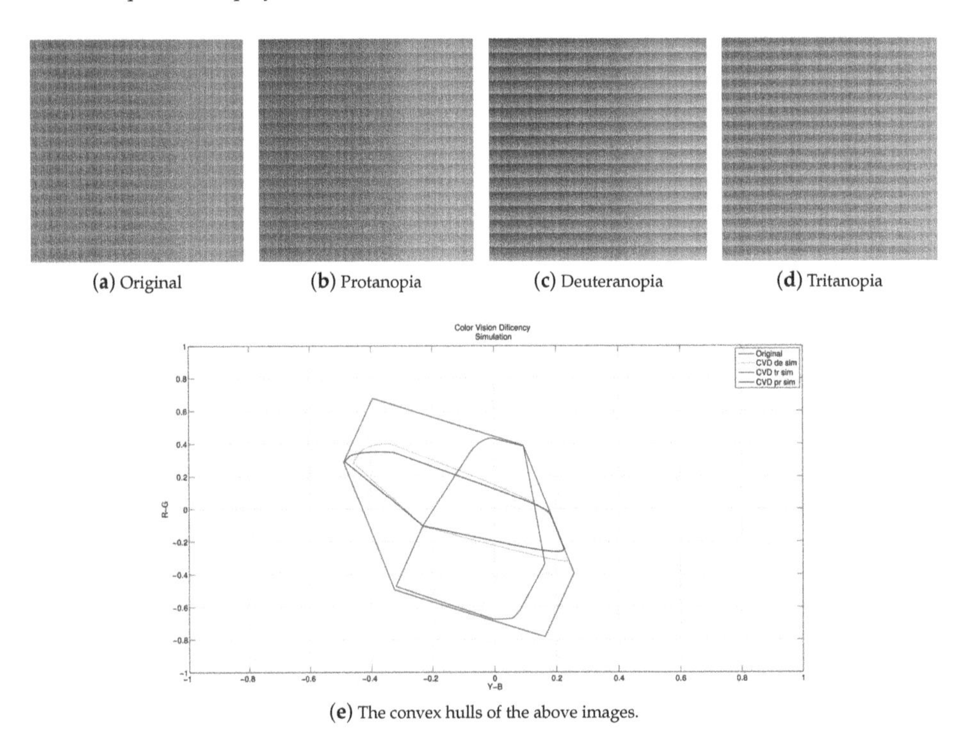

Figure 9. (**a**) Three types of color vision deficiency simulation using a full-color image with 256^3 pixels. The image resolution is 4096×4096. (**e**) The convex hulls of the images in (a–d). All types of CVD simulation cover only a part of the convex hull of the original full color imag.

3. Experiments

The proposed method has been tested on natural images including flowers, fruits, pedestrians and landscape, as well as synthetic images such as patterns with pure colors (see Figure 10). The experiments were carried out on both simulation view and human tests. Figure 11 shows the images of protanopia color vision deficiency with different sensitive from 0.3 to 0.9 after our re-coloring technique. For the color vision deficiency view simulation, we compared the results of the proposed approach with the methods presented by Kuhn et al. [37], Rasche et al. [17] and Huang et al. [12]. Figure 12 shows the results of the deuteranopia color vision deficiency simulation and re-coloring using different algorithms. While all methods are able to separate the flower from the leaves, our result is more distinguishable and much closer to original color.

(**a**) Flower 1 (**b**) Flower 2 (**c**) Natural

(**d**) Fruit (**e**) Pencil (**f**) Pedestrian 1

(**g**) Pedestrian 2

Figure 10. The test images used to evaluate the re-coloring techniques for color vision deficiency.

Figure 11. Enhancement sensitive of protanopia color vision deficiency, (**a**) with sensitivity 0.3, (**b**) with sensitivity 0.5, (**c**) with sensitivity 0.7, (**d**) with sensitivity 0.9.

Figure 12. The comparison of deuteranopia simulation of the flower image in Figure 9a. (**a**) Machado's CVD simulation. (**b**) Our re-coloring technique after Machado's CVD simulation. (**c**) Brettel's CVD simulation. (**d**) Kuhn's re-coloring technique after Brettel's CVD simulation. (**e**) Rasche's CVD simulation. (**f**) Rasche's re-coloring after CVD simulation. (**g**) Huang's CVD simulation. (**h**) Huang's re-coloring after CVD simulation.

3.1. Root Mean Square

We use the root mean square (RMS) value to measure the difference between two images. That is, to evaluate how far between the CVD view simulation and the image processed after our re-coloring algorithm. We calculate the RMS value with k-neighborhood defined by

$$RMS_i = \frac{1}{N}\sqrt{\frac{1}{2}\sum_{j=-k}^{k}[(a^r_{i+j} - a^t_{i+j})^2 + (b^r_{i+j} - b^t_{i+j})^2]}$$

where a^r_{i+j} and b^r_{i+j} are a^*b^* in $L^*a^*b^*$ of the reference image, a^t_{i+j} and b^t_{i+j} are a^*b^* in $L^*a^*b^*$ of the target image, and N is the number of elements in k-neighbor.

An example of tritanopia CVD simulation and the re-coloring results is shown in Figure 13. Compared to the results obtained from Kuhn's and Huang's methods, our approach provides better contrast between the colors. Figure 14 shows the comparison of the RMS values on several test images using the proposed technique and Kuhn's method. The higher RMS value is displayed in dark blue, and the lowest value is shown in white. The figures indicate that, although the distributions of our and Kuhn's results are similar, the RMS values of ours are higher than Kuhn's, which implies a better separation in colors. Additional results of various types of test images are shown in Figure 15. The results of CVD simulation, re-coloring using the proposed technique and CVD simulation on the re-colored images are shown in the first, second and third column, respectively.

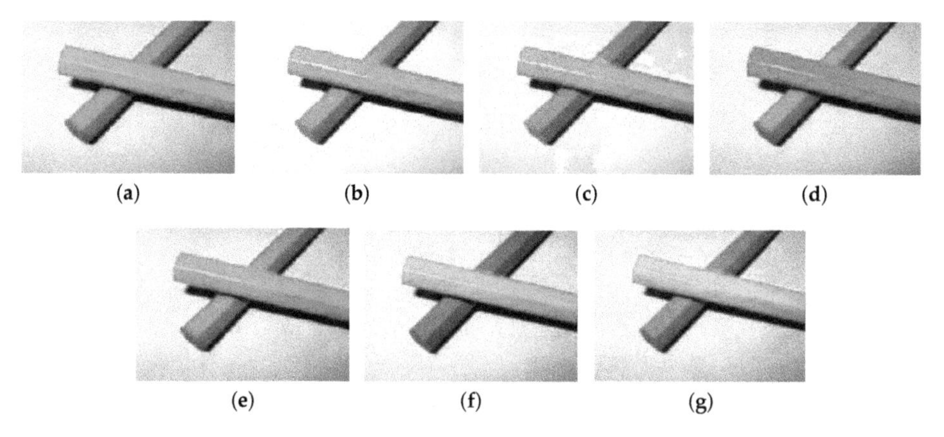

Figure 13. The comparison of tritanopia simulation of the pencil image. (**a**) The original image. (**b**) The CVD simulation using Machado's method. (**c**) Machado's CVD simulation on the image processed by the proposed re-coloring technique. (**d**) The CVD simulation using Brettel's method. (**e**) Brettel's CVD simulation on the image processed by the Kuhn's re-coloring technique. (**f**,**g**) CVD simulation and re-coloring using Huang's approach.

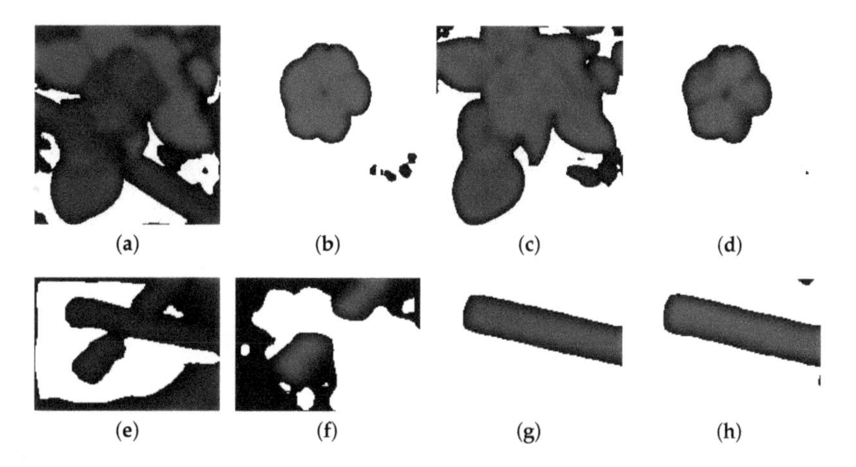

Figure 14. The comparison of RMS values between our method and Kuhn's method. (**a**) The RMS value between Figures 6a and 12a. (**b**) The RMS value between Figure 12b and 12a. (**c**) The RMS value between Figures 6a and 12c. (**d**) The RMS value between Figure 12d and 12c. (**e**) The RMS value between Figure 13a and 13b. (**f**) The RMS value between Figure 13c and 13b. (**g**) The RMS value between Figure 13a and 13d. (**h**) The RMS value between Figure 13e and 13d.

(a) CVD simulation. (b) Our re-coloring. (c) CVD simulation.

(d) CVD simulation. (e) Our re-coloring. (f) CVD simulation.

(g) CVD simulation. (h) Our re-coloring. (i) CVD simulation.

(j) CVD simulation. (k) Our re-coloring. (l) CVD simulation.

(m) CVD simulation. (n) Our re-coloring. (o) CVD simulation.

Figure 15. The results of CVD simulation, re-coloring using the proposed technique and CVD simulation on the re-colored images for some test images in Figure 10 (the first two columns) and 15 (the third column).

3.2. *Visual Difference Predictor*

HDR-VDP is a visual metric that compares a pair of images and predicts their visibility (the probability of the differences between the images) and quality (the quality degradation with the respect to the reference image). In this paper, Mantiuk et al.'s HDR-VDP-2 [23] is adopted to evaluate our re-coloring technique and Kuhn's method. HDR-VDP-2 is a major revision which improves the accuracy of the prediction and changes the metric to predict the visibility (detection/discrimination) and image quality (mean-opinion-score). The new metric also models Long-, Middle-, Short-cone and rod sensitivities for different spectral characteristics of the incoming light. As shown in Figure 16, the first and fourth rows are two test images and their CVD simulation. The images from the left to the right are the original images, CVD simulation results using our re-coloring technique and the results obtained from Kuhn's method. The second and fifth rows are the RMS values with $k = 11$. The higher value of RMS is displayed in a deeper blue color, and the low value is displayed in a white color. The third and sixth rows are the visibility test results using HDR-VDP-2. The probability of detection map tells us how likely we will notice the difference between two images. Red color denotes the high probability and green color indicates a low probability. Finally, as shown in the second and fifth rows, the distribution of our results and Kuhn's are almost the same. However, the third and sixth rows indicate that our re-coloring technique is able to provide more distinguishable colors on the CVD simulation results.

3.3. *Human Subjective Evaluation*

The color sensation is commonly considered as a subjective feeling of human beings. In this work, a human test is carried out to evaluate our color enhancement approach. The procedure consists of an image re-coloring stage to produce the CVD-friendly output, and an evaluation stage to analyze the responses collected from the volunteers. In the human test approved by an IRB (institutional review board) [38], people with color vision deficiency were asked to give judgments for the images enhanced by the re-coloring algorithms. We first let the subjects understand the purpose and process of this color test clearly. Three types of color vision deficiencies: protanopia, deuteranopia and tritanopia are considered, and the subjects are classified to groups for testing. Four different approaches, M_1, M_2, M_3, M_4 are then evaluated as follows.

- M_1: The input image is converted to the $L^*u^*v^*$ color space, projected to u^*v^* and equalized the u^* and v^* coordinates.
- M_2: The input image is used to simulate the CVD view, and find the (R, G, B) difference between input and simulation images. A matrix is then used to enhance the color difference regions.
- M_3: The input image is converted to the $L^*u^*v^*$ color space, and rotated to the non-confused color position.
- M_4: The input image is used to simulate the CVD view, and the distances among the colors are used to obtain the discrepancy. The image is then converted to the λ, *Y-B*, *R-G* color space, and rotated the color difference regions.

We collected 55 valid subjects in the test. The results are tabulated in Table 4. In the table, i is the method of different research stages, j is the index of the test image, the letters are the feeling level of the pros and cons (denoted by A, B, C, D) for the subjects. The summary indicates the proportion of the method M_i for the test image F_j chosen by subjects is over 1/3. As shown in Table 4, 83.64% (marked in blue) of 55 subjects selected level A for the method M_2 and the test image F_1. The numbers marked in red indicate the proportion of method M_i in levels A, B, C, D with higher percentages, and the associated methods are more representative in the level. Thus, each level (A, B, C, D) is represented by the methods: M_2, M_4, M_3, and M_1. It also shows that the best to the worst for color vision deficiency feeling of the four different methods are given by M_2, M_4, M_3, M_1.

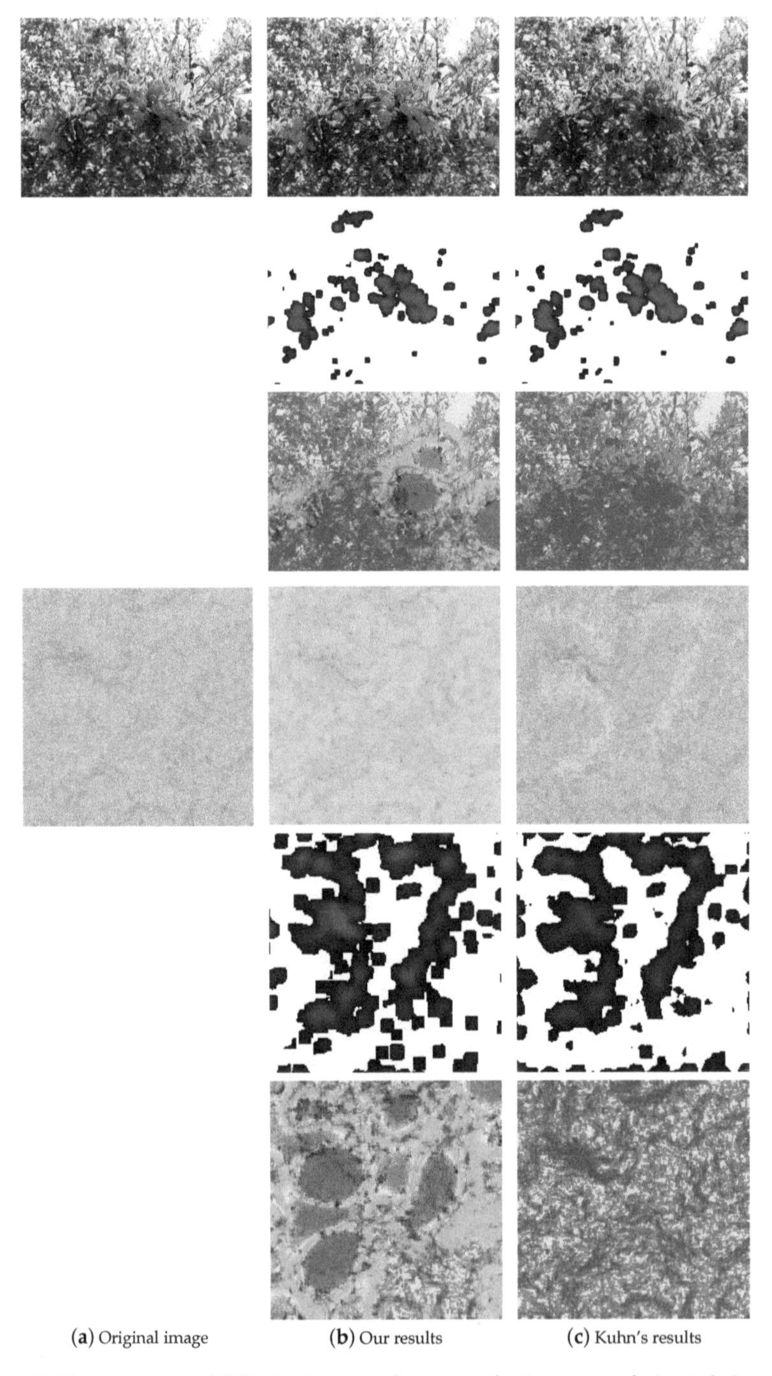

(a) Original image (b) Our results (c) Kuhn's results

Figure 16. The comparison of CVD simulation results processed using our re-coloring technique and Kuhn's method. The first and fourth rows are two test images and their CVD simulation results. The second and fifth rows are the visualized RMS values, and HDR-VDP evaluation is shown in the third and sixth rows.

Table 4. The human test on 55 valid subjects with four different methods. The number are shown in percentage. The numbers marked in red indicate the proportion of method M_i in levels A, B, C, D with higher percentages, and the associated methods are more representative in the level.

Level	A				B			
	M_1	M_2	M_3	M_4	M_1	M_2	M_3	M_4
F_1	5.45	83.64	5.45	5.45	45.45	9.09	9.09	36.36
F_2	5.45	90.91	0.00	3.64	3.64	5.45	21.82	69.09
F_3	1.82	72.73	14.55	10.90	20.00	3.64	25.45	50.91
F_4	1.82	94.55	1.82	1.82	9.09	1.82	25.45	63.64
e F_5	1.89	91.67	0.00	8.33	0.00	5.45	20.00	74.55
F_6	0.00	41.82	12.73	45.45	5.36	8.93	44.64	41.07
F_7	49.09	9.09	3.64	38.18	16.36	27.27	30.91	25.45
F_8	14.81	20.37	16.67	48.15	12.37	9.09	50.91	27.27
Summary	7.14	64.29	3.57	25.00	11.11	3.70	33.33	51.85
Level	**C**				**D**			
	M_1	M_2	M_3	M_4	M_1	M_2	M_3	M_4
F_1	9.09	5.45	54.55	30.91				
F_2	0.00	0.00	76.36	23.64	90.91	3.64	3.64	1.82
F_3	9.09	14.55	43.64	32.73	69.09	9.09	16.36	5.45
F_4	18.18	0.00	49.09	32.73	70.91	3.64	23.64	1.82
F_5	14.55	1.82	63.64	20.00	88.89	0.00	7.41	3.70
F_6	21.82	50.91	14.55	12.73	75.47	0.00	24.53	0.00
F_7	16.36	30.91	32.73	20.00	18.18	32.73	32.73	16.36
F_8	41.82	16.00	32.00	20.00	31.48	51.85	5.56	11.11
Summary	9.68	12.90	58.06	19.35	76.00	12.00	8.00	4.00

4. Conclusions

In this paper, we present an image enhancement approach to assist colorblind people with a better viewing experience. An image re-coloring method based on eigenvector processing is proposed for robust color separation under color deficiency transformation. It is shown that the eigenvector of color vision deficiency is distorted by an angle in the λ, *Y-B*, *R-G* color space. The proposed method represents clearly subjective image quality and the objective evaluation. Compared to the existing techniques, our results of natural images with CVD simulation work very well in terms of RMS, HDR-VDP-2 and IRB-approved human test. Both the objective comparison with previous works and the subjective evaluation on human tests validate the effectiveness of the proposed technique.

Author Contributions: H.-Y.L. proposed the idea, formulated the model, conducted the research and wrote the paper. L.-Q.C. developed the software programs, performed experiments and data analysis, and wrote the paper. M.-L.W. helped with the human test experiments.

Funding: The support of this work is in part by the Ministry of Science and Technology of Taiwan under Grant MOST 106-2221-E-194-004 and the Advanced Institute of Manufacturing with High-tech Innovations (AIM-HI) from The Featured Areas Research Center Program within the framework of the Higher Education Sprout Project by the Ministry of Education (MOE) in Taiwan.

Conflicts of Interest: The authors declare no conflict of interest.

References

1. Wong, B. Points of view: Color blindness. *Nat. Methods* **2011**, *8*, 441. [CrossRef] [PubMed]
2. Ishihara, S. *Ishihara's Tests for Color-Blindness*, 38th ed.; Kanehara, Shuppan: Tokyo, Japan, 1990.
3. Hunt, R. Colour Standards and Calculations. In *The Reproduction of Colour*; John Wiley and Sons, Ltd.: Hoboken, NJ, USA, 2005; pp. 92–125. [CrossRef]
4. Nathans, J.; Thomas, D.; Hogness, D.S. Molecular genetics of human color vision: The genes encoding blue, green, and red pigments. *Science* **1986**, *232*, 193–202. [CrossRef]

5. Michael, K.; Charles, L. Psychophysics of Vision: The Perception of Color. Available online: https://www.ncbi.nlm.nih.gov/books/NBK11538/ (accessed on 30 April 2019).
6. Colblindor Web Site. Available online: https://www.color-blindness.com/category/tools/ (accessed on 30 April 2019).
7. Neitz, M.; Neitz, J. Numbers and ratios of visual pigment genes for normal red-green color vision. *Science* **1995**, *267*, 1013–1016. [CrossRef]
8. Graham, C.; Hsia, Y. Color Defect and Color Theory Studies of normal and color-blind persons, including a subject color-blind in one eye but not in the other. *Science* **1958**, *127*, 675–682. [CrossRef]
9. Fairchild, M. *Color Appearance Models*; The Wiley-IS&T Series in Imaging Science and Technology; Wiley: London, UK, 2013.
10. Dolgin, E. Colour blindness corrected by gene therapy. *Nature* **2009**, *2*, 66–69. [CrossRef]
11. Hunt, R.W.G.; Pointer, M.R. *Measuring Colour*; John Wiley & Sons: Hoboken, NJ, USA, 2011.
12. Huang, J.B.; Wu, S.Y.; Chen, C.S. Enhancing Color Representation for the Color Vision Impaired. In Proceedings of the Workshop on Computer Vision Applications for the Visually Impaired, Marseille, France, 12–18 October 2008.
13. Huang, J.B.; Chen, C.S.; Jen, T.C.; Wang, S.J. Image recolorization for the colorblind. In Proceedings of the IEEE International Conference on Acoustics, Speech and Signal Processing, Taipei, Taiwan, 19–24 April 2009; pp. 1161–1164.
14. Huang, C.R.; Chiu, K.C.; Chen, C.S. Key Color Priority Based Image Recoloring for Dichromats. In *Advances in Multimedia Information Processing, Proceedings of the 11th Pacific Rim Conference on Multimedia, Shanghai, China, 21–24 September 2010*; Springer: Berlin/Heidelberg, Germany, 2010; pp. 637–647. [CrossRef]
15. Chen, Y.S.; Hsu, Y.C. Computer vision on a colour blindness plate. *Image Vis. Comput.* **1995**, *13*, 463–478. [CrossRef]
16. Rasche, K.; Geist, R.; Westall, J. Re-coloring Images for Gamuts of Lower Dimension. *Comput. Graph. Forum* **2005**, *24*, 423–432. [CrossRef]
17. Rasche, K.; Geist, R.; Westall, J. Detail preserving reproduction of color images for monochromats and dichromats. *IEEE Comput. Graph. Appl.* **2005**, *25*, 22–30. [CrossRef]
18. Lau, C.; Heidrich, W.; Mantiuk, R. Cluster-based color space optimizations. In Proceedings of the 2011 IEEE International Conference on Computer Vision (ICCV), Barcelona, Spain, 6–13 November 2011; pp. 1172–1179.
19. Lee, J.; Santos, W. An adaptative fuzzy-based system to evaluate color blindness. In Proceedings of the 17th International Conference on Systems, Signals and Image Processing (IWSSIP 2010), Rio de Janeiro, Brazil, 17–19 June 2010.
20. Poret, S.; Dony, R.; Gregori, S. Image processing for colour blindness correction. In Proceedings of the 2009 IEEE Toronto International Conference Science and Technology for Humanity (TIC-STH), Toronto, ON, Canada, 26–27 September 2009; pp. 539–544.
21. CIE Web Site. Available online: http://cie.co.at/ (accessed on 30 April 2019).
22. Wright, W.D. Color Science, Concepts and Methods. Quantitative Data and Formulas. *Phys. Bull.* **1967**, *18*, 353. [CrossRef]
23. Mantiuk, R.; Kim, K.J.; Rempel, A.G.; Heidrich, W. HDR-VDP-2: A Calibrated Visual Metric for Visibility and Quality Predictions in All Luminance Conditions. *ACM Trans. Graph.* **2011**, *30*, 40:1–40:14. [CrossRef]
24. Moroney, N.; Fairchild, M.D.; Hunt, R.W.; Li, C.; Luo, M.R.; Newman, T. The CIECAM02 Color Appearance Model. *Color Imaging Conf.* **2002**, *2002*, 23–27.
25. Brettel, H.; Viénot, F.; Mollon, J.D. Computerized simulation of color appearance for dichromats. *J. Opt. Soc. Am. A* **1997**, *14*, 2647–2655. [CrossRef]
26. Wild, F. Outline of a Computational Theory of Human Vision. In Proceedings of the KI 2005 Workshop 7 Mixed-Reality as a Challenge to Image Understanding and Artificial Intelligence, Koblenz, Germany, 11 September 2005; p. 55.
27. Busin, L.; Vandenbroucke, N.; Macaire, L. Color spaces and image segmentation. *Adv. Imaging Electron Phys.* **2008**, *151*, 65–168.
28. Vrhel, M.; Saber, E.; Trussell, H. Color image generation and display technologies. *IEEE Signal Process. Mag.* **2005**, *22*, 23–33. [CrossRef]
29. Sharma, G.; Trussell, H. Digital color imaging. *IEEE Trans. Image Process.* **1997**, *6*, 901–932. [CrossRef] [PubMed]

30. Marguier, J.; Süsstrunk, S. Color matching functions for a perceptually uniform RGB space. In Proceedings of the ISCC/CIE Expert Symposium, Ottawa, ON, Canada, 16–17 May 2006.
31. Huang, J.B.; Tseng, Y.C.; Wu, S.I.; Wang, S.J. Information preserving color transformation for protanopia and deuteranopia. *IEEE Signal Process. Lett.* **2007**, *14*, 711–714. [CrossRef]
32. Ballard, D.H.; Brown, C.M. *Computer Vision*; Prentice Hall: Upper Saddle River, NJ, USA, 1982.
33. Swain, M.J.; Ballard, D.H. Color indexing. *Int. J. Comput. Vis.* **1991**, *7*, 11–32. [CrossRef]
34. Ingling, C.R.; Tsou, B.H.P. Orthogonal combination of the three visual channels. *Vis. Res.* **1977**, *17*, 1075–1082. [CrossRef]
35. Machado, G.M.; Oliveira, M.M.; Fernandes, L.A. A physiologically-based model for simulation of color vision deficiency. *IEEE Trans. Vis. Comput. Graph.* **2009**, *15*, 1291–1298. [CrossRef]
36. Smith, V.C.; Pokorny, J. Spectral sensitivity of the foveal cone photopigments between 400 and 500 nm. *Vis. Res.* **1975**, *15*, 161–171. [CrossRef]
37. Kuhn, G.R.; Oliveira, M.M.; Fernandes, L.A. An efficient naturalness-preserving image-recoloring method for dichromats. *IEEE Trans. Vis. Comput. Graph.* **2008**, *14*, 1747–1754. [CrossRef] [PubMed]
38. Wikipedia. Institutional Review Board—Wikipedia. The Free Encyclopedia. Available online: http://en.wikipedia.org/wiki/Institutional_review_board (accessed on 1 July 2013).

Article

Continuous Distant Measurement of the User's Heart Rate in Human-Computer Interaction Applications

Jaromir Przybyło

AGH University of Science and Technology, 30 Mickiewicza Ave., 30-059 Krakow, Poland; przybylo@agh.edu.pl

Received: 9 August 2019; Accepted: 25 September 2019; Published: 27 September 2019

Abstract: In real world scenarios, the task of estimating heart rate (HR) using video plethysmography (VPG) methods is difficult because many factors could contaminate the pulse signal (i.e., a subjects' movement, illumination changes). This article presents the evaluation of a VPG system designed for continuous monitoring of the user's heart rate during typical human-computer interaction scenarios. The impact of human activities while working at the computer (i.e., reading and writing text, playing a game) on the accuracy of HR VPG measurements was examined. Three commonly used signal extraction methods were evaluated: green (G), green-red difference (GRD), blind source separation (ICA). A new method based on an excess green (ExG) image representation was proposed. Three algorithms for estimating pulse rate were used: power spectral density (PSD), autoregressive modeling (AR) and time domain analysis (TIME). In summary, depending on the scenario being studied, different combinations of signal extraction methods and the pulse estimation algorithm ensure optimal heart rate detection results. The best results were obtained for the ICA method: average *RMSE* = 6.1 bpm (beats per minute). The proposed ExG signal representation outperforms other methods except ICA (*RMSE* = 11.2 bpm compared to 14.4 bpm for G and 13.0 bmp for GRD). ExG also is the best method in terms of proposed success rate metric (*sRate*).

Keywords: video pletysmography; image processing; heart rate estimation; human-computer interaction; biomedicine; healthcare; assisted living

---

1. Introduction

Photopletysmography (PPG) is a non-invasive, low-cost optical technique used to detect volumetric changes in blood in the peripheral circulation. It has many medical applications, including clinical physiological monitoring: blood oxygen saturation and heart rate (HR) [1], respiration [2]; vascular assessment: arterial disease [3], arterial ageing [4], venous assessment [5], microvascular blood flow and tissue viability [6]; autonomic function: blood pressure and heart rate variability [7], neurology [8], and telehealth applications [9].

The PPG sensor has to be applied directly to the skin, which limits its practicality in situations such as freedom of movement is required [10]. Among the various contactless methods for measuring cardiovascular parameters [11], video plethysmography (VPG) have recently become popular. One of the first approaches was proposed by Verkruysse et al. [12], who showed that plethysmographic signals can be remotely measured from a human face in normal ambient light using a simple digital, consumer level photo camera. The advantages of this approach, compared to standard photopletysmography (PPG) techniques, are that it does not require uncomfortable wearable accessories and allows easy adaptation to different requirements in various applications, such as: monitoring the driver's vital signs in the automotive industry [13], optimization of training in sport [14] and emotional communication in the field of human-machine interaction [15].

Since then, there has been a rapid development of literature on VPG techniques. A summary of 69 studies related to VPG can be found in [16]. Poh et.al [17,18] introduced a new methodology for non-contact, automatic and motion tolerant cardiac pulse measurements from video images based on blind source separation. They used a basic webcam embedded in a laptop to record videos for analysis. To detect faces in video frames and locate the region of interest (ROI) for each video frame, an automatic face detection algorithm was used.

In [19], the authors proposed a framework that uses face tracking to solve the problem of rigid head movements and use the green background value as a reference to reduce the interference from illumination changes. To reduce the impact of sudden non-rigid facial movements, noisy signal segments are excluded from the analysis. Also, several temporal filters were used to reduce the slow and non-stationary trend of the HR signal.

A complementary method for extracting heart rate from video by analyzing subtle skin color changes due to blood circulation has been proposed in [20]. This algorithm is based on the measurement of subtle head movement caused by Newtonian reaction to the influx of blood inflow with each beat. Thus, the method is effective even when the skin is not visible. A typical procedure for extracting a HR signal from a video frame sequence consists of the following stages [21]: selection and tracking of the region of interest (ROI), pre-processing, extraction and post-processing of the VPG signal, pulse rate estimation. Many different published articles present various improvements of one or several stages. For example, in [22] the author proposed using a new signal extraction method: green-red-difference (GRD) as a robust alternative to G. However, a large proportion of them presents the results of tests carried out under controlled conditions (i.e., lighting, short term monitoring, limited or not natural person movements).

In realistic situations, the task of estimating HR is difficult because many factors can contaminate the pulse signal. For example, the movement of a subject consists of a combination of rigid (head tilts, change of position) and non-rigid movements (facial actions, eye blinking). This can affect pixel values of the face region. Fluctuations in lighting caused by changes in the environment include various forms of noise, such as the blinking of indoor lights or computer screen, a flash of reflected light, and the internal noise of a digital camera.

In this article, we propose a video pulse measurement system designed for continuous monitoring of the user's heart rate (HR) during typical human-computer interaction (HCI) scenarios, i.e., working at the computer. Since physiological activities and changes are a direct reflection of processes in the central and autonomic nervous systems, these signals can be used in an affective computing scenarios (i.e., recognition of human emotions), Assisted Living or healthcare applications (contactless monitoring of cardiovascular parameters). The contribution of this article is following:

- To our knowledge we are the first to systematically study the impact of human activities during various HCI scenarios (i.e., reading text, playing games) on the accuracy of the HR algorithm,
- As far as we know, we are the first to propose the use of new image representation (excess green ExG), which provides acceptable accuracy and at the same time is much faster to compute than other state of the art methods (i.e., blind-source separation—ICA),
- We used the state-of-the art real-time face detection and tracking algorithm, and evaluated four signal extraction methods (preprocessing), and three different pulse rate estimation algorithms,
- To our knowledge we are the first to propose a method of correcting information delay introduced by the algorithm when comparing results with reference data.

The article has the following structure: the next Section 2 describes the experimental setup as well as the algorithmic details. The results and discussion are presented in Section 3, the paper is summarized in Section 4.

2. Materials and Methods

The primary goal of this research was to check the effectiveness of the HR algorithm during typical human-computer interaction (HCI) scenarios. Thus, we evaluated four signal extraction methods and implemented three different HR estimation algorithms. We evaluated the effectiveness of selected algorithms using recorded video sequences of participants performing various HCI tasks. The implementation of the proposed methods can be easily adapted to running in real-time framework, however implementation details are not included in this paper.

2.1. Experimental Setup

An experimental setup consisted of a RealSense™ SR300 camera (model Creative BlasterX Senz3D, Intel, Santa Clara, CA, USA) that can provide RGB video streams with the following parameters: resolution up to 1280 × 720 pixels at 60 FPS (frames per second). To focus on assessing the impact of noise factors on the results of HR detection, we used RGB channel with a resolution of 640 × 480 pixels and a frame rate of 60 FPS. The camera was located 0.5 to 0.6 m from the volunteers (depending on the experiment).

Various extrinsic factors affect the reliability of VPG HR measurement [23]. One of the factors is change in lighting conditions. This factor requires special attention when the user works with a computer exposed to variable illumination caused by the content displayed on the monitor. Another factor that can affect the accuracy of the HR measurement is the sudden user's movements, caused for example by emotions while playing computer games. To estimate the impact of these factors on remote HR measurements, we recorded additional signals using a SimpleLink™ multi-standard SensorTag CC2650 (Texas Instruments, Dallas, TX, USA). It is a low energy Bluetooth device, that includes 10 low-power MEMS sensors of which we used ambient light and motion tracking sensors. The SensorTag was placed on the chest of the subject near the neck and face. To measure the ground truth HR, we used the ECG-based H7 Heart Rate Sensor (Polar Electro OY, Kempele, Finland) connected via Bluetooth).

2.2. Region of Interest (ROI) Selection and Tracking

There are many sources of changes in the appearance of the face. They can be categorized [24] into two groups—intrinsic factors related to the physical nature of the face (identity, age, sex, facial expression) and extrinsic factors resulting from the scene and lightning conditions (illumination, viewing geometry, imaging process, occlusion, shading). All these factors make face detection and recognition a difficult task. Therefore, in recent years there have been many approaches to detecting faces in natural conditions. Surveys of those methods are presented in articles [25,26].

A fast and reliable implementation of the face detection algorithm can be found in Dlib C++ library [27]. It is based on Histograms of Oriented Gradients (HoG) algorithm proposed in [28], combined with Max-Margin Object Detection (MMOD) [29] which produces high quality detectors from relatively small amounts of training data.

In present work, we combined the Dlib's frontal face detector with the KLT tracking algorithm [30] to effectively follow faces in a video sequence. The outline of the algorithm is presented in Figure 1. The face detector implemented in Dlib library appears to be faster and more robust than the Viola-Jones detector [31]. It shows a low ratio of false positive results, which is essential assumption of our system. However, one of the limitations of this implementation is that the face model was trained using frontal images with the face size at least of 80 × 80 pixels. This means that finding smaller faces requires up-sampling the image (which increases processing time) or re-training the model. Detection of non-frontal faces also requires a different model.

The face detector is applied in each of the consecutive image frames. The resulting bounding box is then used by a heart rate estimation algorithm. If the face is not detected by the Dlib detector, the KLT tracking algorithm is used to track a set of feature points from the previous frame and estimate correct bounding box on the current frame. A feature points (corners) are detected inside the face rectangle using the minimum eigenvalue algorithm [32]. The use of a tracking algorithm minimizes the impact of rigid head movements typical in human-computer interaction scenarios. In case of the Dlib detector fails to detect a face, the system automatically switches to the Viola-Jones detector for a single frame. This allows to correctly reinitialize the tracker.

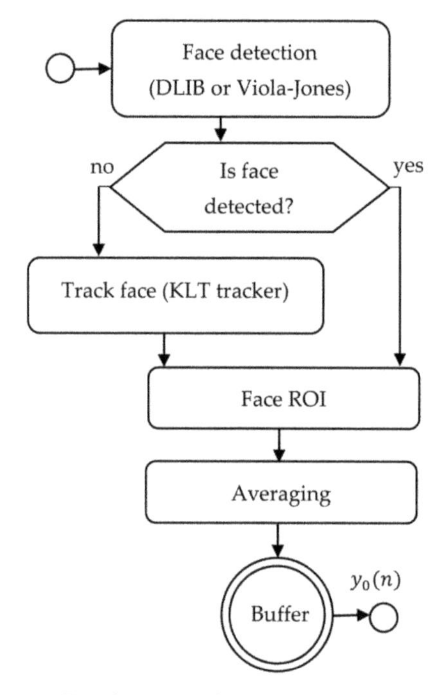

Figure 1. Face detection and tracking—algorithm outline.

The calculated bounding box can include not only skin-color pixels (where the pulse signal is expected), but also objects outside the face. To exclude these regions from the HR estimation, a facial landmark detector [33] is used on the cropped part of the image. Based on detected landmark points, a proper region of interest (ROI) is selected for further analysis (Figure 2).

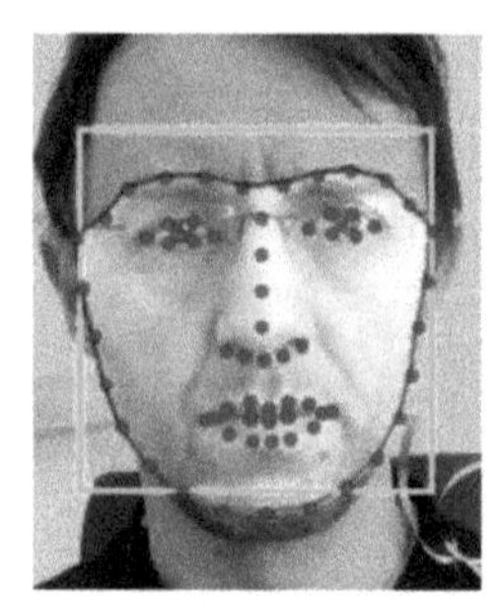

Figure 2. Example image frame with the region of interest (ROI) superimposed.

2.3. *Preprocessing and VPG Signal Extraction*

The selected region of interest is then used to calculate the average color intensities over the ROI for each subsequent image frame. These values are stored in a circular buffer of length N, forming the raw VPG signal $\mathbf{y_0}(n) = [\,R_0(n), G_0(n), B_0(n)]^T$. Then the raw VPG signal is detrended using a simple method consisting of mean-centering and scaling [21] (Equation (1)):

$$\mathbf{y}(n) = \frac{\mathbf{y_0}(n) - \boldsymbol{\mu}(n, L)}{\boldsymbol{\mu}(n, L)} \tag{1}$$

where $\mu(n, L)$ is an L-point running mean vector of VPG signal and $y(n) = [\,R(n), G(n), B(n)]^T$.

The strongest VPG signal can be observed in the green (G) channel. Because the camera's RGB color sensors pick up a mixture of reflected VPG signal along with other sources of fluctuations, such as motion and changes in ambient lighting conditions, various approaches to overcome this problem have been reported in the literature. In [22] a robust alternative to G method has been presented—green-red difference (GRD) which minimizes the impact of artifacts (Equation (2)):

$$GRD(n) = G(n) - R(n) \tag{2}$$

Some authors utilize the fact that each color sensor registers a mixture of original source signals with slightly different weights and uses the independent component analysis (ICA) [17,34]. The ICA model assumes that the observed signals $y(n)$ are linear mixtures of sources $\mathbf{s}(n)$. The aim of ICA is to find the separation matrix $\boldsymbol{W}$ whose output (Equation (3):

$$\hat{\mathbf{s}}(n) = \boldsymbol{W} \cdot \boldsymbol{y}(n) \tag{3}$$

is an estimate of the vector $s(n)$ containing the underlying source signals. The order in which ICA returns the independent components is random. Thus, the component whose power spectrum contained the highest peak can be selected for further analysis. In this work, we used FastICA implementation [35] and calculated power spectrum in the range 35–180 bpm (which corresponds to 0.583–3.00 Hz).

In our research, we found that method for greenness identification [36] utilizing the excess green image component (ExG), amplify the pulse signal and it is faster to compute than the ICA while reducing the impact of noise. The ExG image representation is computed as follows. First, the normalized components r, g and b are calculated using Equation (4):

$$r(n) = \frac{R(n)}{R(n) + G(n) + B(n)} \quad g(n) = \frac{G(n)}{R(n) + G(n) + B(n)} \quad b(n) = \frac{B(n)}{R(n) + G(n) + B(n)} \tag{4}$$

The excess green component ExG is defined by Equation (5):

$$ExG(n) = 2 \cdot g(n) - r(n) - b(n) \tag{5}$$

The refined VPG signal (G, GRD, ICA or ExG) is then band-limited by a zero-phase digital filter (Bartlet-Hamming) yielding the signal VPG(n). The summary of the pre-processing, VPG signal extraction and heart rate estimation steps is provided in Figure 3.

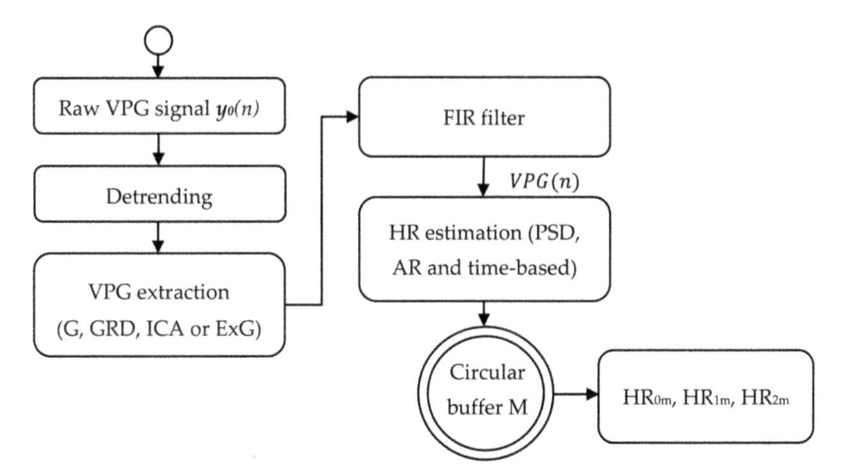

Figure 3. HR estimation algorithm outline.

2.4. Heart Rate Estimation Algorithm

To estimate the heart rate we used three different algorithms. The first algorithm was based on the calculation of the power spectral density (PSD) estimate of the signal VPG(n), using the Welch algorithm and the filter bank approach. To find the pulse frequency, the highest frequency peak was located in the PSD, as a result of which the heart rate was estimated (named as HR_0 in this paper). An important aspect of this classic frequency-based approach is that the frequency resolution *fres* depends on the length of the signal buffer (Equation (6)):

$$f_{res} = \frac{F_s}{N} \tag{6}$$

where: N is the length of the signal observation and Fs is the sampling frequency (frame rate of the video).

We also used a second algorithm based on autoregressive (AR) modelling. In the AR model, the input signal can be expressed by Equation (7):

$$y'(n) = -\sum_{k=1}^{p} a_k \cdot y'(n-k) + e(n) \tag{7}$$

where: p is the model order, a_k are the model coefficients, and $e(n)$ is the white noise.

Using the Yule-Walker method we fit the AR model to the input signal VPG(n) and obtain an estimate of the AR system parameters *ak*. Then, the frequency response of this filter was used to calculate the pulse rate (named as HR_1 in this paper). The HR_1 value was estimated by detecting the highest frequency peak in the filter frequency responsein the selected range (50–180 bpm).

The third approach was time-based (depicted as TIME in the article). On the filtered signal VPG(n), peaks were located using only the peak detection algorithm. Then the intervals between successive peaks were calculated and their median value was used to obtain the heart rate value (HR_2).

To minimize false detections, caused by head movements and other sources of image variations, the estimated HR has been further post-processed. A second heart rate buffer of length M was used to store the latest HR_0, HR_1 and HR_2 values. Then the average value of each HR buffer content was calculated and used as a new estimate of the current heart rate (named as HR_{0m}, HR_{1m} and HR_{2m} respectively).

2.5. Evaluation Methodology

Different kinds of metrics were proposed in other articles for evaluating the accuracy of HR (heart rate) measurement methods. The most common is the root mean squared error denoted as *RMSE* (Equation (8)):

$$\text{RMSE} = \sqrt{\frac{1}{n}\sum\nolimits_{i=1}^{n}[HR_{error}[i]]^2} \tag{8}$$

$$HR_{error} = HR_{video} - HR_{gt} \tag{9}$$

where: HR_{video}– the HR estimated from video, HR_{gt}—the ground truth HR values.

Because *RMSE* is sensitive to extreme values or outliers, we additionally propose using a metric that allows to assess how long the accuracy of a given algorithm is within the assumed error tolerance (Equation (10)). This is particularly important in medical applications where measurement reliability is important:

$$\text{sRate} = \frac{100}{n} \cdot \sum_{i=1}^{n} (|HR_{error}|i|| \leq tolerance) \tag{10}$$

Little or no attention has been given in literature regarding the effect of information delay introduced by the algorithm on the error metrics. Assuming that the algorithm introduces a delay

t_0 and the measured ground truth HR values are also delayed by t_1 (due to acquisition and device measurement method), HR_{error} is biased. Therefore, direct comparison of HR values using HR_{error} is not accurate (a systematic error is introduced). In addition, HR_{video} and HR_{gt} usually are sampled at different frequencies. For example, our camera sampling frequency was 60 FPS and the Polar H7 heart rate sensor provides measurements every approximately three seconds.

To minimize the impact of delays and different sampling frequencies on the results of the HR comparison, we propose the following method. First, HR_{gt} values are interpolated to match the sampling frequency of HR_{video} using simple linear interpolation, resulting in HR_{gt2}. An example of ground truth and measured HR time series is given in Figure 4. All results are available online at [37].

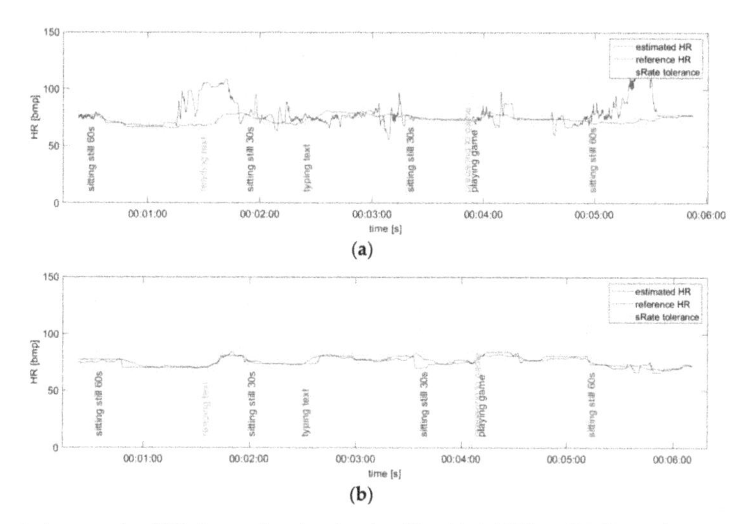

Figure 4. An example of HR time-series plots for algorithm No.1 (PSD) and ExG signal representation: (**a**) video No.5; (**b**) video No.9.

The delay introduced by the algorithm was estimated using the generated artificial signal of known frequency and time of change. Here, we used a signal that changes from 80 to 120 bpm and has a similar amplitude as VPG(n). The resulting delays t_0 do not include the delay t_1 introduced by the Polar H7 device. We have adopted a constant delay introduced by the measuring device. Assuming that the delay introduced by the algorithm is constant for a given algorithm and its parameters, the estimated delay t_2 can be used to correctly evaluate the remaining sequences. Although this is a strong assumption, it improves the accuracy of the results. An estimation of the algorithm delay can also be performed using cross-correlation. However, this analysis is not included in the article, because the estimated delays strongly depended on the shape of the signal and the selected fragment. The results are summarized in Table 1.

Table 1. Results of the delay estimation for selected algorithms.

Algorithm	t_0 [s]	t_1 [s]	$t_2 = t_0 - t_1$[s]
No.1 (PSD)	13.4	3	10.4
No.2 (AR)	6.6	3	3.6
No.3 (TIME)	5.1	3	2.1

It is also worth mentioning that delay correction is useful for correctly positioning the beginnings of individual parts of the experiment. For example—the impact of a user's head movements may be visible only after some time (equal to the algorithm delay) on the estimated pulse signal.

2.6. Details of Experiments

The algorithm parameters have been set to:

- Algorithm No. 1 (PSD, Welch's estimator): the window length N = 1024 samples (which gives a frequency resolution of 3.52 bpm/bin and temporal buffer window of length 21 s),
- Algorithm No. 2 (AR modelling): the order of AR model was equal to 128, the AR model frequency response computed for FFT length of 1024, the window length of N = 600 samples (which gives a frequency resolution of 3.52 bpm/bin and temporal buffer window of length 10 s),
- Algorithm No. 3 (time-based peak detection, depicted as TIME): the buffer length N = 600 samples (which gives a temporal buffer window of length 10 s).

Common parameters for all algorithms were: the bandpass filter of order = 128 and bandwidth = (35–180) bpm (which is equivalent to 0.583–3.00 Hz), the HR postprocessing buffer length M was equivalent to 1 s.

Several video sequences of participants performing HCI tasks were recorded using lossless compression (Huffman codec) and 24-bits-per-pixel format (RGB stream), image resolution of 640 × 480 pixels and frame rate of 60 FPS. Each sequence was approximately 5 min long. The RealSense camera was positioned in such a way that the face of the monitored participant was in the frontal position. All participants were asked to perform various tasks reflecting typical user-computer interaction scenarios. Thus, each video sequence consists of the following parts:

- Part 1—the participant sits still (60 s) without head movements and minimal facial actions,
- Part 2—the participant reads text (short jokes) displayed on the computer screen in front of him, and can express emotions,
- Part 3—the participant sits still (30 s),
- Part 4—the participant rewrites text from the paper located on the left or right side of the desk using the keyboard (which results in head movements),
- Part 5—the participant sits still (30 s),
- Part 6—after the short mental preparation the participant plays the arkanoid game using the mouse and the keyboard,
- Part 7—the participant sits still (60 s).

Only selected parts (1, 2, 4 and 6) were included in the study. The video sequences were recorded in different places and under different conditions (illumination, distance, and if possible similar camera parameters). A description of these videos is provided in Table 2. Examples of video frames are shown in Figure 5. Duration, average illumination values and standard deviation of accelerations for sequences are given in Tables A1 and A2 (Appendix A).

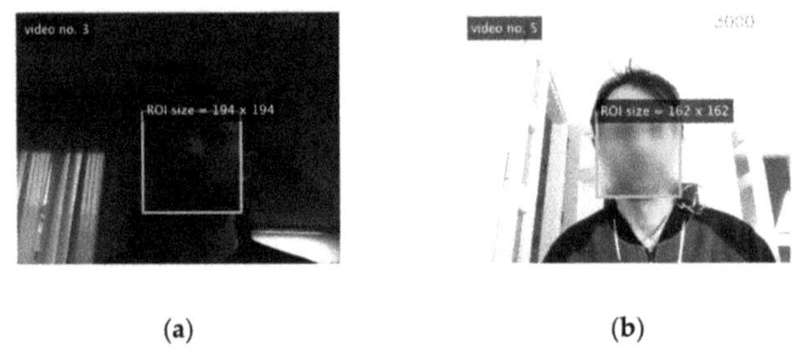

(a) (b)

Figure 5. An example of video frames: (**a**) video No.5; (**b**) video No.9.

Table 2. Recorded video sequences covered by the study.

Video No.	Room Settings	Participant's Details	Camera Parameters
1	room 1: artificial ceiling fluorescent light + natural light (dusk, medium lighting) from a one window on the left side + light from the one computer screen	participant 1: male, ~34 years old	camera-to-face distance ~50 cm, gain = 128, white balance off
2	room 1: artificial ceiling fluorescent light + natural light (dusk, medium lighting) from a one window on the left side + light from the one computer screen	participant 2: male, ~22 years old	camera-to-face distance ~50 cm, gain = 128, white balance off
3	room 2: daylight (cloudy, poor lighting): a one roof window on the left, and a second window in the back on the right + fluorescent lamps in the back (2 m) + ceiling fluorescent lamps + right-side table lamp + light from two computer screens	participant 3: male, ~44 years old	camera-to-face distance ~50 cm, gain = 128, white balance off
4	room 2: daylight (cloudy, medium lighting): a one roof window on the left, and a second window in the back on the right + fluorescent lamps in the back (2 m) + ceiling fluorescent lamps + light from two computer screens	participant 3: male, ~44 years old	camera-to-face distance ~50 cm, gain = 128, white balance on
5	room 3: daylight (sunny, strong lighting): a one window in the front + light from the one computer screen;	participant 3: male, ~44 years old	camera-to-face distance ~60 cm (computer screen slightly lower – user has to gaze slightly downwards), gain = 100, white balance on
6	room 4: nighttime, artificial light only (ceiling lamps, table lamps, led curtain lamps + light from the one computer screen);	participant 3: male, ~44 years old	camera-to-face distance ~50 cm (computer screen slightly lower – user has to gaze slightly downwards), gain = 128, white balance on
7	room 3: daylight (cloudy, medium lighting): a one window in the front + light from the one computer screen;	participant 4: female, ~42 years old	camera-to-face distance ~60 cm (computer screen slightly lower – user has to gaze slightly downwards), gain = 128, white balance on
8	room 2: daylight (cloudy, poor lighting): a one roof window on the left, and a second window in the back on the right + fluorescent lamps in the back (2 m) + light from two computer screens;	participant 3: male, ~44 years old	camera-to-face distance ~50 cm, gain = 100, white balance off
9	room 5: artificial ceiling fluorescent light + natural light (dusk, medium lighting) from a one window on the right side + right side bulb lamp + light from the one computer screen;	participant 5: male, ~23 years old	camera-to-face distance ~60 cm, gain = 128, white balance on

3. Result

3.1. Comparison of the VPG Signal Extraction Methods (G, GRD, ICA and ExG)

To select the appropriate statistical methods to compare the results, a Shapiro-Wilk parametric hypothesis test of composite normality can be used. However, with a small sample size (9 videos), the impact of outliers can be significant. Therefore, median and IQR were used as statistical measures.

Tables A3–A5 (Appendix A) show the results of HR estimation for various signal extraction methods and selected algorithms. The results were calculated for entire video sequences (including all participant activities). The *sRate* value is given for a threshold of 3.52 bpm (equal to the algorithm frequency resolution). Box plots (Figures 6–8) are also included to better illustrate *sRate* and *RMSE* distributions.

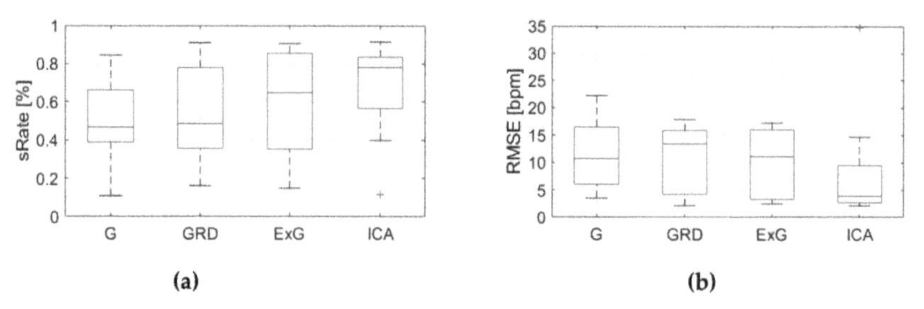

Figure 6. Comparison of signal extraction methods, algorithm No.1 (PSD): (**a**) box plots for *sRate*; (**b**) box plots for *RMSE*. Blue lines—IQR range, red line—median value.

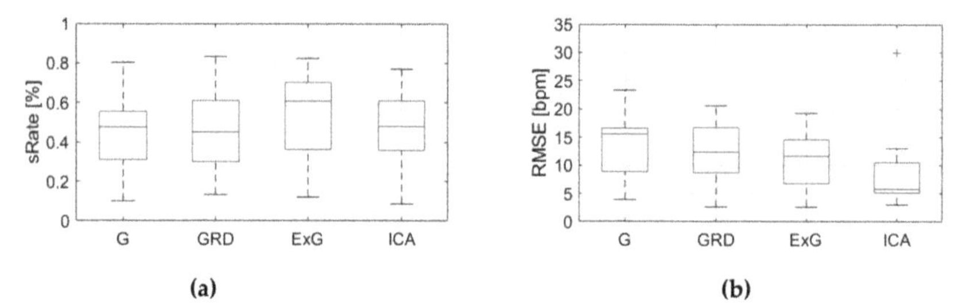

Figure 7. Comparison of signal extraction methods, algorithm No.2 (AR): (**a**) box plots for *sRate*; (**b**) box plots for *RMSE*. Blue lines—IQR range, red line—median value.

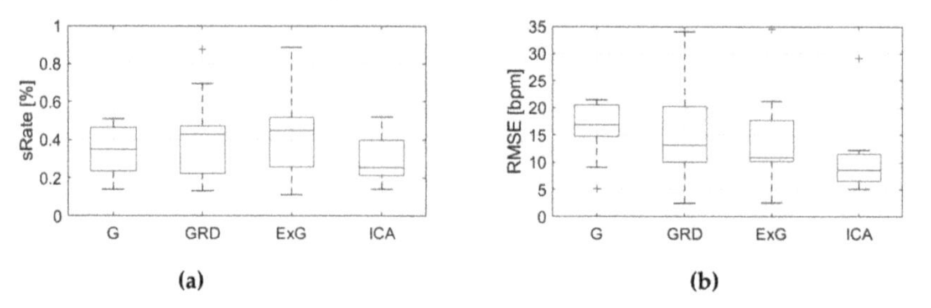

Figure 8. Comparison of signal extraction methods, algorithm No.3 (TIME): (**a**) box plots for *sRate*; (**b**) box plots for *RMSE*. Blue lines—IQR range, red line—median value.

Considering algorithm No. 1 (PSD), the lowest median *RMSE* with low interquartile range (IQR) value is for the ICA signal extraction method. The second lowest *RMSE* values relate to the G and ExG representations. The worst results are for the video No. 9. However, this video was recorded under artificial lighting conditions with lights visible in the scene, which could have a negative effect on the results. Also, the actual heart rate was low (about 50 bpm), which is close to the limit of the measured range (results below 50 bpm are considered incorrect). The *sRate* measure shows similar results—it is the highest for ICA signal extraction method. The ExG method has the highest IQR values.

Looking at the algorithm No. 2 (AR), and *RMSE* - the results are similar to the PSD algorithm. However, all IQR values are lower, which means that this algorithm gives more similar outcome for videos acquired under different conditions. As for *sRate*, the highest value is for ExG signal extraction method but with a large IQR. Given algorithm No. 3 (TIME), the lowest median *RMSE* value with a small interquartile range (IQR) value is for ICA, followed by ExG signal extraction method. All errors are higher for this algorithm than for PSD and AR. The *sRate* is the highest for ExG and then GRD. However, the lowest *sRate* IQR values relate to the ICA and G signal representation.

To compare the medians between groups (signal extraction methods) for statistical differences, a two-sided Wilcoxon rank sum test was used. The Wilcoxon rank sum test is a nonparametric test for the equality of population medians of two independent samples. It is used when the outcome is not normally distributed and the samples are small. The results are shown in Table A6 (Appendix A). The p-values of almost all combinations of signal extraction methods indicate that there is not enough evidence to reject the null hypothesis of equal medians at a default significance level of 5%. This means that all methods provide similar results statistically. The exception is the comparison of G and ICA for algorithm No. 3 (TIME), but only for the *RMSE* metric.

3.2. Comparison of the VPG Signal Extraction Methods for Various Activities

To see how individual activities affect the results of heart rate detection, the *RMSE* and *sRate* values of the following video parts have been compared:

- part 1 (the participant sits still for a minimum of 60 seconds),
- part 2 (the participant reads text),
- part 4 (the participant rewrites text using the keyboard and the mouse),
- and part 6 (the participant plays a game).

Because, *RMSE* and *sRate* can be regarded as a small sample size (nine videos) and the effect of outliers can be significant, the median and IQR were used as statistical measures. Figures 9–20 show the results of the HR estimation and comparison of the signal extraction methods and selected algorithms for selected parts.

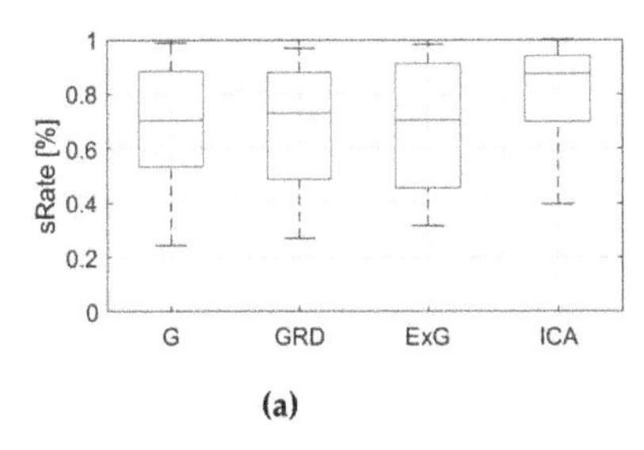

(a)

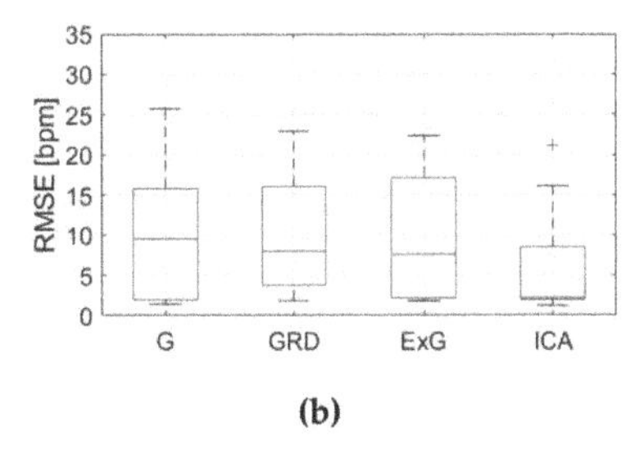

(b)

Figure 9. Comparison of signal extraction methods, algorithm No.1 (PSD), part 1: (**a**) box plots for *sRate*; (**b**) box plots for *RMSE*. Blue lines—IQR range, red line—median value.

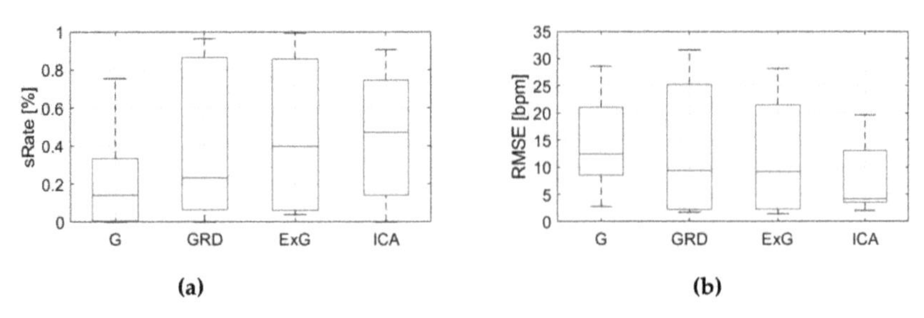

Figure 10. Comparison of signal extraction methods, algorithm No.1 (PSD), part 2: (**a**) box plots for *sRate*; (**b**) box plots for *RMSE*. Blue lines—IQR range, red line—median value.

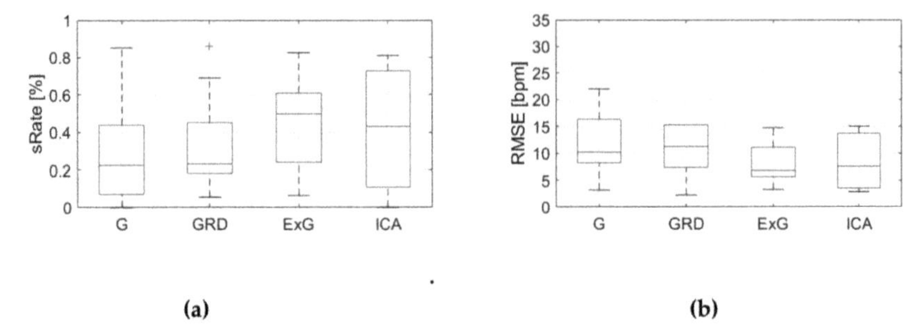

Figure 11. Comparison of signal extraction methods, algorithm No.1 (PSD), part 4: (**a**) box plots for *sRate*; (**b**) box plots for *RMSE*. Blue lines—IQR range, red line—median value.

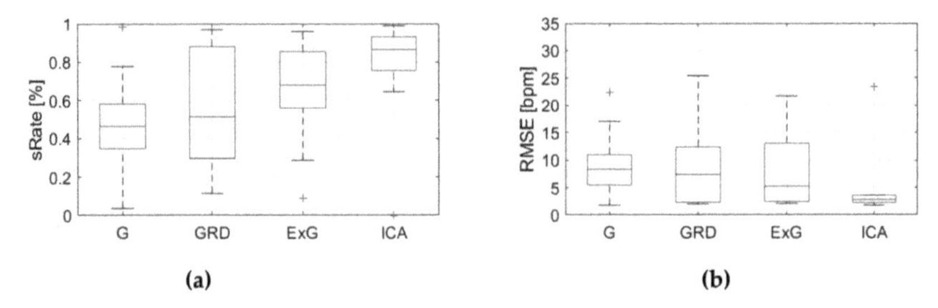

Figure 12. Comparison of signal extraction methods, algorithm No.1 (PSD), part 6: (**a**) box plots for *sRate*; (**b**) box plots for *RMSE*. Blue lines—IQR range, red line—median value.

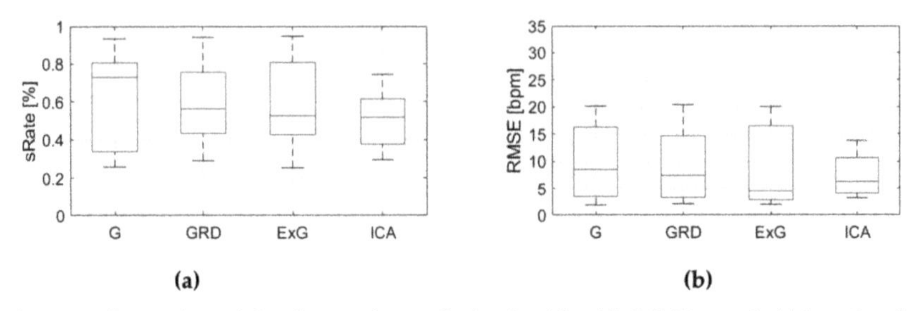

Figure 13. Comparison of signal extraction methods, algorithm No.2 (AR), part 1: (**a**) box plots for *sRate*; (**b**) box plots for *RMSE*. Blue lines—IQR range, red line—median value.

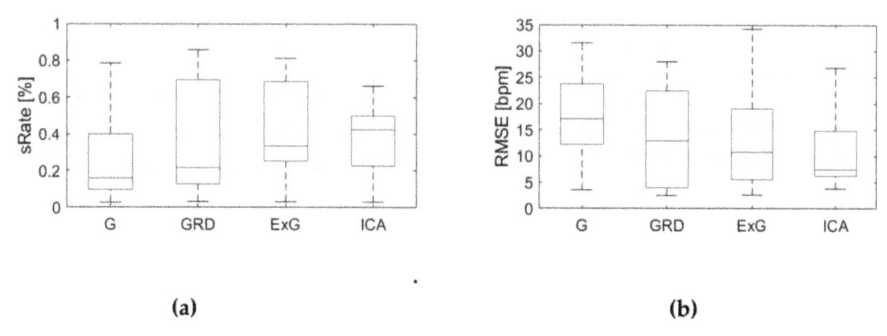

Figure 14. Comparison of signal extraction methods, algorithm No.2 (AR), part 2: (**a**) box plots for *sRate*; (**b**) box plots for *RMSE*. Blue lines—IQR range, red line—median value.

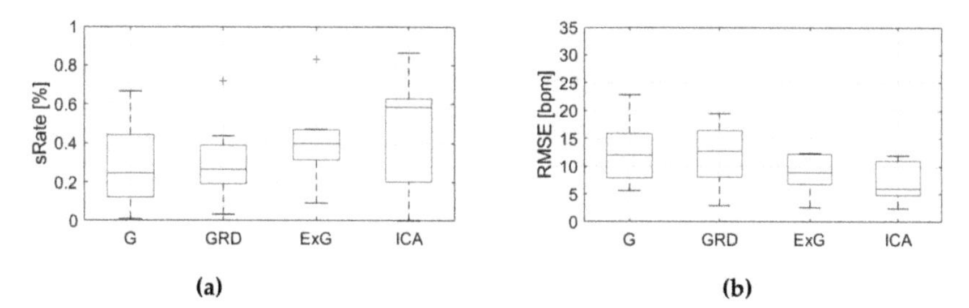

Figure 15. Comparison of signal extraction methods, algorithm No.2 (AR), part 4: (**a**) box plots for *sRate*; (**b**) box plots for *RMSE*. Blue lines—IQR range, red line—median value.

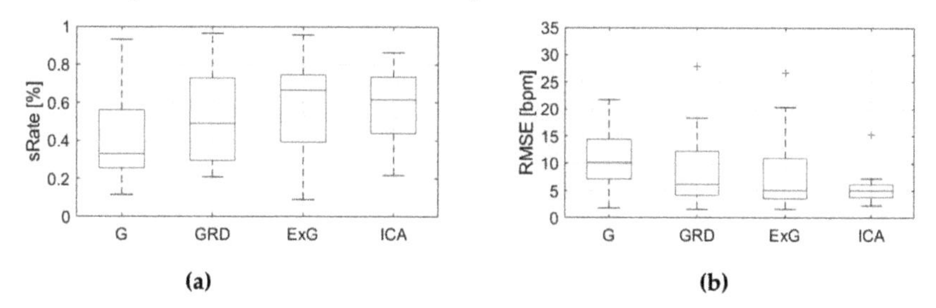

Figure 16. Comparison of signal extraction methods, algorithm No.2 (AR), part 6: (**a**) box plots for *sRate*; (**b**) box plots for *RMSE*. Blue lines—IQR range, red line—median value.

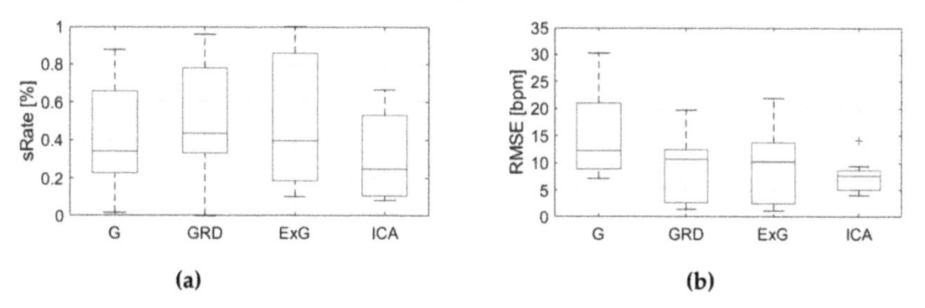

Figure 17. Comparison of signal extraction methods, algorithm No.3 (TIME), part 1: (**a**) box plots for *sRate*; (**b**) box plots for *RMSE*. Blue lines—IQR range, red line—median value.

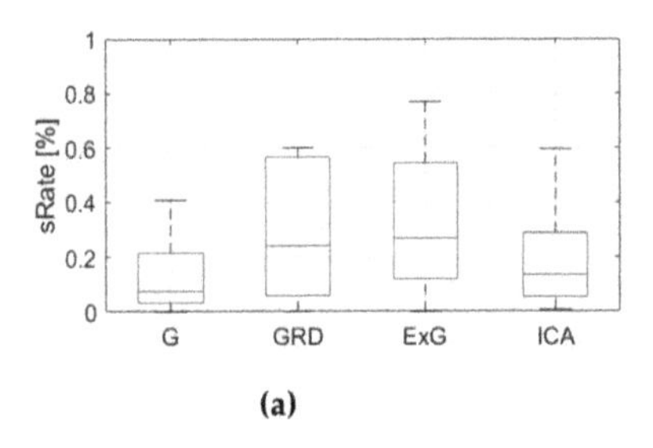

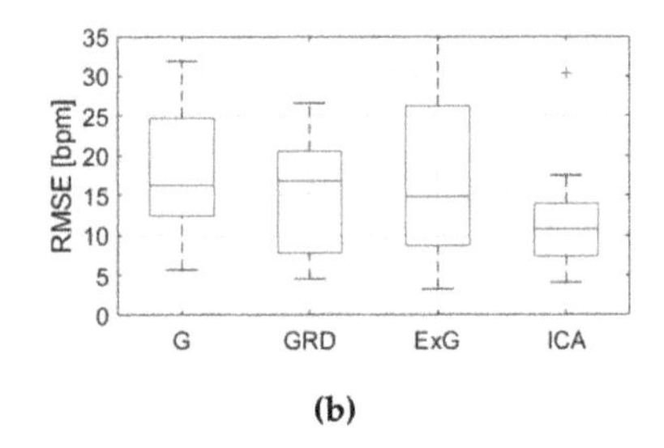

(a) **(b)**

Figure 18. Comparison of signal extraction methods, algorithm No.3 (TIME), part 2: (**a**) box plots for *sRate*; (**b**) box plots for *RMSE*. Blue lines—IQR range, red line—median value.

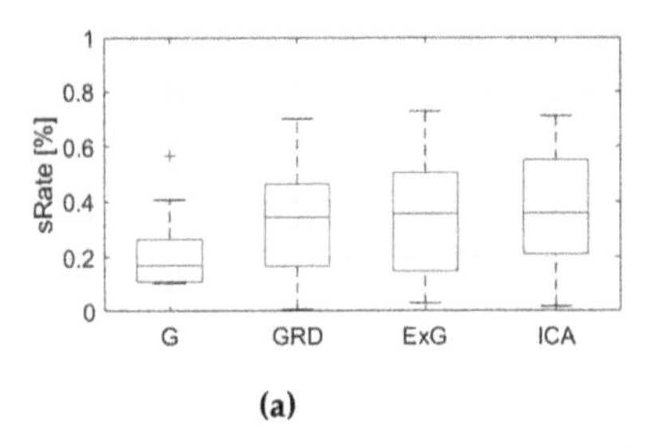

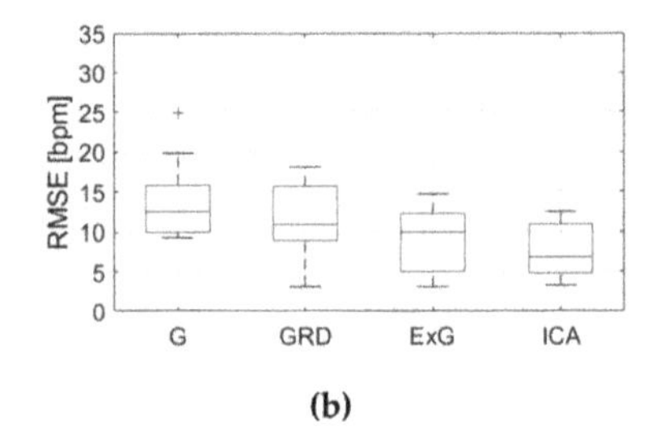

(a) **(b)**

Figure 19. Comparison of signal extraction methods, algorithm No.3 (TIME), part 4: (**a**) box plots for *sRate*; (**b**) box plots for *RMSE*. Blue lines—IQR range, red line—median value.

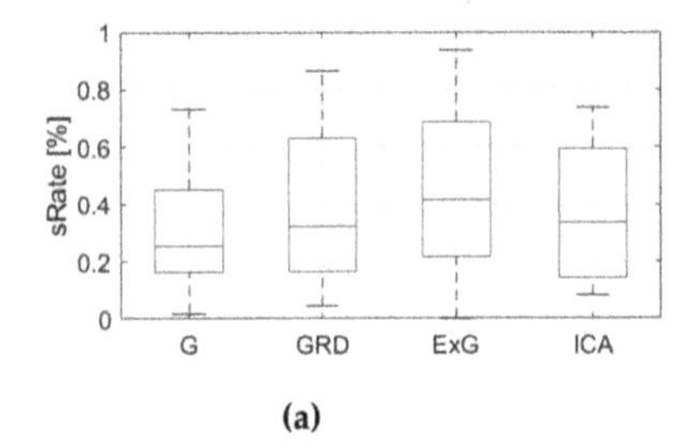

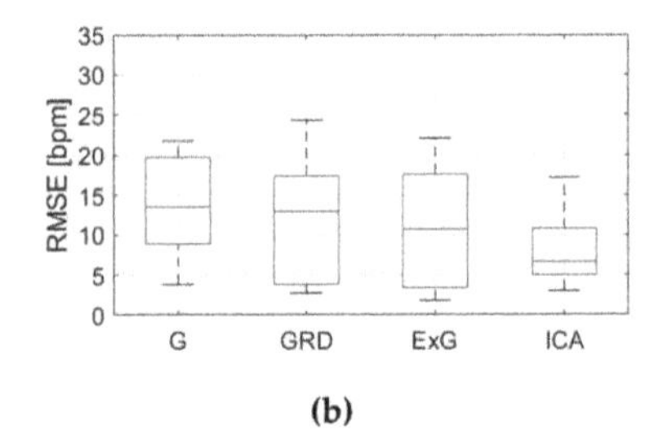

(a) **(b)**

Figure 20. Comparison of signal extraction methods, algorithm No.3 (TIME), part 6: (**a**) box plots for *sRate*; (**b**) box plots for *RMSE*. Blue lines—IQR range, red line—median value.

Considering algorithm No. 1 (PSD), *RMSE* and IQR values are lowest for the ICA for parts 1, 2 and 6 (sitting still, reading text and playing game). For the part 4 (rewriting text) the lowest *RMSE* value applies to the ExG signal representation. Given *sRate*, the best representation is ICA for parts 1,2 and 6, but part 4, where the highest *sRate* is for ExG. However, the IQR values are the lowest for ICA only for parts 1 and 6. For parts 2 and 4 the lowest IQR is for G and GRD representations respectively.

The lowest *RMSE* are for parts 1 and 6 (sitting still and playing a game), in which facial actions and head movements were small. Part 2 (reading text) has the highest IQR values. This means that facial actions in some cases have a negative impact on the accuracy of HR estimation. The large head movements present in part 4 (rewriting text) have the least impact on the accuracy of the ExG signal extraction method.

Considering algorithm No.2 (AR), *RMSE* are the lowest for ICA for parts No. 2, 4 and 6 (reading text, rewriting text and playing game). However, IQR values are not always the lowest for ICA. For part 1 (sitting still) the lowest *RMSE* value applies to the ExG representation, but with a high IQR value. Given *sRate*, it is highest for ICA and parts No. 2, 4 (reading and rewriting text). For part No. 6 (playing game) the best signal extraction method is ExG, and for part No.1 (sitting still) the G image representation.

Given algorithm No. 3 (TIME), *RMSE* values are lowest for ICA for all parts. However, *sRate* is highest for the ExG signal extraction method (parts No. 2 and 6) and GRD for part No.1. This means that there are outliers present because *RMSE* is sensitive to extreme values. The IQR of *sRate* is the lowest for G representation and almost all parts.

To compare the medians between groups (signal extraction methods) for statistical differences, a two-sided Wilcoxon rank sum test was used. The results are shown in Tables A7–A9 (Appendix A). The p-values of almost all combinations of signal extraction methods indicate that there is not enough evidence to reject the null hypothesis of equal medians at a default significance level of 5%. This means that all methods provide similar results for different activities statistically. The exceptions are: comparison between G and ICA for PSD and part 6, G and ICA for AR and part 6 (*RMSE* only), and G and ICA for TIME and parts 1, 4 (*RMSE* only).

3.3. Comparison of the Different Algorithms and Activities

The results of comparing different algorithms (PSD, AR, TIME) are shown in Table 3. Statistics were calculated for entire video sequences (including all participant activities).

Table 3. The median *sRate* and *RMSE* for selected algorithms and signal extraction methods.

Algorithm	*RMSE* [bpm]				*sRate* [%]			
	G	GRD	ExG	ICA	G	GRD	ExG	ICA
PSD	10.7	13.5	11.1	4.0	47%	49%	65%	78%
AR	15.6	12.3	11.6	5.8	48%	45%	61%	48%
TIME	16.8	13.1	10.9	8.6	35%	43%	45%	26%
average	14.4	13.0	11.2	6.1	43%	46%	57%	51%

Considering the median values, the best results (highest *sRate* and lowest *RMSE*) can be observed for algorithm No. 1 based on power spectral density (PSD). The second best algorithm is based on autoregressive modeling (algorithm No. 2). The worst results are for direct analysis of the VPG signal in the time domain (algorithm No. 3). It is worth noting that video No. 9 has a significant impact on results. ICA is the best signal extraction method in terms of *RMSE* values. However, in the case of *sRate* the best results are for ExG.

To compare the medians between groups (algorithms) for statistical differences, a two-sided Wilcoxon rank sum test was used. The results are shown in Table A10 (Appendix A). The p-values of almost all combinations of algorithms and signal extraction methods indicate that there is insufficient evidence to reject the null hypothesis of equal medians at a default significance level of 5%. The only exceptions are: ICA and G for PSD vs TIME, where p-values indicate the rejection of the null hypothesis of equal medians at a default significance level of 5%. This means that the most important issue for the ICA signal extraction method is choosing the right estimation algorithm.

3.4. Analysis of the Impact of Average Lighting and User's Movement on the Results of Pulse Detection.

To assess the effect of the scene illumination on the pulse detection accuracy, a Pearson's correlation coefficient between the median *sRate* and the average scene lighting was calculated for all video sequences (Table A11 in Appendix A). The results show only one strong positive correlation (0.71) for algorithm No. 3 (TIME) and the GRD signal extraction method. There are no medium and strong correlations present, with a significance level of less than 0.05 for other combination of algorithms and signal extraction methods. This may be due to similar and poor lighting for most video sequences.

Similarly—to assess whether the user's movements affect the results, correlation coefficients were calculated between the median *sRate* and the standard deviation of the accelerations (measured by SensorTag) for the entire video sequences (Table A12 in Appendix A). The results show strong positive correlations (> 0.6) for:

- algorithm No. 1 (PSD), and GRD, ExG
- algorithm No. 2 (AR), and GRD
- algorithm No. 3, all except ICA

Counterintuitively, *sRate* raises as the standard deviation of the accelerations increases. This might suggest that ballistocardiographic head movements generated by the flow of blood through the carotid arteries has strongest impact than subtle skin color variations caused by circulating blood. Only the ICA image representation is not sensitive to acceleration. It is also worth noting that this might be the effect of the location of the sensor (chest). However, further investigation of this hypothesis is required. Also, the Pearson's correlation coefficient with a small sample size might lead to inaccurate results. However, it can still provide useful information.

4. Discussion

The main purpose of this research was to investigate the impact of human activity on the accuracy of the VPG heart rate algorithm. We focused on activities performed during typical human-computer interaction (HCI) scenarios (i.e., reading text, rewriting text, playing game). Thus, the evaluation of the continuous HR estimation accuracy was carried out on several video sequences recorded in different places and under different conditions (illumination, person identity, distance from the computer screen and camera). We have used state of the art face detection and tracking algorithm, and compare various signal extraction methods, including (to our knowledge) first time used the ExG image representation. It is worth noting that the scene lighting for most of the videos was very poor, which corresponds to the typical computer work conditions.

For the entire video sequence and taking into account the *RMSE* metric, the ICA signal extraction method results in smallest errors. However, when it comes to reliability of measurements and maintaining the accuracy of a given algorithm within the accepted error tolerance (*sRate* metric), the ExG representation seems to be a promising method. This is especially important in medical applications. It is also worth mentioning that the ExG method is much faster to calculate than ICA (about four times—MATLAB implementation on an Intel i7 machine).

To check how individual activities affect the results of heart rate detection, the following activities were compared: the participant sits still for a minimum of 60 seconds, the participant reads text, the participant rewrites text using the keyboard and the mouse, the participant plays game. In conclusion, considering algorithm No.1 (PSD), the ICA signal extraction method works better in sequences where there are no large head movements (sitting still and playing a game). For large head movements, the ExG representation gives better results. Facial actions (part 2 – reading text) have a negative impact on the accuracy of HR estimation. Given algorithm No.2 (AR), it is difficult to indicate the best signal extraction method. In general, ICA works better on parts with facial actions and head movements. For other parts, the ExG method works well, but for part in which the participant was sitting still, the simplest signal representation (G) is the best. Interestingly, these are the opposite results than in the case of the PSD algorithm, in which the ICA signal extraction method works better in cases where there are no large head movements. Considering algorithm No.3 (TIME), the ExG signal representation method provides better reliability of measurements (*sRate*). The smallest *RMSE* is for ICA, but the *RMSE* metric is more sensitive to extreme values and outliers found in the collected data.

Based on the Wilcoxon rank sum test, almost all signal extraction methods provide similar results statistically with the exception of G and ICA comparisons. This means that for the tested videos it is impossible to indicate the best method that works in all scenarios and lighting conditions. Collecting more data can help indicate a better method. Comparing the results obtained from different algorithms, we found that algorithm No. 1 (PSD) gives the best results, followed by the algorithm No. 2 (AR). The accuracy of the algorithm No. 3 (time-based) is significantly different from other algorithms. In addition, based on the Wilcoxon rank sum test, for the ICA signal extraction method the most important is the selection of the appropriate estimation algorithm.

Taking into account individual activities, the highest average *sRate* applies to the activity in which participants sat still. The second highest average *sRate* is for the activity in which users were playing game. The lowest *sRate* value applies to: reading and typing text respectively. Although, the ICA method seems to provide better results, this is not always the case. There are several combinations of estimation algorithm and signal extraction method in which the ExG is better (i.e., part No.1 and TIME).

The presented analysis and results pave the way for other studies. The following directions of future research remain open:

- Further analysis which external or internal factors influence the results of HR estimation, i.e., Image parameters (saturation, hue), type of user's movements, ROI size, etc.),
- Evaluation of selected algorithms on a larger amount of data,
- Development a metric to detect moments when measurement is correct and reliable,
- Evaluating whether the use of depth and IR channels (provided by the Intel RealSense SR300 camera) as additional sources of pulse signal information increases accuracy.

5. Conclusions

Reliable non-contact cardiovascular parameters monitoring can be difficult because many factors can contaminate the pulse signal, e.g. a subject movement and illumination changes. In this article we examined the accuracy of HR estimation for various human activities during typical HCI scenarios (sitting still, reading text, typing text and playing game). We tested three different heart rate estimation algorithms and four signal extraction methods. The results show that the proposed signal extraction method (ExG) provides acceptable results (65% *sRate* for PSD), while being much faster to calculate that the ICA method. We have found that, depending on the scenario being studied, a different combination of signal extraction methods and pulse estimation algorithm ensures optimal heart rate detection results. We also noticed that the choice of signal representation has a greater impact on accuracy than the choice of estimation algorithm.

Funding: This research was funded by the AGH University of Science and Technology in year 2019 from the subvention granted by the Polish Ministry of Science and Higher Education.

Conflicts of Interest: The authors declare no conflict of interest.

Appendix A

Table A1. Duration of the recorded video sequences and selected parts (mm: ss).

Video No.	Entire Video	Part 1	Part 2	Part 4	Part 6
video 1	05:33	01:00	00:25	00:32	01:01
video 2	04:53	00:50	00:19	00:31	01:01
video 3	05:30	00:55	00:24	00:55	01:04
video 4	05:19	00:48	00:18	00:58	01:02
video 5	05:48	01:00	00:26	01:07	01:02
video 6	05:46	01:00	00:28	01:02	01:03
video 7	05:56	00:57	00:32	01:07	01:01
video 8	05:56	01:00	00:20	01:02	01:01
video 9	05:44	01:01	00:34	01:12	00:20

Table A2. The average illumination and standard deviation of accelerations for recorded video sequences and selected parts.

Video No.	Entire Video	Part 1	Part 2	Part 4	Part 6	Entire Video	Part 1	Part 2	Part 4	Part 6
	Average Illumination [lux]					std of Accelerations [G]				
video 1	72	78	77	62	69	0.134	0.021	0.010	0.011	0.006
video 2	86	93	91	63	76	0.123	0.007	0.006	0.007	0.007
video 3	40	44	43	39	35	0.112	0.011	0.010	0.009	0.008
video 4	54	60	59	58	46	0.109	0.010	0.008	0.008	0.007
video 5	950	1271	1048	806	816	0.118	0.007	0.008	0.008	0.008
video 6	27	27	30	23	25	0.110	0.007	0.009	0.008	0.007
video 7	152	146	140	113	170	0.125	0.013	0.018	0.011	0.008
video 8	49	57	56	53	38	0.099	0.008	0.007	0.008	0.009
video 9	106	108	107	103	99	0.102	0.009	0.011	0.017	0.011

Table A3. Results of HR estimation, algorithm No. 1 (PSD)—comparison of signal extraction methods.

Video No.	*RMSE* [bpm]				*sRate* [%]			
	G	GRD	ExG	ICA	G	GRD	ExG	ICA
video 1	10.7	4.5	3.2	7.8	48%	76%	85%	71%
video 2	16.5	15.2	15.7	14.7	27%	41%	39%	40%
video 3	16.5	13.9	12.9	2.8	47%	49%	57%	83%
video 4	10.1	11.6	7.3	3.2	45%	55%	65%	78%
video 5	3.9	2.2	2.5	2.8	84%	91%	91%	87%
video 6	14.1	13.5	11.1	7.2	43%	42%	65%	62%
video 7	3.5	3.7	3.4	2.2	80%	84%	86%	91%
video 8	6.8	17.9	17.2	4.0	61%	19%	25%	79%
video 9	22.2	36.7	35.8	34.8	11%	16%	15%	12%

Table A4. Results of HR estimation, algorithm No. 2 (AR)—comparison of signal extraction methods.

Video No.	*RMSE* [bpm]				*sRate* [%]			
	G	GRD	ExG	ICA	G	GRD	ExG	ICA
video 1	9.5	4.6	3.5	6.3	45%	62%	74%	55%
video 2	17.0	11.1	11.6	13.0	30%	45%	44%	37%
video 3	15.6	15.3	12.2	5.0	48%	45%	61%	68%
video 4	8.6	12.3	7.0	5.7	57%	57%	69%	47%
video 5	3.9	2.6	2.6	3.0	80%	83%	82%	77%
video 6	19.4	17.2	15.4	11.7	23%	27%	46%	35%
video 7	5.4	7.9	6.7	5.8	76%	73%	71%	48%
video 8	15.6	20.6	19.3	4.8	53%	28%	34%	63%
video 9	23.3	35.3	35.5	30.0	10%	13%	12%	9%

Table A5. Results of HR estimation, algorithm No. 3 (TIME)—comparison of signal extraction methods.

Video No.	*RMSE* [bpm]				*sRate* [%]			
	G	GRD	ExG	ICA	G	GRD	ExG	ICA
video 1	16.8	4.3	2.8	7.4	49%	70%	83%	51%
video 2	21.3	11.3	10.9	12.2	38%	47%	53%	30%
video 3	20.7	18.4	17.7	6.3	28%	25%	36%	43%
video 4	14.2	13.1	10.4	8.6	35%	43%	45%	21%
video 5	16.7	2.5	2.6	5.5	51%	88%	89%	52%
video 6	20.0	21.4	17.8	12.2	22%	21%	22%	26%
video 7	9.1	9.6	10.9	7.8	50%	47%	48%	18%
video 8	21.4	20.9	21.2	9.7	14%	18%	20%	23%
video 9	16.2	34.0	34.5	29.1	18%	13%	11%	14%

Table A6. The Wilcoxon rank sum test results (p-values) for comparing different signal extraction methods.

Comparison	*RMSE* p-Value			*sRate* p-Value		
	PSD	AR	TIME	PSD	AR	TIME
G vs GRD	1.00	0.93	0.49	0.93	1.00	0.93
G vs ExG	0.73	0.49	0.34	0.39	0.55	0.49
G vs ICA	0.14	0.19	**0.01**	0.16	0.86	0.86
GRD vs ExG	0.67	0.67	0.80	0.67	0.60	0.73
GRD vs ICA	0.22	0.30	0.34	0.34	0.86	0.67
ExG vs ICA	0.30	0.39	0.39	0.55	0.60	0.34

Table A7. The Wilcoxon rank sum test results (p-values) for comparing different signal extraction methods and activities, algorithm No.1 (PSD).

Comparison	*RMSE* p-Value				*sRate* p-Value			
	Part1	Part2	Part4	Part6	Part1	Part2	Part4	Part6
G vs GRD	0.80	0.49	1.00	0.93	1.00	0.44	0.60	0.80
G vs ExG	0.93	0.39	0.30	0.60	0.93	0.25	0.16	0.22
G vs ICA	0.44	0.09	0.34	**0.05**	0.30	0.16	0.45	**0.04**
GRD vs ExG	0.67	0.86	0.44	1.00	0.80	1.00	0.30	0.86
GRD vs ICA	0.22	0.73	0.39	0.22	0.26	0.75	0.67	0.26
ExG vs ICA	0.30	0.73	0.86	0.19	0.45	0.93	1.00	0.30

Table A8. The Wilcoxon rank sum test results (p-values) for comparing different signal extraction methods and activities, algorithm No.2 (AR).

Comparison	*RMSE* p-Value				*sRate* p-Value			
	Part1	Part2	Part4	Part6	Part1	Part2	Part4	Part6
G vs GRD	1.00	0.44	0.80	0.39	0.93	0.55	0.73	0.55
G vs ExG	1.00	0.34	0.39	0.22	1.00	0.19	0.22	0.26
G vs ICA	0.67	0.22	0.09	**0.05**	0.30	0.30	0.26	0.14
GRD vs ExG	0.93	0.73	0.30	0.67	0.86	0.49	0.14	0.60
GRD vs ICA	0.73	0.67	0.14	0.44	0.44	0.73	0.34	0.49
ExG vs ICA	1.00	0.80	0.26	0.73	0.34	0.73	0.60	1.00

Table A9. The Wilcoxon rank sum test results (p-values) for comparing different signal extraction methods and activities, algorithm No.3 (TIME).

Comparison	*RMSE* p-Value				*sRate* p-Value			
	Part1	Part2	Part4	Part6	Part1	Part2	Part4	Part6
G vs GRD	0.30	0.49	0.67	0.49	0.49	0.14	0.22	0.55
G vs ExG	0.34	0.80	0.22	0.49	0.60	0.17	0.39	0.34
G vs ICA	**0.01**	0.08	**0.03**	0.06	0.44	0.67	0.22	0.67
GRD vs ExG	0.86	0.80	0.55	0.80	0.86	0.93	1.00	0.73
GRD vs ICA	0.49	0.44	0.22	0.60	0.22	0.49	0.80	0.93
ExG vs ICA	0.44	0.30	0.60	0.86	0.22	0.30	0.86	0.55

Table A10. The Wilcoxon rank sum test results (p-values) for comparing different algorithms.

Comparison	*RMSE* p-Value				*sRate* p-Value			
	G	GRD	ExG	ICA	G	GRD	ExG	ICA
PSD vs AR	0.67	0.80	0.93	0.44	0.93	0.80	0.73	0.05
PSD vs TIME	**0.05**	0.73	0.67	0.11	0.26	0.44	0.26	**0.01**
AR vs TIME	0.16	0.80	0.86	0.26	0.22	0.49	0.55	0.06

Table A11. The Pearson's correlation values between the *sRate* and the average scene lighting.

Algorithm	Correlation Value				p-Value			
	G	GRD	ExG	ICA	G	GRD	ExG	ICA
PSD	0.57	0.57	0.45	0.27	0.11	0.11	0.23	0.49
AR	0.56	0.61	0.46	0.47	0.11	0.08	0.22	0.20
TIME	0.50	**0.71**	0.62	0.51	0.17	0.03	0.08	0.16

Table A12. The Pearson's correlation values between the *sRate* and the standard deviation of the accelerations.

Algorithm	Correlation Value				p-Value			
	G	GRD	ExG	ICA	G	GRD	ExG	ICA
PSD	0.27	**0.75**	**0.70**	0.24	0.49	0.02	0.04	0.53
AR	0.32	**0.68**	0.66	0.22	0.41	0.04	0.06	0.56
TIME	**0.87**	**0.72**	**0.78**	0.53	0.00	0.03	0.01	0.14

References

1. Aoyagi, T.; Miyasaka, K. Pulse oximetry: Its invention, contribution to medicine, and future tasks. *Anesth. Analg.* **2002**, *94*, S1–S3. [PubMed]
2. Nilsson, L.; Johansson, A.; Kalman, S. Respiration can be monitored by photoplethysmography with high sensitivity and specificity regardless of anaesthesia and ventilatory mode. *Acta Anaesthesiol. Scand.* **2005**, *49*, 1157–1162. [CrossRef] [PubMed]
3. Kvernebo, K.; Megerman, J.; Hamilton, G.; Abbott, W.M. Response of skin photoplethysmography, laser Doppler flowmetry and transcutaneous oxygen tensiometry to stenosis-induced reductions in limb blood flow. *Eur. J. Vasc. Surg.* **1989**, *3*, 113–120. [CrossRef]
4. Loukogeorgakis, S.; Dawson, R.; Phillips, N.; Martyn, C.N.; Greenwald, S.E. Validation of a device to measure arterial pulse wave velocity by a photoplethysmographic method. *Physiol. Meas.* **2002**, *23*, 581–596. [CrossRef] [PubMed]
5. Incze, A.; Lazar, I.; Abraham, E.; Copotoiu, M.; Cotoi, S. The use of light reflection rheography in diagnosing venous disease and arterial microcirculation. *Rom. J. Intern. Med.* **2003**, *41*, 35–40. [PubMed]
6. Jones, M.E.; Withey, S.; Grover, R.; Smith, P.J. The use of the photoplethysmograph to monitor the training of a cross-leg free flap prior to division. *Br. J. Plast. Surg.* **2000**, *53*, 532–534. [CrossRef] [PubMed]
7. Imholz, B.P.; Wieling, W.; van Montfrans, G.A.; Wesseling, K.H. Fifteen years experience with finger arterial pressure monitoring: Assessment of the technology. *Cardiovasc. Res.* **1998**, *38*, 605–616. [CrossRef]
8. Avnon, Y.; Nitzan, M.; Sprecher, E.; Rogowski, Z.; Yarnitsky, D. Different patterns of parasympathetic activation in uni- and bilateral migraineurs. *Brain* **2003**, *126*, 1660–1670. [CrossRef]
9. Gregoski, M.J.; Mueller, M.; Vertegel, A.; Shaporev, A.; Jackson, B.B.; Frenzel, R.M.; Sprehn, S.M.; Treiber, F.A. Development and validation of a smartphone heart rate acquisition application for health promotion and wellness telehealth applications. *Int. J. Telemed. Appl.* **2012**, *2012*, 696324. [CrossRef]
10. Allen, J. Photoplethysmography and its application in clinical physiological measurement. *Physiol. Meas.* **2007**, *28*, R1–R39. [CrossRef]
11. Kranjec, J.; Beguš, S.; Geršak, G.; Drnovšek, J. Review. *Biomed. Signal Process. Control* **2014**, *13*, 102–112. [CrossRef]

12. Verkruysse, W.; Svaasand, L.O.; Nelson, J.S. Remote plethysmographic imaging using ambient light. *Opt. Express* **2008**, *16*, 21434–21445. [CrossRef] [PubMed]
13. Zhang, Q.; Wu, Q.; Zhou, Y.; Wu, X.; Ou, Y.; Zhou, H. Webcam-based, non-contact, real-time measurement for the physiological parameters of drivers. *Measurement* **2017**, *100*, 311–321. [CrossRef]
14. Wang, W.; den Brinker, A.C.; Stuijk, S.; de Haan, G. Robust heart rate from fitness videos. *Physiol. Meas.* **2017**, *38*, 1023–1044. [CrossRef] [PubMed]
15. McDuff, D.J.; Hernandez, J.; Gontarek, S.; Picard, R.W. COGCAM: Contact-free Measurement of Cognitive Stress During Computer Tasks with a Digital Camera. In Proceedings of the 2016 CHI Conference on Human Factors in Computing Systems, San Jose, CA, USA, 7–12 May 2016; pp. 4000–4004.
16. Sun, Y.; Thakor, N. Photoplethysmography Revisited: From Contact to Noncontact, From Point to Imaging. *IEEE Trans. Biomed. Eng.* **2016**, *63*, 463–477. [CrossRef] [PubMed]
17. Poh, M.Z.; McDuff, D.J.; Picard, R.W. Non-contact, automated cardiac pulse measurements using video imaging and blind source separation. *Opt. Express* **2010**, *18*, 10762–10774. [CrossRef] [PubMed]
18. Poh, M.Z.; McDuff, D.J.; Picard, R.W. Advancements in noncontact, multiparameter physiological measurements using a webcam. *IEEE Trans Biomed. Eng.* **2011**, *58*, 7–11. [CrossRef]
19. Li, X.; Chen, J.; Zhao, G.; Pietikainen, M. Remote heart rate measurement from face videos under realistic situations. In Proceedings of the IEEE Conference on Computer Vision and Pattern Recognition, Columbus, OH, USA, 24–27 June 2014; pp. 4264–4271.
20. Balakrishnan, G.; Durand, F.; Guttag, J. Detecting Pulse from Head Motions in Video. In Proceedings of the IEEE Conference on Computer Vision and Pattern Recognition (CVPR), Portland, OR, USA, 23–28 June 2013.
21. Unakafov, A.M. Pulse rate estimation using imaging photoplethysmography: Generic framework and comparison of methods on a publicly available dataset. *Biomed. Phys. Eng. Express* **2018**, *4*, 045001. [CrossRef]
22. Hülsbusch, M. An image-based functional method for opto-electronic detection of skin-perfusion. Ph.D. Thesis, RWTH Aachen University, Aachen, Germany, 2008.
23. Przybyło, J.; Kańtoch, E.; Jabłoński, M.; Augustyniak, P. Distant Measurement of Plethysmographic Signal in Various Lighting Conditions Using Configurable Frame-Rate Camera. *Metrol. Meas. Syst.* **2016**, *23*, 579–592. [CrossRef]
24. Gong, S.; McKenna, S.J.; Psarrou, A. *Dynamic Vision: From Images to Face Recognition*, 1st ed.; Imperial College Press: London, UK, 2000; ISBN 1-86094-181-8.
25. Zafeiriou, S.; Zhang, C.; Zhang, Z. A Survey on Face Detection in the Wild. *Comput. Vis. Image Underst.* **2015**, *138*, 1–24. [CrossRef]
26. Zhang, C.; Zhang, Z. A Survey of Recent Advances in Face Detection. Available online: https://www.microsoft.com/en-us/research/publication/a-survey-of-recent-advances-in-face-detection/ (accessed on 27 September 2019).
27. King, D. Dlib c++ library. Available online: http://dlib.net (accessed on 22 January 2018).
28. Dalal, N.; Triggs, B. Histograms of Oriented Gradients for Human Detection. In Proceedings of the 2005 IEEE Computer Society Conference on Computer Vision and Pattern Recognition (CVPR'05), San Diego, CA, USA, 20–25 June 2005; Volume 1, pp. 886–893.
29. King, D.E. Max-Margin Object Detection. *arXiv* **2015**.
30. Tomasi, C.; Kanade, T. Detection and Tracking of Point Features. Available online: https://www2.cs.duke.edu/courses/fall17/compsci527/notes/interest-points.pdf (accessed on 27 September 2019).
31. Viola, P.; Jones, M.J. Robust Real-Time Face Detection. *Int. J. Comput. Vis.* **2004**, *57*, 137–154. [CrossRef]
32. Shi, J.; Tomasi, C. *Good Features to Track*; Cornell University: Ithaca, NY, USA, 1993.
33. Kazemi, V.; Sullivan, J. One Millisecond Face Alignment with an Ensemble of Regression Trees. In Proceedings of the 2014 IEEE Conference on Computer Vision and Pattern Recognition, Columbus, OH, USA, 24–27 June 2014; pp. 1867–1874.
34. Yu, Y.P.; Raveendran, P.; Lim, C.L. Dynamic heart rate measurements from video sequences. *Biomed. Opt. Express* **2015**, *6*, 2466–2480. [CrossRef] [PubMed]
35. Hyvärinen, A.; Oja, E. Independent Component Analysis: Algorithms and Applications. *Neural Netw.* **2000**, *13*, 411–430. [CrossRef]

36. Yang, W.; Wang, S.; Zhao, X.; Zhang, J.; Feng, J. Greenness identification based on HSV decision tree. *Inf. Process. Agric.* **2015**, *2*, 149–160. [CrossRef]
37. Jaromir Przybyło. Available online: http://home.agh.edu.pl/~{}przybylo/download_en.html (accessed on 26 September 2019).

Article

Target-Specific Action Classification for Automated Assessment of Human Motor Behavior from Video

Behnaz Rezaei [1], Yiorgos Christakis [2], Bryan Ho [3], Kevin Thomas [4], Kelley Erb [2], Sarah Ostadabbas [1] and Shyamal Patel [2,*]

1 Augmented Cognition Lab (ACLab), Department of Electrical and Computer Engineering, Northeastern University, Boston, MA 02115, USA; brezaei@ece.neu.edu (B.R.); ostadabbas@ece.neu.edu (S.O.)
2 Digital Medicine & Translational Imaging Group, Pfizer, Cambridge, MA 02139, USA; Yiorgos.Christakis@pfizer.com (Y.C.); MichaelKelley.Erb@pfizer.com (K.E.)
3 Neurology Department, Tufts University School of Medicine, Boston, MA 02111, USA; bho@tuftsmedicalcenter.org
4 Department of Anatomy & Neurobiology, Boston University School of Medicine, Boston, MA 02118, USA; kipthoma@bu.edu
* Correspondence: Shyamal.Patel@pfizer.com

Received: 30 August 2019; Accepted: 28 September 2019; Published: 1 October 2019

Abstract: Objective monitoring and assessment of human motor behavior can improve the diagnosis and management of several medical conditions. Over the past decade, significant advances have been made in the use of wearable technology for continuously monitoring human motor behavior in free-living conditions. However, wearable technology remains ill-suited for applications which require monitoring and interpretation of complex motor behaviors (e.g., involving interactions with the environment). Recent advances in computer vision and deep learning have opened up new possibilities for extracting information from video recordings. In this paper, we present a hierarchical vision-based behavior phenotyping method for classification of basic human actions in video recordings performed using a single RGB camera. Our method addresses challenges associated with tracking multiple human actors and classification of actions in videos recorded in changing environments with different fields of view. We implement a cascaded pose tracker that uses temporal relationships between detections for short-term tracking and appearance based tracklet fusion for long-term tracking. Furthermore, for action classification, we use pose evolution maps derived from the cascaded pose tracker as low-dimensional and interpretable representations of the movement sequences for training a convolutional neural network. The cascaded pose tracker achieves an average accuracy of 88% in tracking the target human actor in our video recordings, and overall system achieves average test accuracy of 84% for target-specific action classification in untrimmed video recordings.

Keywords: action classification; human motor behavior; computer vision; deep learning; pose tracking

1. Introduction

Clinical assessment of human motor behavior plays an important role in the diagnosis and management of medical conditions like Parkinson's Disease (PD) [1]. However, such assessments can only be performed intermittently by trained clinical examiners, which limits the quantity and quality of information that can be collected to understand the impact of disease in the real-world setting. To address these limitations, significant efforts have been made to develop wearable sensing technologies that can be used for continuously monitoring various types of motor symptoms and behaviors [2–4]. While data collected using wearable sensors are well suited for detecting and measuring basic movements (e.g., arm or leg movements, tremor) and actions (e.g., sitting, standing, walking), they are ill-suited when it comes to complex activities (e.g., cooking, grooming) and behaviors (e.g., personal habits, routines)—particularly if they involve the interpretation of environmental interactions (e.g., with other humans, animals, or objects). Understanding the various factors that influence physical behavior can help clinicians better understand the impact of motor and non-motor symptoms on the daily life of patients with PD [5].

Recently, artificial intelligence (AI) assisted classification of human behavior using computer vision has received newfound attention among researchers in machine learning and pattern recognition communities for applications spanning from automatic recognition of daily life activities in smart homes to monitoring the health and safety of elderly and patients with mobility disorders in their homes/hospitals [6–12]. However, in contrast to wearable devices, vision-based approaches pose a greater risk to privacy and security of an individual [13]. Vision-based assessment of human behavior enables us to automate the detection and measurement of the full range of human behaviors. As illustrated in Figure 1, the taxonomy of human behaviors can be viewed as a four-level hierarchical framework with basic movements at the bottom (e.g., movement of body segments) and complex behaviours (e.g., personal habits and routines) at the top. Automatic recognition at any level requires that actions and/or behaviors at the level below it are also recognized. For example, in order to recognize walking, we first need to assess if the pose is upright, the arms are swinging and legs are moving. At the first level (motion), recognition deals with tasks such as movement detection or background extraction/segmentation in video recordings of the target [14–16]. These techniques try to locate the moving objects in a scene by extracting a silhouette of the object in a single frame or over a few consecutive frames. However, segmentation algorithms without any further processing provide only very basic pose estimation of the object with little to no temporal information. At the second level (action), human movements along with environmental interactions are classified in order to recognize what the target is doing over a period of seconds or minutes [17]. At the third level (activity), the recognition task is focused on identifying activities as a combination of sequence of actions and environmental interactions over a period of minutes to hours. Finally, at the fourth level (behavior), sequence of activities and environmental interactions along with information about their temporal dependencies are used to recognize complex human behaviors.

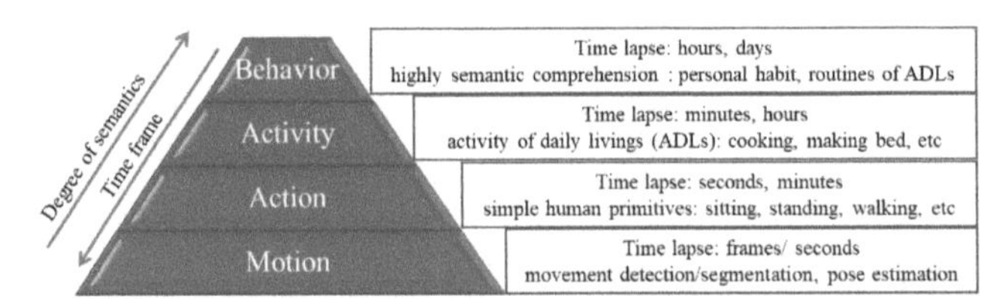

Figure 1. Taxonomy of human behaviors with different levels of semantics and complexity. Recognition of each level requires most of the underlying tasks to be recognized [6].

1.1. Our Contributions

Automated assessment of human behavior in multi-person video (i.e., when several people are present in the video) requires the tracking and classification of a sequence of actions performed by a target (e.g., patient). Therefore, accurate temporal tracking of the target is an essential requirement for this application, along with robust feature extraction that can be used for classifying human behaviors at different levels of complexity. In this paper, we present a hierarchical target-specific action classification method, which is illustrated as a block diagram in Figure 2. Detection of different actions performed by the target is done using pose evolution feature representation. We define pose evolution as a low-dimensional embedding of a sequence of posture movements that are required to perform an action (e.g., walking). In order to find the pose evolution feature representation corresponding to the target, we present a cascaded target pose tracking algorithm that receives multi-person pose estimation results from an earlier stage and tracks the target pose throughout the video. Our main contributions in this paper are: (1) development of a robust hierarchical multiple-target pose tracking method to facilitate action recognition in videos recorded in uncontrolled environments in the presence of multiple human actors; (2) introducing pose evolution, an explicit body movement representation, as complementary information to the appearance and motion cues for robust action recognition; and (3) a novel target-specific action classification architecture applied to untrimmed video recordings of patients with PD.

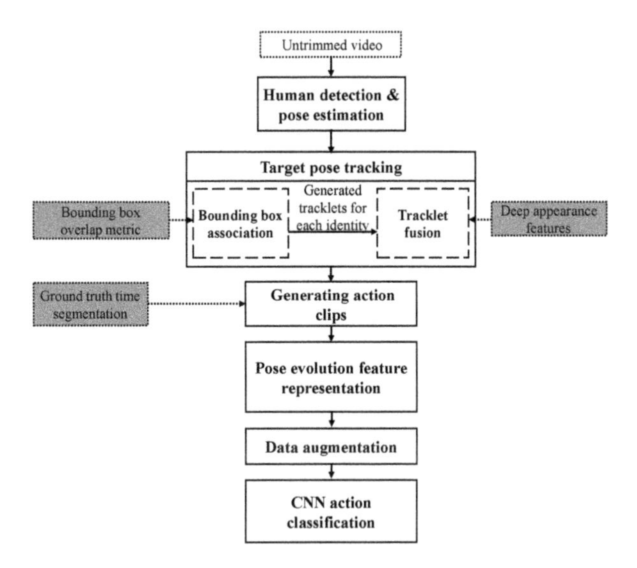

Figure 2. Overview of the proposed multi-stage method for human behavior phenotyping in untrimmed videos. At the first stage, human detection and pose estimation are applied to the recorded video. At the second stage, the regressed bounding boxes for each detected person and corresponding keypoints are used for tracking the identities in the video. Tracking is done in an incremental process incorporating both appearance and time information. Outputs of tracking the target identity along with ground-truth time segmentation are used for generating a compact representation of the target actor pose evolution in time for each action clip. Finally, the augmented pose evolution representation is fed to a convolutional neural network (CNN)-based action classification network to recognize actions of interest.

1.2. Related Works

Assigning a single action label to a multi-person video clip dilutes the specificity of information and makes it less meaningful. For many real-world applications such as video-based assessment of human behavior, there is a need for person-centric action recognition, which assigns an action label to each person in a multi-person video clip. One of the challenges in person-centric action recognition is robust tracking of the target in long-term videos. Tracking is challenging because there are many sources of uncertainty, such as clutter, occlusions, target interactions, and camera motion. However, most of the research studies on human activity classification have typically dealt with videos with a single human actor or video clips with ground-truth tracking provided [18], with the exception of few that performed human-centeric action recognition [8,19]. Girdhar, et al. [19] re-purposed an action transformer network to exclude non-target human actors in the scene and aggregated spatio–temporal features around the target human actor. Chen, et al. [8] presented human activity classification using skeleton motions in videos with interference from non-target objects aimed at supporting applications in monitoring frail and elderly individuals. However, neither work provided details on how they addressed non-target filtering in their human action classification pipelines.

Beside the importance of dealing with non-target objects in providing a well-performing real-world human action recognition system, creating robust and discriminating feature representations for each video action clip plays an important role in detecting different human activities [20]. Most of the state-of-the-art action recognition architectures process appearance and motion cues in two independent streams of information, which are fused right before the classification phase or a few stages before the classification stage in a merge and divide scheme [21,22]. Others have used 3D spatio-temporal convolutions to directly extract relevant spatial and temporal features [23–25]. However, human pose cues, which can provide low-dimensional interpretations for different activities, have been overlooked in these studies. Most recently, Choutas, et al. [26] and Mengyuan, et al. [27] used temporal changes of pose information with two different representations for boosting action recognition performance. In [27], authors claim that if there are multiple people in the scene, pose motion representation does not need the time associations of the joints to work but they did not address how their proposed method can handle multiple human actors in a video.

In general, convolutional neural network (CNN) based action recognition approaches can be divided into three different categories based on their underlying architecture: (1) spatio-temporal convolutions (3-dimensional convolutions), (2) recurrent neural networks, and (3) two stream convolutional networks. The benefit of multi-stream networks is that different modalities can be aggregated in the network to improve performance of the final action classification task. In this paper, we addressed the problem of person-centric action recognition by long-term tracking of the target human actor. In addition, our method provides a novel pose evolution representation of the target human actor rather than the common spatio-temporal features extracted from raw video frames to the classification network. It is worth mentioning that our pose-based action recognition stream can be used to augment the current multi-stream action classification networks.

The rest of the paper is organized as follows. In Section 2, we describe the proposed method for tracking target human actor in untrimmed videos in order to extract appropriate pose evolution features from actions performed in a video. In Section 3, we describe the subsequent stages for action classification (illustrated in Figure 2), which include pose evolution feature representation and classification network. We present our experimental setup and performance evaluation results of the proposed method in Section 4. Finally, we discuss the results in Section 5 and conclude our paper in Section 6.

2. Target Pose Tracking

Diverging from the common approach of learning spatio-temporal features from videos for action classification, pose-based action classification methods have shown promising results by providing a compact representation of human pose evolution in videos [26–29]. The temporal evolution of pose can be used as the only discriminating feature information for classification of human actions that involve different pose transitions (e.g., walking). This approach can further be combined with spatio-temporal features to improve the performance of context-aware action classification in the case of more complex behaviors (e.g., moving an object from one place to another).

The primary task in pose-based action classification in untrimmed videos is locating the target. This requires a robust estimation and tracking of human body poses by addressing the challenges associated with long-term videos recorded for assessment of human motor behavior. These challenges include partial to complete occlusion, change of scene, and camera motion. In this section, we propose a cascaded multi-person pose tracking method using both time and appearance features, which will be used in later steps to generate pose evolution feature representations for action classification.

2.1. Human Pose Estimation

In order to extract human pose information in each video frame along with their associated bounding boxes as the first step in our system, we used a 2D version of the state-of-the-art human pose estimation method proposed in [30]. The pre-trained model performs efficient frame-level multi-person pose estimation in videos using the Mask R-CNN network [31]. This model was initialized on ImageNet [32] and then trained on the COCO keypoint detection task [33]. The Mask R-CNN network was then fine-tuned on the PoseTrack dataset [34]. The architecture of this pose estimation network is illustrated in Figure 3. The network uses ResNet-101 [35] as the base convolutional network for extracting image features. Extracted features are then fed to a region proposal network (RPN) trained to highlight regions that contain object candidates [36]. Candidate regions of the output feature map are all aligned to a fixed resolution via a spatial region of interest (ROI)-align operation. This operation divides feature maps that may have different sizes depending on the size of detected bounding boxes to a fixed number of sub-windows. The value for each sub-window is calculated by finding a bi-linear interpolation of four regularly sampled locations inside the sub-window. The aligned features are then fed into two heads, a *classification head* responsible for person detection and bounding box regression, and a *keypoint head* for estimating the human body joints defined as a human pose in each detected bounding box. The outputs of this pose estimation network are seventeen keypoints associated with various body joints and a bounding box surrounding each person.

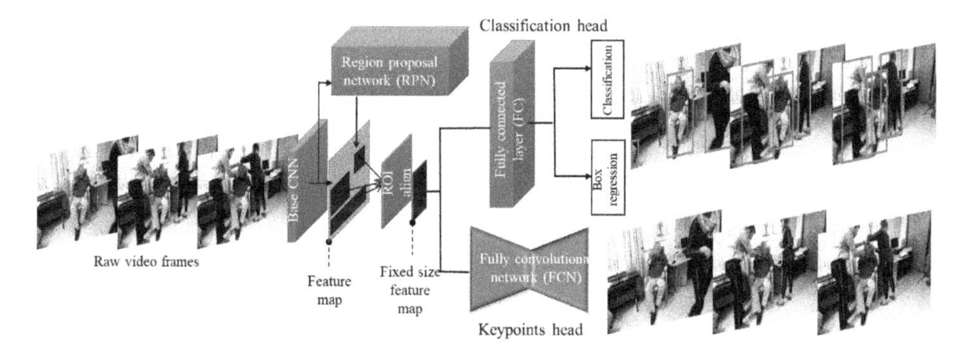

Figure 3. Architecture of the pose estimation network. Each video frame is fed separately to the base network (ResNet 101) for feature extraction. A region proposal network is applied on the output feature map to find the areas with the highest objectness probability. The fixed size features for proposed regions are then given to the classification and pose estimation heads to find the human bounding boxes and their corresponding keypoints.

2.2. Cascaded Pose Tracking

In many real-world settings where a person has to be tracked across videos recorded from different cameras located in different environments, a single tracker is unable to track the person throughout the video and all of them fail when the target leaves one environment and appears into another environment or is occluded from the camera view and then reappears in the camera's field of view [37]. In order to address this problem of tracking people in videos recorded in multiple environments (in our case different rooms and hallways) various person, re-identification methods have been proposed [38–40]. Most of the existing re-identification (re-id) methods are supervised with the assumption of availability of large manually labeled matching identity pairs. This assumption does not hold in many practical scenarios (such as our dataset) where the model has to be generalizable for any person and providing manually labeled identity matches is not feasible. Unsupervised learning for person re-id has become important in various scenarios where the system needs to be adapted to new identities such as video surveillance applications [41,42]. In this work, we have adapted the idea of person re-id, which is used for the matching the identities among non-overlapping cameras for tracking the target throughout the non-overlapping videos. This would address challenges such as changing environments or turning away from the camera, which can be treated as the case of re-identification across different non-overlapping cameras. In the traditional re-id problem, we typically have a gallery of images containing the images taken using different cameras for different identities. Given a probe image, the aim is to match the probe identity with the images in the gallery that belong to the same identity as the probe. In our problem of long-term tracking of the target human (patient) in videos, we have a set of tracklets and a given probe (an image of the target) and the aim is to fuse all the tracklets in the set which belong to the same identity as the probe in order to find the single track of the patient throughout the video. In contrast to the re-identification problem, multiple tracklets are generated because of the failure in the tracking of the target throughout the video because of the occlusions, change of environment and abrupt camera motions.

In order to continuously track the pose of the target (i.e., the subject in our dataset) in video recordings, we propose a two-step procedure based on the estimated bounding boxes and keypoints provided by the pose estimation network in Section 2.1. As illustrated in Figure 4, in the first stage (short-term tracking, Section 2.2.1) we use a lightweight data association approach to link the detected bounding boxes in consecutive frames into tracklets. Tracklets are a series of bounding boxes in consecutive frames associated with the same identity (person). In the next stage (long-term tracking, Section 2.2.2), we fuse tracklets of the same identity using their learned appearance features to provide continuous tracking of the target actor across the entire video recording. The implementation details are described in Section 2.2.2.

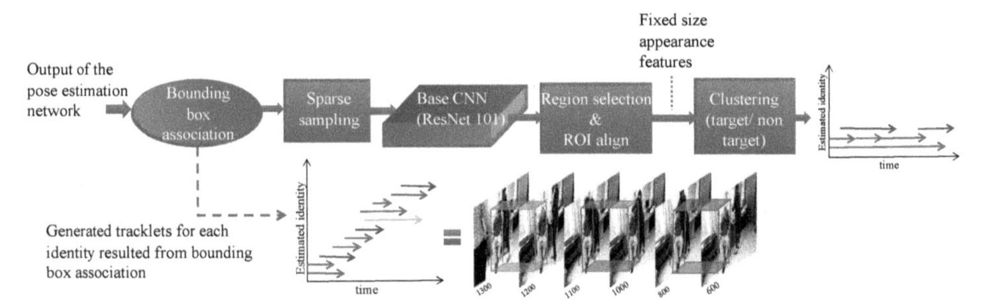

Figure 4. Hierarchical pose tracking using temporal and appearance features. Tracking starts by associating detected bounding boxes in each pair of consecutive frames using the intersection over union metric. Output of this step is a number of different tracklets for each identity. At the next step generated tracklets are pruned based on their length, and pose estimation confidence followed by sparse sampling. Finally, generated tracklets which belong to the target identity are merged according to their appearance similarity to create the endpoint track for the target human actor (best viewed in color).

2.2.1. Short Term Tracking Based on Temporal Association

Given the detected bounding boxes for each person in the video, we link the bounding boxes that belong to the same identity in time to create pose tracklets. Assuming that there is no abrupt movement in the video, tracklets are generated by solving a data association problem with similarity measurement defined as the intersection over union between the currently detected bounding boxes and the bounding boxes from the previous frame. Like [30,43], we formulate the task as a bipartite matching problem, and solve it using the Hungarian algorithm [44]. We initialize tracklets on the first frame and propagate the labels forward one frame at a time using the matches. Any box that does not get matched to an existing tracklet instantiates a new tracklet. This method is computationally efficient and can be adapted to any video length or any number of people. However, tracking can fail due to challenges such as abrupt camera motion, occlusions and change of scene, which can result in multiple tracklets for the same identity. For instance, as illustrated in Figure 4, short term tracking generates 3 distinct tracklets for the target in just 700 consecutive frames (23 s).

2.2.2. Long Term Tracking using Appearance based Tracklet Fusion

Given the large number of tracklets generated from the previous stage (i.e., short term tracking), we fuse tracklets that belong to the same identity to generate a single long-term track for the target. As illustrated in Figure 4, in order to merge the generated tracklets belonging to the same identity throughout the video, we first apply sparse sampling by pruning the tracklets based on their length and the number of estimated keypoints, and then selecting the highest confidence bounding box from each tracklet. Finally, we merge the tracklets into a single track based on their similarity to the reference tracklet. The affinity metric between the tracklet T_i, and the reference tracklet T_{ref}, is calculated as:

$$P_a(T_i, T_{ref}) = ||f_i^{t'} - f_{ref}^{t}||_2, \quad (1)$$

where $f_i^{t'}$ is feature vector of the sampled detection in tracklet T_i at time t', and f_{ref}^{t} is feature vector of the sampled detection in reference tracklet T_{ref} at time t. Affinity metric, $P_a(.)$ is the Euclidean distance between the above feature vectors. In order to extract deep appearance features, we feed every sampled detection of each tracklet to the base network of a Mask R-CNN (i.e., ResNet-101), which has been trained on the PoseTrack dataset for pose estimation [31,35]. The extracted feature map is then aligned spatially to a fixed resolution via ROI-align operation. It is worth mentioning that we do not pay an extra computational cost for learning the features for merging the associated tracklets of the target into one track. In order to show the importance of the target tracking in the performance of the action classification network we trained the action classification network on the pose evolution maps without any tracking involved, more details are provided in Section 4.3.

3. Action Classification Based on Pose Evolution Representation

After locating the target, providing a compact yet discriminative pose evolution feature representation for each action clip plays an essential role in recognizing different actions. To achieve this, we first provide a compact spatio-temporal representation of the target's pose evolution in Section 3.1 for each video clip inspired by PoTion pose motion representation introduced in [27]. Then, we use the tracked pose evolution to recognize five categories of human actions: sitting, sit-to-stand, standing, walking, and stand-to-sit in Section 3.2.

3.1. *Pose Evolution Representation*

By using pose of the target for each frame of the video clip provided by the pose tracking stage, we create a fixed-size pose evolution representation by temporally aggregating these pose maps. Pose tracking in preceding stages gives us locations of the body joints of the target (i.e., the subject in our case) in each frame of the video clip. We first generate joint heatmaps from given keypoint positions by locating a Gaussian kernel around each keypoint. These heatmaps are gray scale images showing the probability of the estimated location for each body joint. The pose evolution representations are created based on these joint heatmaps.

As illustrated in Figure 5, in order to capture the temporal evolution of pose in a video clip, after generating pose heatmaps for the target actor in a video frame, we colorize them according to their relative time in the video. In other words, each gray scale joint heatmap of dimension $H \times W$ generated for the current frame at time t is transformed into a C-channel color image of $C \times H \times W$. As indicated in Equation (2), this transformation is done by replicating the original heatmaps C times and multiplying values of each channel with a linear function of the relative time of the current frame in the video clip.

$$Je_i(j,x,y) = \frac{\sum_{t=0}^{T-1} JH_i^t(x,y) \times oc_j(t)}{max_{x,y} \sum_{t=0}^{T-1} JH_i^t(x,y) \times oc_j(t)} \quad (2)$$
$$\text{for } i \in \{1,2,...,14\}, \ \ j \in \{1,...,C\}$$

where $JH_i^t(x,y)$ designates the estimated joint heatmap for joint number i of the target in a given frame number t. $oc_j(t)$ is the linear time encoding function for channel j evaluated at time t. Je_i is the joint evolution representation for each joint i. The final pose evolution representation, Pe is derived by concatenating all calculated joint evolutions, as $Pe = concatenate(Je_1, Je_2, ..., Je_{14})$, where we have 14 joints given by reducing the head keypoints to one single keypoint.

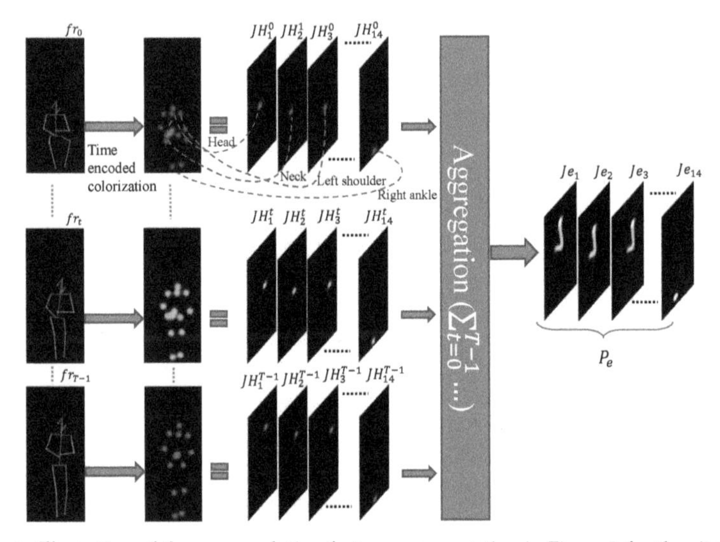

Figure 5. Illustration of the pose evolution feature representation in Figure 2 for the sit-to-stand task. Given the estimated keypoints of the target human actor from preceding stages in the first column, colorized joint heatmaps in the second column are generated using the time encoding function represented in Figure 6. The final pose evolution representation is generated by aggregating and normalizing the colorized joint heatmaps in time (best viewed in color).

In order to calculate the time encoding function for a C-channel pose evolution representation, the video clip time length T is divided into $C-1$ intervals with duration $l = \frac{T}{C}$ each. For each given frame at time t that sits in kth interval which $k = \lceil \frac{t}{T} \rceil$, $oc_j(t)$ is defined as follows:

$$oc_j(t) = \begin{cases} \frac{(-t+\frac{kT}{C-1})}{l}, & \text{for } j = k \\ \frac{(t-\frac{T(k-1)}{C-1})}{l}, & \text{for } j = k+1 \\ 0, & \text{otherwise.} \end{cases} \tag{3}$$

Figure 6 illustrates the time encoding functions that are defined based on the Equation (3) for 3-channel colorization used in our pose evolution representation. After creating the pose evolution representations, we augment them by adding white noise to our representation to train the action classification network.

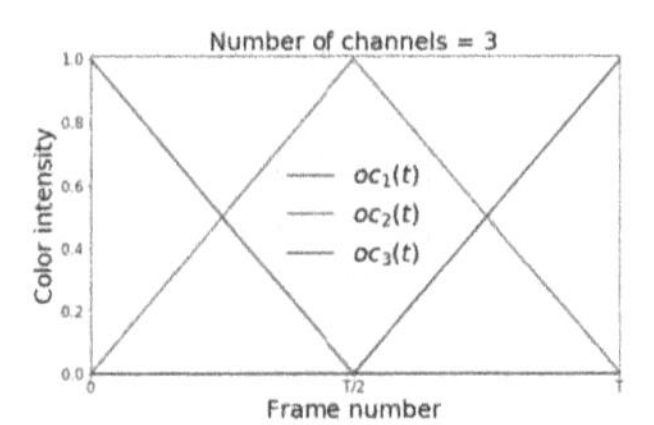

Figure 6. Demonstration of the time encoded colorization method utilized for creating body pose motion map representation. $oc_1(t)$, $oc_2(t)$, and $oc_3(t)$ show the time encoding function for each color channel.

3.2. *Classification Network*

We trained a CNN for classifying different actions using the pose evolution representations. Since pose evolution representations are very sparse and have no contextual information of the raw video frames, the network does not need to be very deep or pre-trained to be able to classify actions. We used the network architecture illustrated in Figure 7 consisting of 4 fully convolutional layers (FCN), and one fully connected layer (FC) as the classifier. The input of the first layer is the pose evolution representation of size 14 $C \times H \times W$, where 14 is the number of body joints that are used in our feature representation. In this work, we used $C = 3$ as the number of channels for encoding the time information into our feature representation. In Section 4.3, we explore the effect of number of channels on the performance of the action classification network.

The action classification network includes two blocks of convolutional layers, a global average pooling layer, and a fully connected layer with a Softmax loss function as the classification layer. Each block contains two convolution layers with filter sizes of $3 \times 3 \times 128$, and $3 \times 3 \times 256$, respectively. The first convolution layer in each block is designed with a stride of 2 pixels and a second layer with a stride of 1 pixel. All convolutional layers are followed by a rectified linear unit (ReLU), batch normalization, and dropout. We investigated the performance of several variations of this architecture on action classification in Section 4.

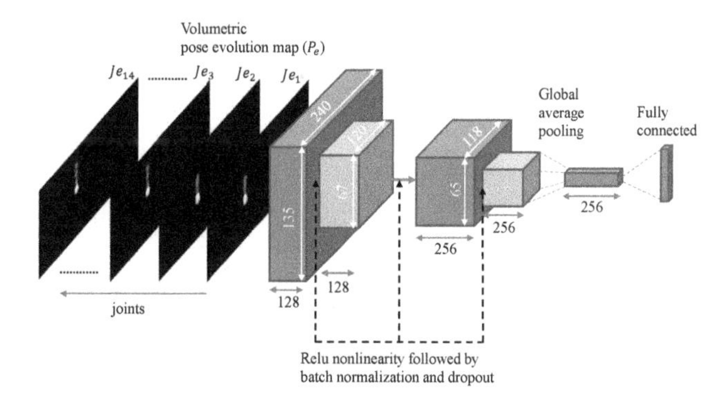

Figure 7. Architecture of the action classification network. This network takes the volumetric pose evolution map of the target human actor from a video clip as the input and classifies occurrence of an action in the video into one of the five predefined actions (best viewed in color).

4. Experiments

To evaluate the performance of the proposed approach, we used a real-world dataset collected in a neurology clinic. We provide an overview of the dataset in Section 4.1, and report on the performance of target tracking and action classification in Section 4.2 and Section 4.3 respectively.

4.1. *Dataset*

Our dataset consists of video recordings of 35 patients with Parkinson's disease (Age: 68.31 ± 8.03 (46–79) years; Sex: 23M/12F; Hoehn & Yahr I/II/III: 2/26/7; MDS-UPDRS III: 52.86 ± 16.03) who participated in a clinical study to assess changes in their motor symptoms before (OFF state) and after (ON state) medication intake. Individuals with a clinical diagnosis of PD between 30–80 years old, able to recognize wearing-off periods, with Hoehn & Yahr stage $\leq$ III and currently on L-dopa therapy were eligible to participate in this study. Exclusion criteria included the presence of other comorbidities (e.g., head injuries, psychiatric illness, cardiac disorders), recent treatment with investigational drugs, pregnant women and allergy to silicone or adhesives. The study had approval from the Tufts Medical

Center and Tufts University Health Sciences Institutional Review Board (study ID: HUBB121601) and all experimental procedures were conducted at Tufts Medical Center [45]. All subjects provided written informed consent.

The study protocol included two visits to the clinic; subjects were randomly assigned to be in the ON (after medication intake) or OFF (before medication intake) state for the first visit, and underwent the second visit in the other state. During each study visit, patients performed a battery of motor tasks including activities of daily living (e.g., dressing, writing, drinking from a cup of water, opening a door, folding clothes) and a standard battery of clinical motor assessments from the Movement Disorder Society's Unified Parkinson's Disease Rating Scale (MDS-UPDRS) [46] administered by a trained clinician with experience in movement disorders. Each visit lasted approximately 1 h and most of the experimental activities were video recorded at 30 frames per second by two Microsoft Kinect™ cameras (1080 × 1920-pixel resolution), one mounted on a mobile tripod in the testing room and another on a wall mount in the adjacent hallway. In total, the dataset consists of 70 video recordings (35 subjects × 2 visits per subject). The video camera was positioned to capture a frontal view of the subject at most times. Besides the subject, there are several other people (e.g., physicians, nurses, study staff) who appear in these video recordings.

Behaviors of interest were identified within each video using structured definitions and, their start and end times annotated using human raters as described elsewhere [47]. Briefly, each video recording was reviewed and key behaviors annotated by two trained raters. To maximize inter-rater agreement, each behavior had been explicitly defined to establish specific, anatomically based visual cues for annotating its start and end times. The completed annotations were reviewed for agreement by an experienced arbitrator, who identified and resolved inter-rater disagreements (e.g., different start times for a behavior). The annotated behaviors were categorized into three classes: postures (e.g., walking, sitting, standing), transitions (e.g., sit-to-stand, turning), and cued behaviors (i.e., activities of daily living and MDS-UPDRS tasks). In this manuscript, we focus on the recognition of postures (sitting, standing and walking) and transitions (sitting-to-standing and standing-to-sitting). Recognizing these activities in PD patients provide valuable context for understanding motor symptoms like tremor, bradykinesia and freezing of gait. Major challenges in recognizing activities of the target (i.e., subject) in this dataset were camera motion (when not on tripod), change of scene as the experimental activities took place in different environments (e.g., physician office, clinic hallway, etc.) and long periods occlusion (around a few minutes) due to interactions between the patient and the study staff.

4.2. Tracking Target Human and Pose

Given that video recordings involved the presence of multiple people, we first detected all human actors along with their associated keypoints in each video frame using the multi-person pose estimation method described in Section 2.1 (illustrated in Figure 3). This pose estimation network was pre-trained on the COCO dataset and fine-tuned on the PoseTrack dataset previously [30,33,34]. As illustrated in Figure 4, the output of this stage is a list of the bounding boxes for human actors detected in each video frame and the estimated locations of keypoints for each person along with a confidence estimate for each keypoint.

In order to recognize activities of the target, we first locate and track the subject (i.e., PD patient) in each frame. This was accomplished by using the hierarchical tracking method described in Section 2.2. Given all detected bounding boxes across all frames from the pose estimation stage, we first generate tracklets for each identity appearing in the video via short-term tracking explained in Section 2.2.1. Each tracklet is a list of detected bounding boxes in consecutive frames that belong to the same identity. In order to find the final patient track for the entire video, we use the long-term tracking method described in Section 2.2.2 to remove non-target tracklets (e.g., study staff, physician, nurse) and fuse the tracklets that belong to the patient using the appearance features. There is no supervision in tracking of the patient during the video except providing a reference tracklet, which is associated to with the target (i.e., subject) in the long-term tracking step.

To evaluate the performance of our target tracking method, we first manually annotated all tracklets generated by short-term tracking and then calculated accuracy of the long-term tracking method with respect to the manually generated ground-truth. Accuracy is calculated by treating the long-term tracker as a binary classifier as it excludes non-patient tracklets and fuses tracklets belonging to the target to find a single final patient track for the entire video recording. Considering patient tracklets as the positive class and non-patient tracklets as the negative class, our tracker achieved an average classification accuracy of 88% across 70 videos on this dataset.

4.3. Action Classification

In the last stage of our multi-stage target-specific action classification system, we trained a CNN to recognize the following five actions of interest: sitting, standing, walking, sitting-to-standing, and standing-to-sitting. After applying the target pose tracking system illustrated in Figure 4, we segmented the resulting long-term video into action clips based on ground-truth annotations provided by human raters. Although the action clips have variable lengths (ranging from a few frames to more than 10 min), each video clip includes only one of the five actions of interest. As a result, we ended up with a highly imbalanced dataset. In order to create a more balanced dataset for training and evaluating the action classification network, we first excluded action clips less than 0.2 s (too short for dynamic activities like walking) and divided the ones longer than four seconds into four-second clips. Assuming that four seconds is long enough for most activities of interest and below 0.2 s (lower than six frames) is too short to be used for recognizing an action [48]. This resulted in a total of 44,580 action clips extracted from video recordings of 35 subjects. We used 29 subjects (39,086 action clips) for training/validation set and the remaining 6 subjects (5494 action clips) were held out for testing. As shown in Figure 8, the resulting dataset is highly imbalanced with a significant skew towards the sitting class, which can result in over-fitting issues. To address this imbalance, we randomly under-sampled the walking, sitting, and standing classes to 4000 video clips each.

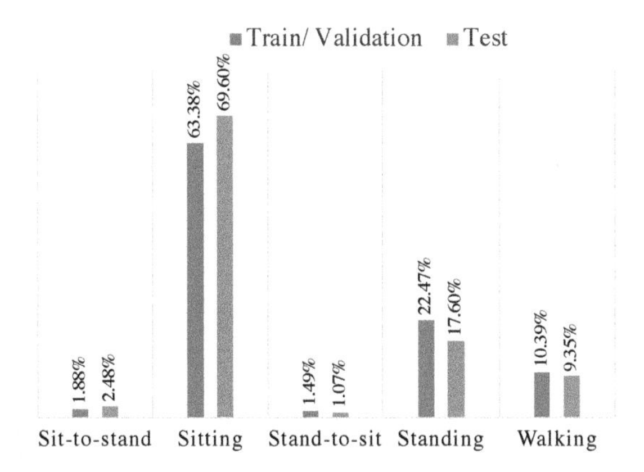

Figure 8. Distribution of the action clips based on the type of the actions for test and train/validation datasets. The distribution of the original set of action clips is highly imbalanced.

To prepare input data for the action classification network, we transformed each action clip into a pose evolution representation as described in Section 3.1. To create the pose evolution maps, we scaled the original size of each video frame (1080 × 1920) by a factor of 0.125 and chose 3 channels to represent the input based on training time and average validation accuracy in diagnostic experiments. The training dataset was also augmented by adding Gaussian noise. In addition, we tried data augmentation techniques like random translation and flipping during our diagnostic experiments, but the classification performance degraded by about 3%. Therefore, we only used additive Gaussian noise to randomly selected video frames as the only type of data augmentation.

We used 90% of the train/validation dataset for training the action classification network with architecture illustrated in Figure 7 and the rest for validation. The network training started with random weight initialization and we used the Adam optimizer with a base learning rate of 0.01, a batch size of 70 and a dropout probability of 0.3. We experimented with several variants of the network architecture proposed in Section 3 by increasing the number of the convolution blocks to three and changing the number of filters in each block to 64, 128, 256, and 512. Based on the performance on the validation set and training loss, Figure 7 provided the best performance while avoiding over-fitting to the training data. In addition, we investigated the impact of using a different number of channels for representing the temporal pose evolution on the performance of action classification. Figure 9 illustrates the accuracy of the classification network with different representations as input. We chose 3 channels for our representation because adding more channels would only increase the computational cost without any significant improvement in accuracy. The trained action classification model achieved a best-case weighted classification accuracy of 83.97% on the test dataset. In order to demonstrate the importance of target tracking on the performance of the action classification network, we conducted another experiment without using any tracking on the recorded videos. The results show that while the best case weighted overall accuracy for the validation set was slightly better (84.04%), it dropped to 63.14% on the test set. This is an indication that the model is not able to generalize well, because the quality of training data degrades without target tracking. More details of the classification performance including per class accuracy in the test and validations phase can be found in Table 1.

Table 1. Best case classification accuracy (%) per action class on the validation and test set with and without long-term tracking. Weighted overall accuracy was calculated to account for the class imbalance. Mean and standard deviation (std) of the weighted average accuracy (last column) were calculated by training the network 10 times using the same hyper-parameters but with different initialization and evaluating it on the validation split.

	Sit	Sit-to-Stand	Stand	Walk	Stand-to-Sit	Weighted Overall Accuracy	Mean ± Std. of Average Accuracy
				With long-term tracking			
Validation	92.8	68.1	81.5	78.9	70.7	82.00	79.85 ± 2.38
Test	91.6	75.0	85.7	81.0	78.6	83.97	-
				Without long-term tracking			
Validation	90.9	88.1	91.0	71.8	75.8	84.04	71.42 ± 10.32
Test	72.6	63.9	81.6	51.7	16.3	63.14	-

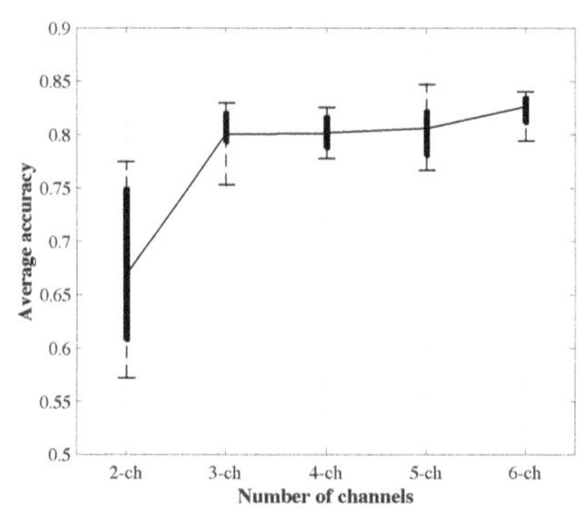

Figure 9. Average classification accuracy with respect to the number of channels of input pose evolution representations.

5. Discussion

Real-world assessment of motor behaviour can provide valuable clinical insights for diagnosis, prognosis and prevention of disorders ranging from psychiatry to neurology [49,50]. In this paper, we propose a new approach for automated assessment of target-specific behavior from video recordings in the presence of other actors, occlusions and changes in scene. This approach relies on using temporal relationships for short-term tracking and appearance-based features for long-term tracking of the target. Short-term tracking based on temporal relationships between adjacent frames resulted in 1466 ± 653 tracklets per video, which were then fused by using appearance-based features for long-term tracking. Using this approach, we were able to identify the target track throughout the video recording with an accuracy of 88% in our dataset of 70 videos belonging to 35 targets (i.e., PD patients). However, one of the limitations of our dataset was that the target's appearance did not change significantly (except for a brief period when the subject put a lab coat on to perform a task) over the duration of the recording. This is unlikely in the real-world as we expect appearance to change on a daily basis (e.g., clothing, makeup) as well as over weeks and months (e.g., age or disease-related changes). Therefore, the proposed method requires further validation on a larger dataset collected during daily life and would benefit from strategies for dealing with changes in appearance.

The second aspect of our work focused on classification of activities of daily living. Activities like sitting, standing, sit-to-stand, stand-to-sit and walking are basic elements of most of the tasks that we perform during daily life. To train the activity classification model, we used pose evolution representations to capture both temporal and spatial features associated with these activities. While this model achieved a classification accuracy of 84%, as we can see in Figure 10, a significant source of error was the misclassification of 18% (25/142) of walking as standing. This could be attributed to two factors. Firstly, video recordings of the walking activity were performed with a frontal view of the subject, which limits the ability of pose evolution representations to capture features associated with spatial displacement during walking. As a result, pose evolution representations of walking and standing look similar. This would be challenging to deal with in real-world scenarios because the camera's field of view is typically fixed. This limitation highlights the need for developing methods that are robust to changes in feature maps associated with different fields of views. Secondly, the activity transition period from standing to walking was labeled as walking during the ground-truth annotation process. As a result, when the action classification network is applied to short action clips, those containing such transitions are more likely to be misclassified as standing. Examples of the aforementioned misclassification are illustrated in Figure 11. In Figure 11a the subject takes a couple of steps to reach for a coat hanging on the rack and in Figure 11b the subject is about to start walking from a standing position. In both cases, the ground-truth annotation was walking but the video clip was classified as standing. This is a potential limitation of a video-based action recognition approach as its performance will be dependent on factors like the camera view. By comparison, approaches using one or more wearable sensors (e.g., accelerometers and gyroscopes on the thigh, chest and ankle [51]) are relatively robust to such problems as their measurements are performed in the body frame of reference, which results in high classification accuracy (>95%) across a range of daily activities.

Another source of error which impacts overall performance is the error propagated from pose tracking (~12%) and pose estimation stages. Pose estimation error can be tolerated to some degree by aggregating the colorized joint heatmaps in the pose evolution feature representation. However, since the output of the pose tracking is directly used for generating pose evolution maps, any error in tracking the patient throughout the video would negatively impact the action classification performance. One approach for tolerating the error from pose tracking stage is to incorporate raw RGB frames as the second stream of information for action classification and using attention maps based on the tracking outcome rather than excluding non-target persons from the input representations.

Vision-based monitoring tools have the distinct advantage of being transparent to the target, which would help with issues of compliance associated with the use of wearable devices. Also, unlike wearable devices, vision-based approaches can capture contextual information, which is necessary for understanding behavior at higher level. However, this also comes at an increased risk to privacy for the target as well as other people in the environment. The proposed approach can potentially mitigate this concern by limiting monitoring to the target (e.g., patient) and transforming data at the source into sparse feature maps (i.e., pose evolution representations).

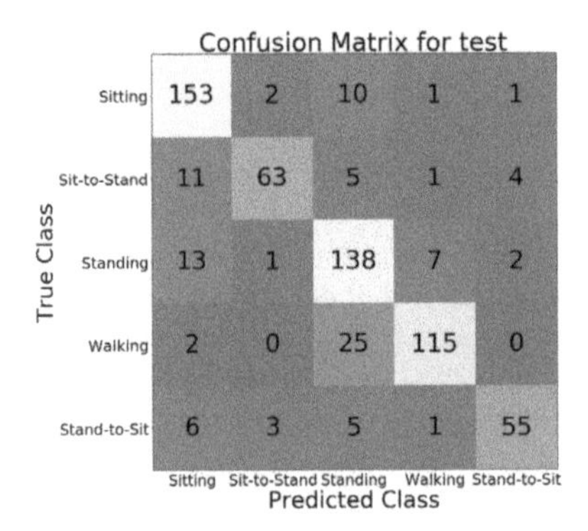

Figure 10. Confusion matrix of the action recognition network evaluated on the test dataset.

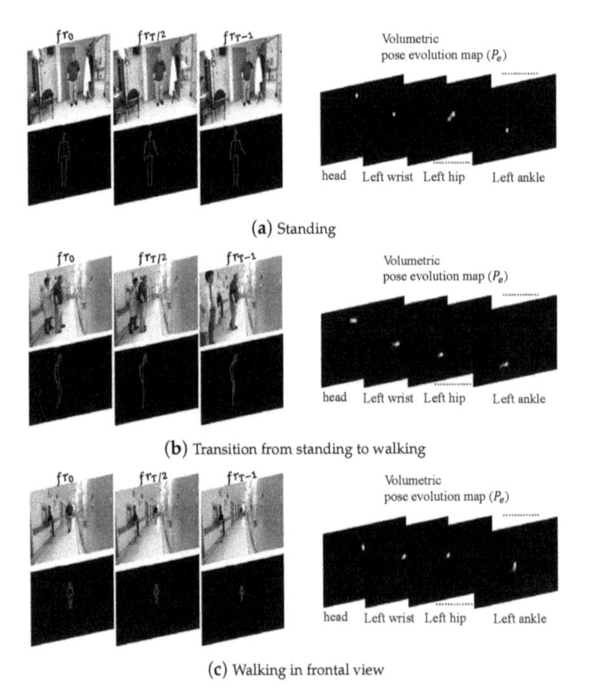

Figure 11. An example of the misclassification of walking as standing. (**a**–**c**) The first, middle, and last frame of three action video clips along with the corresponding pose estimations and pose evolution maps. During the manual annotation process (**a**) was labeled as standing, whereas (**b**,**c**) were labeled as walking. The action classification network classifies (**a**,**b**) as standing because they have a very similar pose evolution map (best viewed in color and zoomed in).

6. Conclusions and Future Work

In this paper, we have presented an AI-assisted method for automatic assessment of human motor behavior from video recorded using a single RGB camera. Results demonstrate that the multi-stage method, which includes pose estimation, target tracking and action classification, provides an accurate target-specific classification of activities in the presence of other human actors and is robust to changing environments. The work presented herein focused on the classification of basic postures (sitting, standing and walking) and transitions (sitting-to-standing and standing-to-sitting), which commonly occur during the performance of many daily activities and are relevant to understanding the impact of diseases like Parkinson's disease and stroke on the functional ability of patients. This has laid the foundation for future research efforts that will be directed towards detecting and quantifying clinically meaningful information like detection of emergency events (e.g., falls, seizures) and assessment of symptom severity (e.g., gait impairments, tremor) in patients with various mobility limiting conditions. The proposed method is mainly intended for offline processing of video recording. To provide real-time detection of serious events (e.g., falls), future research efforts should focus on enabling real-time target tracking and action classification by developing more computationally efficient approaches. In addition, achieving high-resolution temporal localization of actions will be necessary to ensure accurate assessment of clinical events of interest (e.g., duration of a seizure) for certain medical applications. Lastly, the code and models developed during this work are being made available for the benefit of the broader research community (https://github.com/brezaei/PoseTrack_ActionClassification).

Author Contributions: Conceptualization: B.R., S.O. and S.P.; data curation: B.R., Y.C., B.H., K.T. and K.E.; formal analysis: B.R.; investigation, B.H. and K.E.; methodology: B.R., S.O. and S.P.; project administration: K.E.; resources: B.H.; software: B.R.; supervision: S.O. and S.P.; writing—original draft: B.R. and S.P.; writing—review and editing: B.R., Y.C., B.H., K.T., K.E., S.O. and S.P.

Funding: Funding for this work was provided by Pfizer, Inc.

Acknowledgments: The authors would like to acknowledge the BlueSky project team for generating the data that made this work possible. Specifically, we would like to acknowledge Hao Zhang, Steve Amato, Vesper Ramos, Paul Wacnik and Tairmae Kangarloo for their contributions to study design and data collection.

Conflicts of Interest: S.P., Y.C. and M.K.E. are employees of Pfizer, Inc. The remaining authors declare no conflict of interest.

References

1. Post, B.; Merkus, M.P.; de Bie, R.M.; de Haan, R.J.; Speelman, J.D. Unified Parkinson's disease rating scale motor examination: Are ratings of nurses, residents in neurology, and movement disorders specialists interchangeable? *Mov. Disord. Off. J. Mov. Disord. Soc.* **2005**, *20*, 1577–1584. [CrossRef] [PubMed]
2. Espay, A.J.; Bonato, P.; Nahab, F.B.; Maetzler, W.; Dean, J.M.; Klucken, J.; Eskofier, B.M.; Merola, A.; Horak, F.; Lang, A.E.; et al. Movement Disorders Society Task Force on Technology. Technology in Parkinson's disease: Challenges and opportunities. *Mov. Disord.* **2016**, *31*, 1272–1282. [CrossRef] [PubMed]
3. Thorp, J.E.; Adamczyk, P.G.; Ploeg, H.L.; Pickett, K.A. Monitoring Motor Symptoms During Activities of Daily Living in Individuals With Parkinson's Disease. *Front. Neurol.* **2018**, *9*, 1036. [CrossRef] [PubMed]
4. Lara, O.D.; Labrador, M.A. A survey on human activity recognition using wearable sensors. *IEEE Commun. Surv. Tutor.* **2013**, *15*, 1192–1209. [CrossRef]
5. van Nimwegen, M.; Speelman, A.D.; Hofman-van Rossum, E.J.M.; Overeem, S.; Deeg, D.J.H.; Borm, G.F.; van der Horst, M.H.L.; Bloem, B.R.; Munneke, M. Physical inactivity in Parkinson's disease. *J. Neurol.* **2011**, *258*, 2214–2221. [CrossRef] [PubMed]
6. Chaaraoui, A.A.; Climent-Pérez, P.; Flórez-Revuelta, F. A review on vision techniques applied to human behaviour analysis for ambient-assisted living. *Expert Syst. Appl.* **2012**, *39*, 10873–10888. [CrossRef]
7. Vrigkas, M.; Nikou, C.; Kakadiaris, I.A. A review of human activity recognition methods. *Front. Robot. AI* **2015**, *2*, 28. [CrossRef]
8. Chen, Y.; Yu, L.; Ota, K.; Dong, M. Robust Activity Recognition for Aging Society. *IEEE J. Biomed. Health Inform.* **2018**, *22*, 1754–1764. [CrossRef]
9. Li, M.H.; Mestre, T.A.; Fox, S.H.; Taati, B. Vision-based assessment of parkinsonism and levodopa-induced dyskinesia with pose estimation. *J. Neuroeng. Rehabil.* **2018**, *15*, 97. [CrossRef]

10. Brattoli, B.; Buchler, U.; Wahl, A.S.; Schwab, M.E.; Ommer, B. LSTM Self-Supervision for Detailed Behavior Analysis. In Proceedings of the IEEE Conference on Computer Vision and Pattern Recognition (CVPR), Honolulu, HI, USA, 21–26 July 2017; pp. 6466–6475.
11. Song, S.; Shen, L.; Valstar, M. Human behaviour-based automatic depression analysis using hand-crafted statistics and deep learned spectral features. In Proceedings of the 2018 13th IEEE International Conference on Automatic Face & Gesture Recognition (FG 2018), Xi'an, China, 15–19 May 2018; pp. 158–165.
12. Schmitt, F.; Bieg, H.J.; Herman, M.; Rothkopf, C.A. I see what you see: Inferring sensor and policy models of human real-world motor behavior. In Proceedings of the Thirty-First AAAI Conference on Artificial Intelligence, San Francisco, CA, USA, 4–9 February 2017.
13. Chen, A.T.; Biglari-Abhari, M.; Wang, K.I. Trusting the Computer in Computer Vision: A Privacy-Affirming Framework. In Proceedings of the 2017 IEEE Conference on Computer Vision and Pattern Recognition Workshops (CVPRW), Honolulu, HI, USA, 21–26 July 2017; pp. 1360–1367.
14. Rezaei, B.; Ostadabbas, S. Background Subtraction via Fast Robust Matrix Completion. In Proceedings of the 2017 IEEE International Conference on Computer Vision Workshop (ICCVW), Venice, Italy, 22–29 October 2017; pp. 1871–1879.
15. Rezaei, B.; Huang, X.; Yee, J.R.; Ostadabbas, S. Long-term non-contact tracking of caged rodents. In Proceedings of the 2017 IEEE International Conference on Acoustics, Speech and Signal Processing (ICASSP), New Orleans, LA, USA, 5–9 March 2017; pp. 1952–1956.
16. Rezaei, B.; Ostadabbas, S. Moving Object Detection through Robust Matrix Completion Augmented with Objectness. *IEEE J. Sel. Top. Signal Process.* **2018**, *12*, 1313–1323, doi:10.1109/JSTSP.2018.2869111. [CrossRef]
17. Herath, S.; Harandi, M.; Porikli, F. Going deeper into action recognition: A survey. *Image Vis. Comput.* **2017**, *60*, 4–21. [CrossRef]
18. Dawar, N.; Ostadabbas, S.; Kehtarnavaz, N. Data Augmentation in Deep Learning-Based Fusion of Depth and Inertial Sensing for Action Recognition. *IEEE Sens. Lett.* **2018**, *3*, 1–4. [CrossRef]
19. Girdhar, R.; Carreira, J.; Doersch, C.; Zisserman, A. Video action transformer network. In Proceedings of the IEEE Conference on Computer Vision and Pattern Recognition, Long Beach, CA, USA, 16–21 June 2019; pp. 244–253.
20. Zhang, H.B.; Zhang, Y.X.; Zhong, B.; Lei, Q.; Yang, L.; Du, J.X.; Chen, D.S. A comprehensive survey of vision-based human action recognition methods. *Sensors* **2019**, *19*, 1005. [CrossRef] [PubMed]
21. Li, N.; Huang, J.; Li, T.; Guo, H.; Li, G. Detecting action tubes via spatial action estimation and temporal path inference. *Neurocomputing* **2018**, *311*, 65–77. [CrossRef]
22. Simonyan, K.; Zisserman, A. Two-stream convolutional networks for action recognition in videos. In Proceedings of the Advances in Neural Information Processing Systems, Montreal, ON, Canada, 8–13 December 2014; pp. 568–576.
23. Zhou, Y.; Sun, X.; Zha, Z.J.; Zeng, W. MiCT: Mixed 3D/2D Convolutional Tube for Human Action Recognition. In Proceedings of the IEEE Conference on Computer Vision and Pattern Recognition, Salt Lake City, UT, USA, 18–22 June 2018; pp. 449–458.
24. Tran, D.; Wang, H.; Torresani, L.; Ray, J.; LeCun, Y.; Paluri, M. A Closer Look at Spatiotemporal Convolutions for Action Recognition. In Proceedings of the IEEE Conference on Computer Vision and Pattern Recognition, Salt Lake City, UT, USA, 18–22 June 2018; pp. 6450–6459.
25. Tran, D.; Bourdev, L.; Fergus, R.; Torresani, L.; Paluri, M. Learning spatiotemporal features with 3d convolutional networks. In Proceedings of the IEEE International Conference on Computer Vision, Santiago, Chile, 7–13 December 2015; pp. 4489–4497.
26. Liu, M.; Yuan, J. Recognizing Human Actions as the Evolution of Pose Estimation Maps. In Proceedings of the IEEE Conference on Computer Vision and Pattern Recognition, Salt Lake City, UT, USA, 18–22 June 2018.
27. Choutas, V.; Weinzaepfel, P.; Revaud, J.; Schmid, C. PoTion: Pose MoTion Representation for Action Recognition. In Proceedings of the IEEE Conference on Computer Vision and Pattern Recognition, Salt Lake City, UT, USA, 18–22 June 2018.
28. Cherian, A.; Sra, S.; Gould, S.; Hartley, R. Non-Linear Temporal Subspace Representations for Activity Recognition. In Proceedings of the IEEE Conference on Computer Vision and Pattern Recognition, Salt Lake City, UT, USA, 18–22 June 2018; pp. 2197–2206.
29. Zolfaghari, M.; Oliveira, G.L.; Sedaghat, N.; Brox, T. Chained Multi-stream Networks Exploiting Pose, Motion, and Appearance for Action Classification and Detection. In Proceedings of the IEEE International Conference on Computer Vision (ICCV), Venice, Italy, 22–29 October 2017.

30. Girdhar, R.; Gkioxari, G.; Torresani, L.; Paluri, M.; Tran, D. Detect-and-Track: Efficient Pose Estimation in Videos. In Proceedings of the IEEE Conference on Computer Vision and Pattern Recognition, Salt Lake City, UT, USA, 18–22 June 2018; pp. 350–359.
31. He, K.; Gkioxari, G.; Dollár, P.; Girshick, R. Mask r-cnn. In Proceedings of the IEEE International Conference on Computer Vision, Venice, Italy, 22–29 October 2017; pp. 2980–2988.
32. Deng, J.; Dong, W.; Socher, R.; Li, L.J.; Li, K.; Fei-Fei, L. Imagenet: A large-scale hierarchical image database. In Proceedings of the IEEE Conference on Computer Vision and Pattern Recognition, Miami, FL, USA, 20–25 June 2009; pp. 248–255.
33. Lin, T.Y.; Maire, M.; Belongie, S.; Hays, J.; Perona, P.; Ramanan, D.; Dollár, P.; Zitnick, C.L. Microsoft coco: Common objects in context. In *European Conference on Computer Vision*; Springer: Berlin/Heidelberg, Germany, 2014, pp. 740–755.
34. Andriluka, M.; Iqbal, U.; Milan, A.; Insafutdinov, E.; Pishchulin, L.; Gall, J.; Schiele, B. Posetrack: A benchmark for human pose estimation and tracking. In Proceedings of the IEEE Conference on Computer Vision and Pattern Recognition, Salt Lake City, UT, USA, 18–22 June 2018; pp. 5167–5176.
35. He, K.; Zhang, X.; Ren, S.; Sun, J. Deep residual learning for image recognition. In Proceedings of the IEEE Conference on Computer Vision and Pattern Recognition, Las Vegas, NV, USA, 26 June 26–1 July 2016; pp. 770–778.
36. Ren, S.; He, K.; Girshick, R.; Sun, J. Faster R-CNN: Towards Real-Time Object Detection with Region Proposal Networks. *IEEE Trans. Pattern Anal. Mach. Intell.* **2016**, *39*, 1137–1149. [CrossRef] [PubMed]
37. Gou, M.; Wu, Z.; Rates-Borras, A.; Camps, O.; Radke, R.J.; others. A systematic evaluation and benchmark for person re-identification: Features, metrics, and datasets. *IEEE Trans. Pattern Anal. Mach. Intell.* **2018**, *41*, 523–536.
38. Gou, M.; Camps, O.; Sznaier, M. Mom: Mean of moments feature for person re-identification. In Proceedings of the IEEE International Conference on Computer Vision, Venice, Italy, 22–29 October 2017; pp. 1294–1303.
39. Liao, S.; Hu, Y.; Zhu, X.; Li, S.Z. Person re-identification by local maximal occurrence representation and metric learning. In Proceedings of the IEEE Conference on Computer Vision and Pattern Recognition, Boston, MA, USA, 7–12 June 2015; pp. 2197–2206.
40. Ahmed, E.; Jones, M.; Marks, T.K. An improved deep learning architecture for person re-identification. In Proceedings of the IEEE Conference on Computer Vision and Pattern Recognition, Boston, MA, USA, 7–12 June 2015; pp. 3908–3916.
41. Li, M.; Zhu, X.; Gong, S. Unsupervised person re-identification by deep learning tracklet association. In Proceedings of the European Conference on Computer Vision (ECCV), Munich, Germany, 8–14 September 2018; pp. 737–753.
42. Lv, J.; Chen, W.; Li, Q.; Yang, C. Unsupervised cross-dataset person re-identification by transfer learning of spatial-temporal patterns. In Proceedings of the IEEE Conference on Computer Vision and Pattern Recognition, Salt Lake City, UT, USA, 18–22 June 2018; pp. 7948–7956.
43. Pirsiavash, H.; Ramanan, D.; Fowlkes, C.C. Globally-optimal greedy algorithms for tracking a variable number of objects. In Proceedings of the IEEE Conference on Computer Vision and Pattern Recognition, Springs, CO, USA, 20–25 June 2011; pp. 1201–1208.
44. Kuhn, H.W. The Hungarian method for the assignment problem. *Nav. Res. Logist.* **2005**, *52*, 7–21. [CrossRef]
45. Erb, K.; Daneault, J.; Amato, S.; Bergethon, P.; Demanuele, C.; Kangarloo, T.; Patel, S.; Ramos, V.; Volfson, D.; Wacnik, P.; et al. The BlueSky Project: Monitoring motor and non-motor characteristics of people with Parkinson's disease in the laboratory, a simulated apartment, and home and community settings. In Proceedings of the 2018 International Congress, Hong Kong, China, 5–9 October 2018; Volume 33, p. 1990.
46. Goetz, C.G.; Tilley, B.C.; Shaftman, S.R.; Stebbins, G.T.; Fahn, S.; Martinez-Martin, P.; Poewe, W.; Sampaio, C.; Stern, M.B.; Dodel, R.; et al. Movement Disorder Society-sponsored revision of the Unified Parkinson's Disease Rating Scale (MDS-UPDRS): Scale presentation and clinimetric testing results. *Mov. Disord. Off. J. Mov. Disord. Soc.* **2008**, *23*, 2129–2170. [CrossRef] [PubMed]
47. Brooks, C.; Eden, G.; Chang, A.; Demanuele, C.; Kelley Erb, M.; Shaafi Kabiri, N.; Moss, M.; Bhangu, J.; Thomas, K. Quantification of discrete behavioral components of the MDS-UPDRS. *J. Clin. Neurosci.* **2019**, *61*, 174–179. [CrossRef] [PubMed]
48. Barrouillet, P.; Bernardin, S.; Camos, V. Time constraints and resource sharing in adults' working memory spans. *J. Exp. Psychol. Gen.* **2004**, *133*, 83. [CrossRef]

49. Insel, T.R. Digital Phenotyping: Technology for a New Science of Behavior. *JAMA* **2017**, *318*, 1215–1216. [CrossRef] [PubMed]
50. Arigo, D.; Jake-Schoffman, D.E.; Wolin, K.; Beckjord, E.; Hekler, E.B.; Pagoto, S.L. The history and future of digital health in the field of behavioral medicine. *J. Behav. Med.* **2019**, *42*, 67–83. [CrossRef] [PubMed]
51. Attal, F.; Mohammed, S.; Dedabrishvili, M.; Chamroukhi, F.; Oukhellou, L.; Amirat, Y. Physical Human Activity Recognition Using Wearable Sensors. *Sensors* **2015**, *15*, 31314–31338. [CrossRef] [PubMed]

Article

A New Approach for Motor Imagery Classification Based on Sorted Blind Source Separation, Continuous Wavelet Transform, and Convolutional Neural Network

César J. Ortiz-Echeverri [1], Sebastián Salazar-Colores [1], Juvenal Rodríguez-Reséndiz [2,*] and Roberto A. Gómez-Loenzo [2]

[1] Facultad de Informática, Universidad Autónoma de Querétaro, C.P. 76230 Querétaro, Mexico; cortiz08@alumnos.uaq.mx (C.J.O.-E.); ssalazar05@alumnos.uaq.mx (S.S.-C.)

[2] Facultad de Ingniería, Universidad Autónoma de Querétaro, C.P. 76010 Querétaro, Mexico; rob@uaq.mx

* Correspondence: juvenal@uaq.edu.mx; Tel.: +52-442-192-12-00

Received: 21 September 2019; Accepted: 15 October 2019; Published: 18 October 2019

Abstract: Brain-Computer Interfaces (BCI) are systems that allow the interaction of people and devices on the grounds of brain activity. The noninvasive and most viable way to obtain such information is by using electroencephalography (EEG). However, these signals have a low signal-to-noise ratio, as well as a low spatial resolution. This work proposes a new method built from the combination of a Blind Source Separation (BSS) to obtain estimated independent components, a 2D representation of these component signals using the Continuous Wavelet Transform (CWT), and a classification stage using a Convolutional Neural Network (CNN) approach. A criterion based on the spectral correlation with a Movement Related Independent Component (MRIC) is used to sort the estimated sources by BSS, thus reducing the spatial variance. The experimental results of 94.66% using a k-fold cross validation are competitive with techniques recently reported in the state-of-the-art.

Keywords: Brain-Computer Interface; Blind Source Separation; Movement Related Independent Component; Wavelet Transform; Convolutional Neural Network

1. Introduction

The Brain-Computer Interface (BCI) is a method of communication between a user and a system, where the intention of the subject is translated into a control signal by classifying the specific pattern which is characteristic of the imagined task, for example, the movement of the hand and/or foot [1]. The most widely used technique to register the electrical activity for BCI applications is the electroencephalography (EEG), which is a non-invasive and low-cost technique. The recording is done by placing electrodes on the scalp according to the 10–20 system [2], which records electrical impulses associated with neuronal activity in the brain cortex. The BCI can be based on exogenous such as the event-related P300 potential and Visual Evoked Potentials (VEPs), or endogenous potentials, where Motor Imagery (MI) widely used in BCI applications is the dynamic state where a subject evokes a movement or gesture. The event related phenomena represent frequency-specific changes in the ongoing EEG activity and may consist, in general terms, of either decreases or increases of power in given frequency bands [1]. Most of the brain activity is concentrated in electrophysiological bands called: delta δ (0.5–4 Hz), theta θ (4–7.5 Hz), alpha α (8–13 Hz), and beta β (14–26 Hz) [2]. Another important frequency for applications in BCI is the μ or sensorimotor rhythm, with the same frequency bands as α, but located in the motor cortex instead of the visual cortex where α is mainly generated [3]. There are several works which report the importance of μ frequencies for MI detection [3–7], where Pfurtscheller et al. published [8–12]. They demonstrate the changes of EEG activity in μ and β rhythms caused by voluntary movements.

Endogenous MI-BCI-based system does not require external stimuli, hence it is more acceptable to the users [4]. Nonetheless, MI depends on the ability to control the electrophysiological activity, which makes feature extraction and classification for MI-BCI based system more difficult than for exogenous responses. One of the major limitations of EEG records is the low signal-to-noise ratio and the fact that the signals picked up at the electrodes are a mixture of sources that cannot be observed directly by non-invasive methods. Therefore, for endogenous BCI approaches, a preprocessing step is required to identify independent sources of the mixtures observed in the electrodes. A well-known preprocessing method is based on the decomposition of multi-channel EEG data into spatial patterns which are calculated from two classes of MI, known as Common Spatial Patterns (CSP) [13,14]. CSP is a supervised method where class information must be available a priori and its effectiveness relies on the subject-specific frequency bands [7,15].

As an unsupervised alternative to the estimation of independent sources, Blind Source Separation (BSS) algorithms have been incorporated in EEG preprocessing, mainly in medical applications to improve the tasks of diseases diagnosis [16]. BSS algorithms make the source estimations from the mixed observation using statistical information. It has been shown that BSS is especially suitable for removing a wide variety of artifacts in EEG recordings [17] and separating μ rhythms generated in both brain hemispheres [18]. Therefore, BSS is a useful method for constructing spatial filters for preprocessing raw multi-channel EEG data in BCI research [15].

Due to its unsupervised and statistical nature, BSS does not require a priori information about MI classes, nor specific frequency bands, which is an advantage over CSP approaches. Nonetheless, an inherent disadvantage of BSS algorithms is that for each processed trial, the order is not preserved, which limits its direct application in further classifier stages used in BCI, where the order of the input vectors must be conserved to avoid loss of the adjustment parameters for each new data entry. Some automated BSS approaches have been proposed to discern between sources of interest and artifacts, and thus minimize the aforementioned inconvenience making use of statistical concepts [19–21].

In the classification stage, the most widely used approaches are Linear Discriminant Analysis (LDA) [22], Support Vector Machine (SVM) [23], Multilayer Perceptron [24], and Bayesian classifier [25]. A recent approach that has given excellent results, mainly in computer vision is deep learning [26]. However, deep learning techniques have not been widely used for EEG-BCI applications, due to factors such as as noise, the correlation between channels, and the high dimensional EEG data [27]. Some works where deep learning has been used for MI classification have been proposed [27–35]. However, for MI-BCI based paradigm the datasets are small due to the fatigue where the participants are exposed in each session. Therefore, it has been difficult to use deep learning for this purpose [32].

In this research, a fastICA BSS algorithm is used to obtain estimated independent components. A typical spectral profile of Movement Related Independent Components (MRIC) with significant components in the μ and β frequencies is used for sorting in each processed trial, thus ensuring that the sources estimated to be the most active in MI frequencies remain at the beginning of the array, while the artifacts are placed in the final positions. For each estimated source, the Continuous Wavelet Transform (CWT) is calculated for a given time window, generating an image containing temporal, frequential, and spatial information. This process is carried out throughout all trials, forming a set of images to train and test a Convolutional Neural Network (CNN).

A contribution in the present work is the use of BSS instead of the widely used CSP. Even though this has worked well for MI-BCI based, these spatial filters require prior information of the classes to be separated in order to maximize the differences between them. In addition, BSS is an unsupervised approach that does not require prior information about the classes. The problem of large datasets needed for training is minimized using the MRIC criterion to sort the estimated sources. The paper is structured as follows: Section 2 the background of BSS, CWT, and CNN are explained.

Section 3 the proposed methodology to obtain CWT maps from estimated sources is described, along with the details of the CNN architecture. The experimental results and discussion are presented in Section 4. Finally, conclusions and a future work overview are presented in Section 5.

2. Background

In this section, theoretical background about preprocessing, feature extraction, and classification stages are presented.

2.1. Blind Source Separation

Firstly, BSS is an approach to estimate and recover the original sources using only the information of the observed mixture at the recording channels. For the instantaneous and ideal case, where the source signals arrive to the sensors at the same time such as in the EEG, the mathematical model of BSS can be expressed as

$$\mathbf{x}(t) = \mathbf{A}\mathbf{s}(t), \tag{1}$$

where $\mathbf{x}(t)$ is the vector of the mixed signals, $\mathbf{A}$ is the unknown non-singular mixing matrix, and $\mathbf{s}(t)$ is the vector of sources [2]. Then, BSS consists of estimating a separation matrix $\mathbf{B}$ such that in an ideal case $\mathbf{B} = \mathbf{A}^{-1}$, and then computed the sources $\mathbf{s}'(t)$ as

$$\mathbf{s}'(t) = \mathbf{B}\mathbf{x}(t), \tag{2}$$

The BSS algorithms are grouped into two main categories: (1) Second-Order Statistics (SOS), built from correlations and time-delayed windowed signals [36]; and (2) High Order Statistics (HOS) which are based on the optimization of relevant elements in the Probability Density Function (PDF) [37]. The prior SOS algorithms search for the independence of signals based on the criterion of correlation between them. However, uncorrelatedness does not imply independence in all cases. In the HOS algorithms, the assumption to find the estimated sources is that the independent sources have non-Gaussian PDF, while the mixtures present a Gaussian distribution, which is valid for most cases including EEG signals [38]. The classical algebraic approach of ICA is based on negentropy, which is a measure of the entropy of a random variable $H(y)$ with respect to the entropy of a Gaussian distribution variable $H_g(y)$. By definition negentropy is the Kullback-Liebler divergence between a density of probability $p(\mathbf{y})$ and the Gaussian probability density of $g(\mathbf{y})$ of the same mean and same variance. The negentropy J of $\mathbf{y}$ is defined as

$$J(\mathbf{y}) = H_g(\mathbf{y}) - H(\mathbf{y}), \tag{3}$$

According to the definition of mutual information, $I(\mathbf{y})$ can be expressed as

$$I(\mathbf{y}) = J(\mathbf{y}) - \sum_i J(y_i) + \frac{1}{2}\log\frac{\prod_i v_{ii}}{\det V}, \tag{4}$$

where $\mathbf{V}$ is the variance-covariance matrix of $\mathbf{y}$, with diagonal elements v_{ii}. Since maximizing independence means minimizing mutual information, maximizing of the sum of marginal negentropies is to minimize mutual information after whitening of the observations. This method is also similar to those using the notion of kurtosis. Under these conditions, the separation problem consists in the search for a rotation matrix $\mathbf{J}$ such that $\sum_i H(y_i)$ is minimal. The use of second-order moment is not sufficient to decide if non-Gaussian variables are independent. On the other hand, the use of cumulants (cross-referenced) of all kinds makes it possible to show whether variables are independent or not: if all the crossed cumulants of a set of random variables of all kinds are null, then the random variables are independent.

2.2. Wavelet Transform

Second, EEG registers are initially time series, from which it is possible to obtain information related to the temporal evolution of some characteristics. However, in this space, it is not possible to know frequency information. On the other hand, the Fourier transformation makes it possible to identify information in the frequency domain, but the temporal evolution of the frequency components is unknown. The CWT generates 2D maps from 1D time series containing information of time and scale, with a logarithmic relationship with the frequency components. Unlike the Short Time Fourier Transform (STFT), CWT performs a multiresolution analysis [39]. CWT is described by

$$W_x(a,\tau) = \frac{1}{\sqrt{|a|}} \int_{-\infty}^{\infty} x(t)\psi^* \left(\frac{t-\tau}{a}\right) dt, \tag{5}$$

where a is the scale factor, ψ is the mother wavelet, and τ is the shift time of the mother wavelet on the $x(t)$ signal.

2.3. Convolutional Neural Network

The Convolutional Neural Network (CNN) architecture is comprised of a sequence of convolutions and sub-sampling layers in which the content or values of the convolutional kernels is learned via an optimization algorithm. With a fixed number of filters in each layer, each individual filter is convoluted transversely with the width and height of the input figure in the forward transmits. The output of this layer is a two-dimensional feature map of that filter to detect the pattern. This is followed by a Rectified Linear Unit (ReLU) where the non-linearity is increased in the network using a rectified function. The governing equation of the convolution operation is given as

$$h_i^\ell = f(c) = f\left(\sum_{n=1}^{M} W_{ni}^\ell \cdot h_n^{\ell-1}, + b_i^\ell\right) \tag{6}$$

with the ReLU function is defined as

$$f(c) = \begin{cases} 0, & c < 0, \\ c, & c \geq 0, \end{cases} \tag{7}$$

where h_i^ℓ is the i-th output of layer ℓ, W_{ni}^ℓ is the convolutional kernel that is operated on the n-th map of the $\ell-1$-th layer used for the i-th output of the ℓ-th layer, and b_i^ℓ is the bias term; $f(c)$ is an activation function imposed on the output of the convolution (c). The optimization algorithm focuses on optimizing the convolutional kernel W. The output of each convolutional operation or any operation is denoted as a feature map. After convolution, the sub-sampling layer reduces the dimension of the feature maps by representing a neighborhood of values in the feature map by a single value.

Additionally, CNN are also known as shift invariant or space invariant artificial neural networks (SIANN), based on their shared-weights architecture and translation invariance characteristics [40]. This particular feature makes CNN appropriate to deal with the problem described above with the loss of order in each trial processed by BSS algorithms. The parameters of each convolutional layer are:

- Filters: The number of output filters in the convolution.
- kernel size: The height and width of the 2D convolution window.
- Strides: The strides of the convolution along the height and width.

Another important operation applied usually in each convolutional layer is the max-pooling, that is a sample-based discretization process. The objective is to down-sample an input representation, reducing its dimensionality and allowing for assumptions to be made about features contained in the sub-regions binned. This is done to in part to help over-fitting by providing an abstracted form of the representation. Also, it reduces the computational cost by reducing the number of parameters to learn and provides basic translation invariance to the internal representation [41].

3. Methodology

In this section the dataset format, the selected electrodes, and the proposed approach for MI classification are described.

3.1. Dataset

The dataset was provided by Intelligent Data Analysis Group is the IVa from the BCI competition III was used in current study. This consists of five healthy subjects sat in a comfortable chair with arms resting. The subjects are labeled as *aa, al, av, aw*, and *ay* [42]. Visual cues indicated for 3.5 s which of the following two MI the subject should perform: right hand (class1) and foot (class2) MI. The presentation of target cues were intermitted by periods of random length, 1.75 to 2.25 s, in which the subject could relax [43]. In Figure 1 is shown a diagram where it is illustrated the extraction of MI segments inside a trial, where the interest segment of MI occurs in time interval between 4.5 and 8 s. The recording was made using BrainAmp amplifiers and a 128 channel Ag/AgCl electrode cap from ECI. 118 EEG channels were measured at positions of the extended international 10–20-system. Signals were digitized at 1000 Hz with 16 bit (0.1 μV) accuracy.

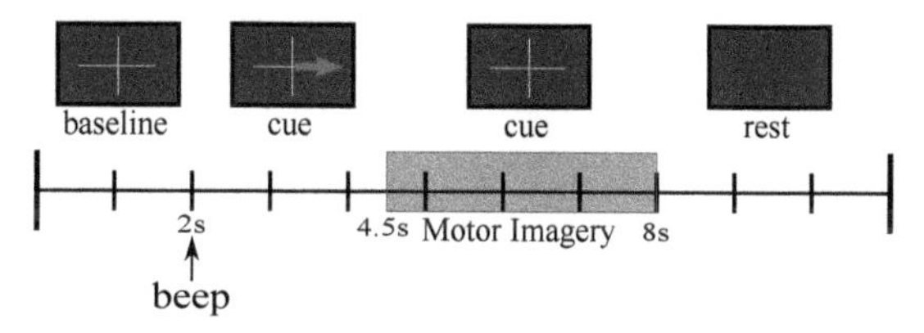

Figure 1. Extraction of MI for each trial.

As mentioned above, μ rhythms are generated in the motor cortex [5], therefore, the selected electrodes for this region are C_3 for the left hemisphere and C_4 for the right hemisphere. In this work, 18 channels were located in both hemispheres around the sensorimotor cortex were selected. FC_5, FC_3, FC_1, C_5, C_3, C_1, CP_5, CP_3, CP_1 in left side, and FC_2, FC_4, FC_6, C_2, C_4, C_6, CP_2, CP_4,CP_6 in right side. The selected electrodes coincide with the regions reported in [21]. The two clusters left and right are filtered using a second-order IIR Butterworth, set as band-pass filter. To preserve the relevant information for MI, the selected cut-off frequencies were adjusted between 0.5 and 90 Hz.

3.2. Proposed Approach

Description of proposed approach is depicted in Figure 2. Clusters $left$ and $right$ are separated in $trials$, from $trial_1$ to $trial_N$. Each trial is preprocessed using a BSS algorithm, which generates equal number of estimated sources $\mathbf{s}'(t)$ from the input channels $\mathbf{x}(t)$. These sources were sorted using as criterion the correlation between their spectral components and the MRIC. This procedure helps to separate the sources and the unwanted artifacts that have low correlations with MRIC. Sorted trials of $\mathbf{s}'(t)$ are passed through a CWT block The CWT is obtained using generalized Morse wavelets. Analytic wavelets are complex-valued wavelets whose Fourier transforms are supported only on the positive real axis. They are useful for analyzing modulated signals, which are signals with time-varying amplitude and frequency [44]. The window size of each CWT is 1.0 s, each window is computed using steps of 0.25 s. CNN architecture has as input the CWT figures and finally a fully connected Multilayer Perceptron (MLP) separates into two classes, *right hand MI* and *right foot MI*. In Figure 3a is shown the CWT of a single estimated source $\mathbf{s}_1'$, in Figure 3b is shown an example of input containing all estimated sources stacked along y axis. Each figure is re-scaled to a size of 128×256.

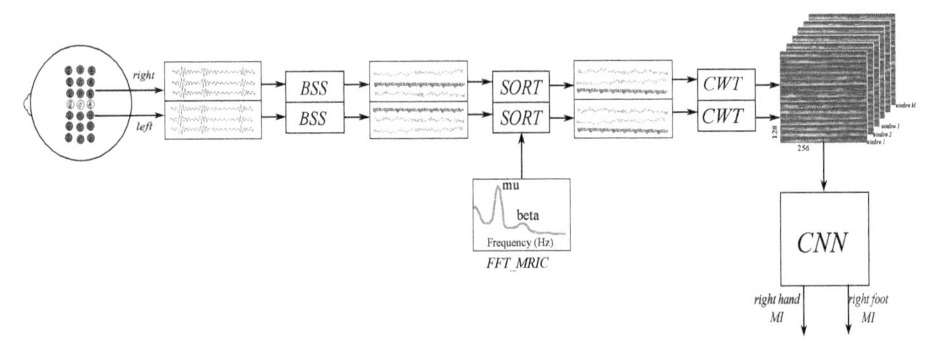

Figure 2. Proposed methodology. The left and right channels are prepossessed using a BSS algorithm, the MRIC sorts the estimated sources, in the CWT stage the images for each time window are obtained, finally the CNN separates the classes.

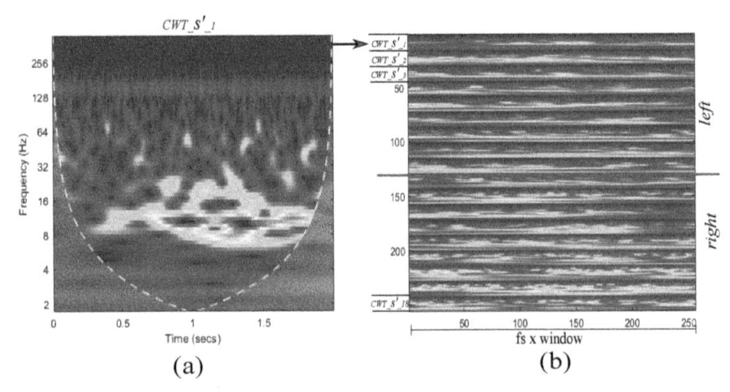

Figure 3. CWT maps for (**a**) one estimated source; (**b**) CWT stacked maps for left and right estimated sources.

An example of CNN architecture is depicted in Figure 4. Each CWT input images are passed through the CNN architecture. Two convolutional layers with respective max-pooling are responsible for obtaining descriptors from CWT maps. In the third layer, the matrices are flattened and passed through a dense layer. Finally, an output layer composed of two neurons is the classifier for two MI classes.

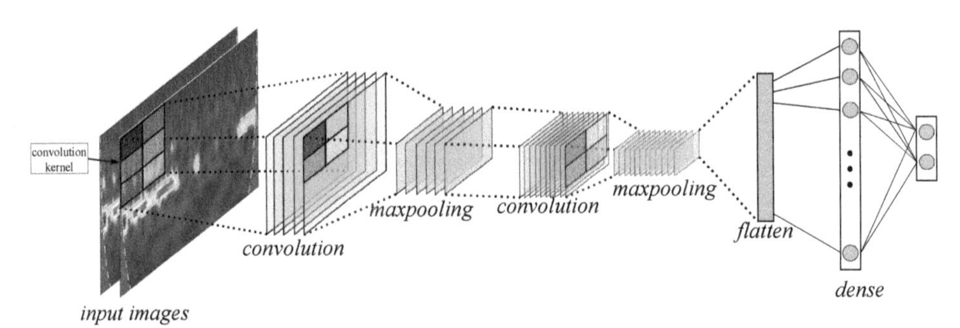

Figure 4. Scheme of CNN architecture used. The CWT input images are pass through two convolutional layers with respective max-pooling. The matrices are flattened and passed by a dense layer.

3.3. Experiment Setup

The experiments were conducted on an Intel Core(TM) i7-7700HQ 2.80 GHz with 16 GB of RAM. Matlab 2017 was used to compute fastICA [38] and CWT maps, while python Tensorflow and Keras libraries were used to compute the CNN architecture.

4. Results and Discussion

In this section a validation of the proposed method is made, some relevant hyper-parameters of the CNN architecture are explored, and the classification results are compared with other reported methods.

4.1. Validation of Proposed Method

The first experiment was carried out to compare the performance between two BSS approaches: SOS and HOS. To evaluate the proposed methodology, an SOS called Robust Second Order Blind Identification (SOBIRO), and the previously mentioned fastICA, belonging to the family of HOS algorithms were tested in one subject. In Figure 5 is showed the train and validation values for 20 epochs in cases Figure 5a without BSS preprocessing, Figure 5b with fastICA without the sort criterion, Figure 5c with sorted SOBIRO, and Figure 5d with sorted fastICA. Considering the structure of the database used in [32], 50% trials were set as the training set, and the remaining 50% were selected as the test set. The CNN architecture initially used was the same proposed in [32]. Nevertheless, the input data is organized in a different way, for which it was necessary to make some adjustments in the convolution stride and max-pooling size. The CNN architecture used in this work is showed in Table 1.

Table 1. CNN modified architecture inspired from [32].

Layer	Operation	Kernel	Stride	Output Shape
1	Conv2D → 250	(3, 1)	(2, 2)	(63,256,250)
	Activation→ *ReLU*			(63,256,250)
	Max-pooling→ (4, 4)			(15,64,250)
2	Conv2D → 150	(1, 2)	(1, 1)	(15,63,150)
	Activation→ *ReLU*			(15,63,150)
	Max-pooling→ (3, 3)			(5,21,150)
3	Flatten			(15750)
	Dense→ 2048			(2048)
	Activation→ *ReLU*			(2048)
	Dropout→ 0.4			(2048)
4	Dense→ 2			(2)
	Activation→ *softmax*			(2)

In the graphs it is possible to observe that in case (a) without preprocessing BSS stage, the maximum accuracy value for train is near 0.8, while the validation accuracy is below 0.6. These results can be explained taking into account the reduced number of data for a deep learning approach, where large amounts of data are required to achieve an end-to-end system, where convolutional layers are able to find the determining patterns that allow for classifying movement intentions; In the case (b), where fastICA is applied in each trial but without the sort step, the training accuracy reaches values close to 0.98, but the validation accuracy is below 0.60 and decrease for each epoch. This can be explained by the disorder of the estimated sources in each trial, where CNN learns training set, but fails to generalize in the test set. This result validates the hypothesis of the need to use a sort criterion for sources estimated through BSS; In case (c), using the second-order Statistic approach (SOS) SOBIRO as BSS algorithm and sorted with the same explained criterion in Figure 2, the test accuracy achieved 0.73 values, improving the (a) and (b) responses. However, differences between train and test are considerably large which indicates overfitting; Finally, in case (d), with the sorted HOS fastICA, both the training data and the validation data achieve values higher than 0.8, reducing the phenomenon of overfitting. This result validates several previous works where the superiority of the HOS algorithms over the SOS for BSS is reported. Results (c) and (d) are in accordance with numerous works reporting superiority of HOS-BSS algorithms over SOS-BSS algorithms for EEG preprocessing [45–47]. Sorted fastICA is then chosen as BSS algorithm before CWT generation and posterior CNN classification stages.

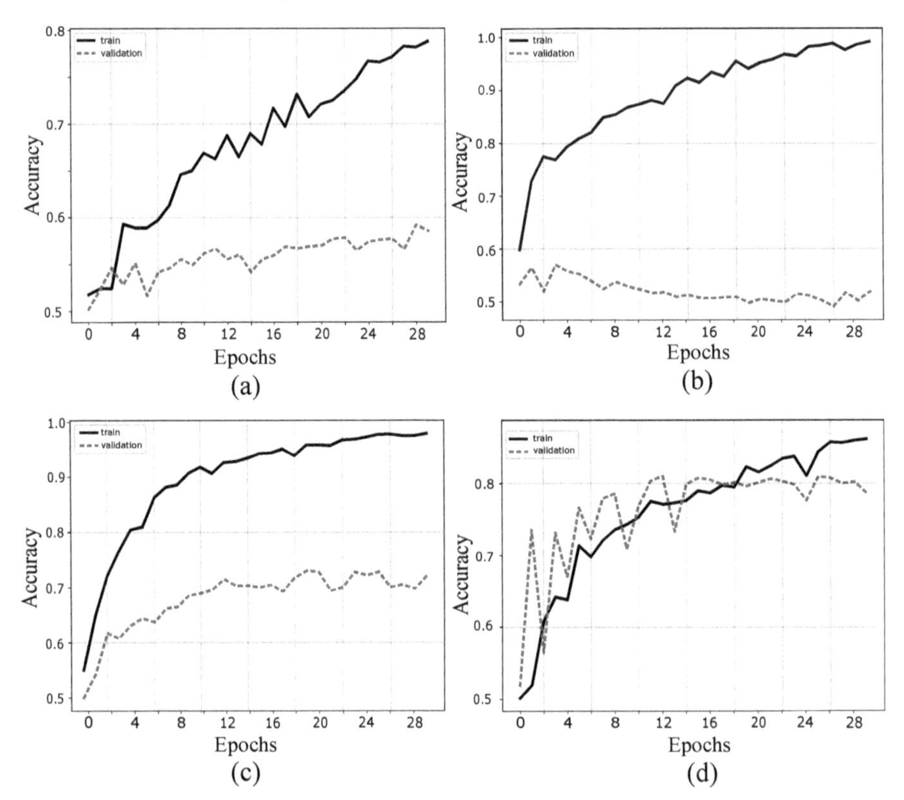

Figure 5. Train and test validation behaviour for subject *aa* in cases. (**a**) without BSS; (**b**) with no sorted fastICA; (**c**) with sorted SOBIRO; and (**d**) with sorted fastICA (30 epochs). with the sorted HOS fastICA, both the training data and the validation data achieve values higher than 0.8, reducing the phenomenon of overfitting in comparison to the other cases.

4.2. Comparison with Other Methods

In Table 2 are shown the validation accuracy for each k validation and each subject of dataset IVa of BCI competition III. The test set was divided into 10 equal parts for each cross validation. The maximum classification value was chosen in each case.

As is shown in Table 2, the average ranking percentage of 94.66%, with a standard deviation σ of 6.46. For the five subjects, the maximum k-fold accuracy average was 97.81% with σ of 3.34 in subject *aa*, and a minimum value of 92.18 with σ of 6.98 in subject *ay*. Table 3 shows some recent work that reports the same used dataset [48–51].

Table 2. 10-fold cross validation accuracy.

Subject	Accuracy										
	k = 1	k = 2	k = 3	k = 4	k = 5	k = 6	k = 7	k = 8	k = 9	k = 10	Average	Std
subject aa	94.79	100.00	96.87	89.58	100.00	100.00	100.00	98.95	98.95	98.95	97.81	3.34
subject al	91.66	94.79	94.79	98.95	85.41	87.50	97.91	100.00	98.95	94.79	94.47	4.96
subject av	95.83	97.91	100.00	98.95	97.91	88.54	68.75	100.00	100.00	100.00	94.78	9.79
subject aw	98.75	92.50	95.00	99.37	99.37	91.87	100.00	98.75	76.87	88.12	94.06	7.26
subject ay	85.41	97.91	85.41	79.16	96.87	95.83	95.83	100.00	96.87	88.54	92.18	6.98
Average											**94.66**	6.46

Table 3. Comparison with other works using the IVa of BCI competition III dataset.

Author	Method	Classifier	Accuracy (%)	Year
Lu et al.	R-CSP with aggregation	R-CSP	83.90	2010
Siuly et al.	CT	LS-SVM	88.32	2011
Zhang et al.	Z-score	LDA	81.10	2013
Siuly et al.	OA	NB	96.36	2016
Kevric et al.	MSPCA, WPD, HOS	k-NN	92.80	2017
Taran et al.	TQWT	LS-SVM	96.89	2018
Proposed	sorted-fastICA-CWT	CNN	**94.66**	2019

4.3. Discussion

One of the main contributions of this work is the criterion of sorting the estimated sources. In Figure 6 are shown the spectral components of estimated sources before and after apply the sort criterion in one trial. The components with more information in μ and β will generally be placed in the top positions, while the components least associated with these frequencies will be at the bottom as for example in Figure 6a, the first spectral component has more energy in frequencies over 20 Hz but without α and β contribution, is placed in the bottom in Figure 6b. In contrast, the component located in the seventh position in Figure 6a, which has the most energy in the region μ, is located in the first position in Figure 6b.

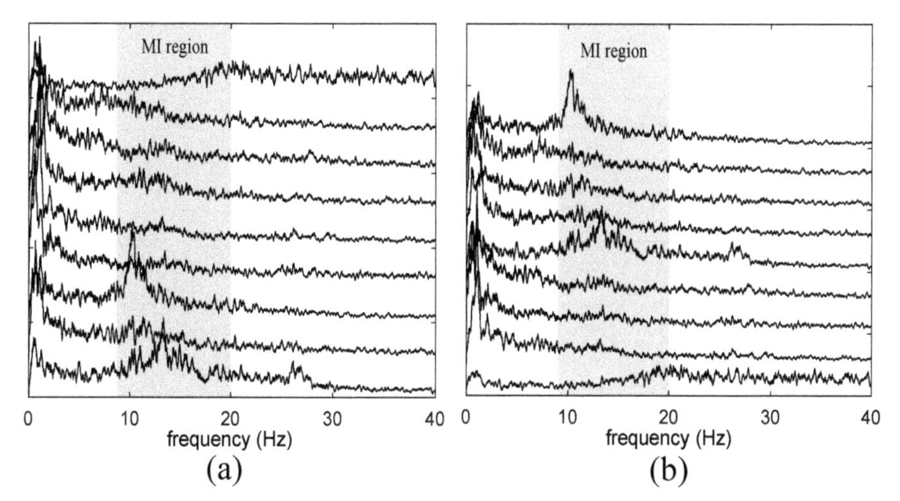

Figure 6. Analysis spectral components: (**a**) before sort; (**b**) after sort. The components with more MRIC frequencies are placed at the top after sorting.

The other distinctive part of proposed method is the use of a CNN architecture, and the adjustment of some relevant hyper-parameters. Taking as initial values those shown in Table 1, and changing the kernel size along height and width for each convolutional layer, the behaviour of validation accuracy for each case is analyzed. The kernel size is (y,x), where y-axis contain frequential and spatial information, while x-axis contain temporal information. In Figure 7 are depicted the validation accuracy for subject *aa*. The kernel sizes selected to the comparison in the first convolutional layer were $(i,1)$, $(i,3)$, $(i,5)$, $(i,7)$, with i taking odd values from 1 to 9.

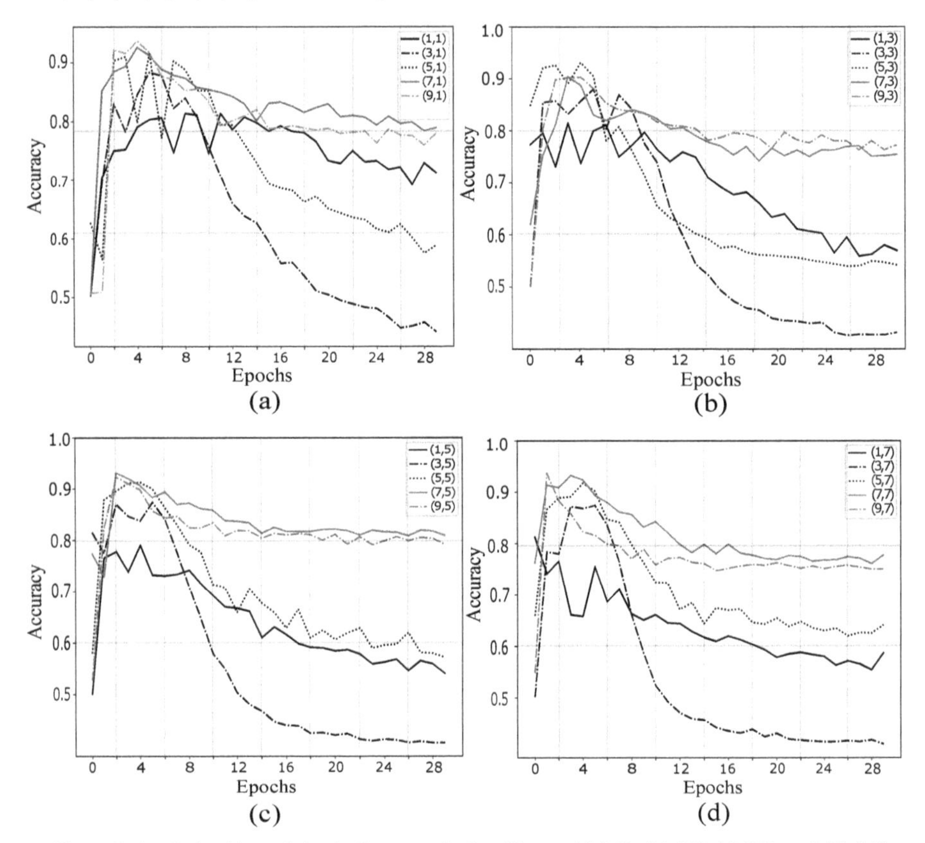

Figure 7. Analysis of kernel size in first convolutional layer: (**a**) (i,1); (**b**) (i,3); (**c**) (i,5); and (**d**) (i,7). The kernel size (7,5) in (**c**) presented less overfitting and major classification accuracy.

According to the analysis of kernel size in the first convolutional layer, the size (7,5) (Figure 7c) presented less overfitting and major classification accuracy. It is also noted that when the kernel has a size of in the size 3 for y-axis, while reaching a maximum value close to the best case, this kernel size generates a high overfitting after certain times. On the other hand, the the size 1 for y-axis generates the lowest maximum values in all combinations of x-axis.

These results are in accordance with other work where the size of the kernel is also studied, where they report that the vertical locations (frequency-space) is of great importance for the classification performance, while, in contrast, the horizontal locations (time) are not as significant [27].

Once fixed the kernel size in first layer, a similar analysis was made for the kernel of the second convolutional layer. In Figure 8 are depicted the validation accuracy for subject *aa*. Taking into account that a (4,4) max-pooling has been previously applied, the kernel sizes selected were (j,1), (j,2), (j,3), and (j,4), with j taking values from 1 to 5.

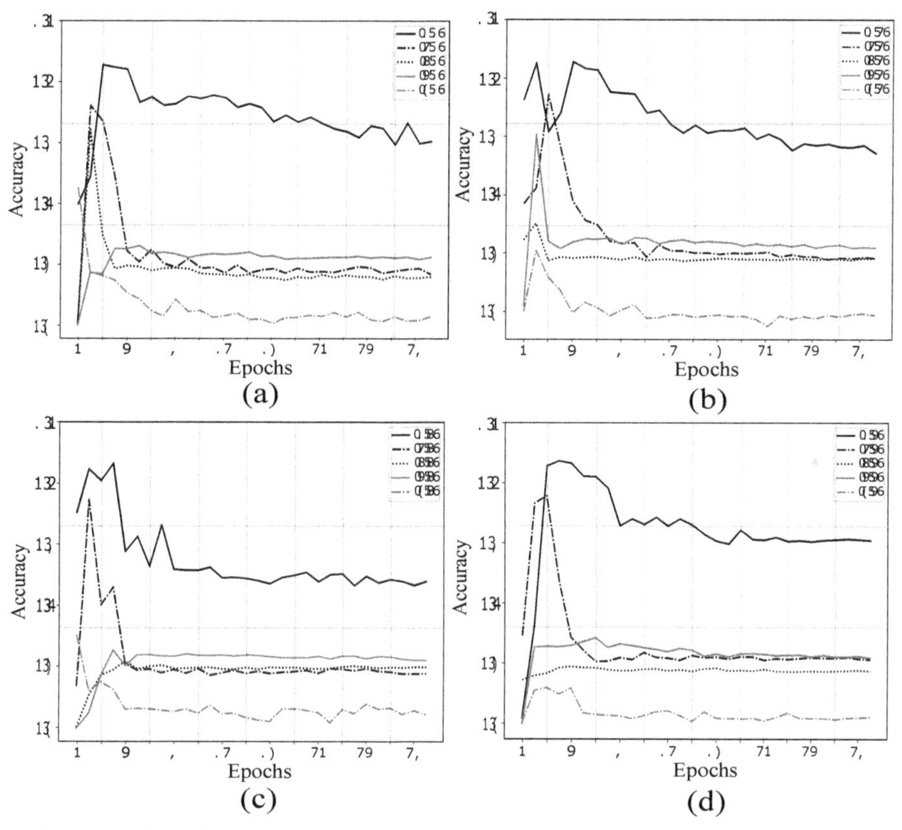

Figure 8. Analysis of kernel size in second convolutional layer: (**a**) (j,1); (**b**) (j,2); (**c**) (j,3); and (**d**) (j,4). The kernel size (1,1) in (**a**) presented less overfitting and major classification accuracy. The only y-axis where the CNN can achieve a stable accuracy throughout the epochs is 1.

At this case, the only y-axis where the CNN can achieve a stable accuracy throughout the epochs is 1 independently of x-axis size. In other cases, the validation accuracy decreases to 60% or even around 50% in case y-axis = 5.

As is well known, currently deep learning approaches are state-of-the-art in many images processing and artificial vision. However, in contrast with two-dimensional static images, the EEG signals are dynamic time series, where the generalizable MI patterns in EEG signals are spatially distributed and mixed in the channels around the motor region. In addition to this, low signal-to-noise ratio could make learning features in an end-to-end fashion more difficult for EEG signals than for images [29]. On the other hand, deep learning approaches require a large amount of training data in order to obtain descriptors that allow discrimination between different classes. In particular case of EEG and MI, this is a limitation since the data must be processed independently for each subject, and due to fatigue, the MI databases are relatively small. To deal with this problem, some works proposed using deep learning have done data augmentation using some criteria to generate simulated data from the training set. This approach has yielded good results. However, the generation of artificial data can be risky without a rigorous methodology and thus generate false data that increases accuracy for a particular dataset.

5. Conclusions

In this work, the estimated independent components were obtained using a fastICA algorithm, separating relevant MI-related independent components from unwanted artifacts. However, the source separation by itself is not sufficient if for each trial, the order of them is not preserved. For this reason, a spectral correlation with MRIC helps to sort the sources by reducing the spatial variance, leaving in the last positions the sources with a more significant influence of artifacts and less μ and β components. These operations help to reduce the complexity in the search for relevant patterns in the posterior extraction and classification stages. The use of CWT maps in the feature extraction stage allows obtaining a 2D representation of time series. In contrast with Short-Time Fourier Transform (STFT), CWT performs a multi-resolution analysis. According to the experimentation carried out, obtained results in this work are competitive with the state-of-the-art with a 94.66% in the k-fold cross validation. Regarding the architecture of CNN, it was found that the hyper-parameters related to the size of the kernel as well as the kernel stride in each convolutional layer have a significant influence on network performance, while the number of convolutions has less impact in final accuracy. Two future works derived: first, the development of a methodology that allows to find the hyper-parameters close to the optimum and then, improve the current results. Second, to replace the BSS stage with some autoencoder architecture, as for example Variational Autoencoder (VAE), to obtain the estimated sources.

Author Contributions: Conceptualization, C.J.O.-E. and S.S.-C.; Methodology, C.J.O.-E. and S.S.-C.; Writing–original draft preparation, C.J.O.-E., S.S.-C., J.R.-R. and R.A.G.-L.; Writing–review and editing, C.J.O.-E., S.S.-C., J.R.-R. and R.A.G.-L.; Supervision, J.R.-R. and R.A.G.-L.

Funding: This research received no external funding.

Acknowledgments: The authors would like to thank the Informatics Faculty and the Department of Engineering of the Universidad Autónoma de Querétaro, as well as Consejo Nacional de Ciencia y Tecnología (CONACyT) for the Doctorate Fellowship support.

Conflicts of Interest: The authors declare no conflict of interest.

References

1. Gandhi, V.; Prasad, G.; Coyle, D.; Behera, L.; McGinnity, T.M. Evaluating Quantum Neural Network filtered motor imagery brain-computer interface using multiple classification techniques. *Neurocomputing* **2015**, *170*, 161–167. [CrossRef]
2. Sanei, S.; Chambers, J.A. *EEG Signal Processing*; Wiley: Hoboken, NJ, USA, 2007; p. 10.
3. Niedermeyer, E.; da Silva, F.L. *Electroencephalography: Basic Principles, Clinical Applications, and Related Fields*; Lippincott Williams & Wilkins: Philadelphia, PA, USA, 2005.
4. Wolpaw, J.R.; McFarland, D.J. Control of a two-dimensional movement signal by a noninvasive brain-computer interface in humans. *Proc. Natl. Acad. Sci. USA* **2004**, *101*, 17849–17854. [CrossRef] [PubMed]
5. Pineda, J.A. The functional significance of mu rhythms: Translating "seeing" and "hearing" into "doing". *Brain Res. Rev.* **2005**, *50*, 57–68. [CrossRef] [PubMed]
6. McFarland, D.J.; Miner, L.A.; Vaughan, T.M.; Wolpaw, J.R. Mu and beta rhythm topographies during motor imagery and actual movements. *Brain Topogr.* **2000**, *12*, 177–186. [CrossRef] [PubMed]
7. Islam, M.R.; Tanaka, T.; Molla, M.K.I. Multiband tangent space mapping and feature selection for classification of EEG during motor imagery. *J. Neural Eng.* **2018**, *15*, 046021. [CrossRef] [PubMed]
8. Pfurtscheller, G.; Aranibar, A. Evaluation of event-related desynchronization (ERD) preceding and following voluntary self-paced movement. *Electroencephalogr. Clin. Neurophysiol.* **1979**, *46*, 138–146. [CrossRef]
9. Pfurtscheller, G. Central beta rhythm during sensorimotor activities in man. *Electroencephalogr. Clin. Neurophysiol.* **1981**, *51*, 253–264. [CrossRef]
10. Pfurtscheller, G.; Neuper, C. Event-related synchronization of mu rhythm in the EEG over the cortical hand area in man. *Neurosci. Lett.* **1994**, *174*, 93–96. [CrossRef]

11. Pfurtscheller, G.; Da Silva, F.L. Event-related EEG/MEG synchronization and desynchronization: Basic principles. *Clin. Neurophysiol.* **1999**, *110*, 1842–1857. [CrossRef]
12. Pfurtscheller, G.; Brunner, C.; Schlögl, A.; Da Silva, F.L. Mu rhythm (de) synchronization and EEG single-trial classification of different motor imagery tasks. *NeuroImage* **2006**, *31*, 153–159. [CrossRef]
13. Koles, Z.; Lind, J.; Soong, A. Spatio-temporal decomposition of the EEG: A general approach to the isolation and localization of sources. *Electroencephalogr. Clin. Neurophysiol.* **1995**, *95*, 219–230. [CrossRef]
14. Liu, A.; Chen, K.; Liu, Q.; Ai, Q.; Xie, Y.; Chen, A. Feature selection for motor imagery EEG classification based on firefly algorithm and learning automata. *Sensors* **2017**, *17*, 2576. [CrossRef] [PubMed]
15. Naeem, M.; Brunner, C.; Leeb, R.; Graimann, B.; Pfurtscheller, G. Seperability of four-class motor imagery data using independent components analysis. *J. Neural Eng.* **2006**, *3*, 208. [CrossRef] [PubMed]
16. Vázquez, R.R.; Velez-Perez, H.; Ranta, R.; Dorr, V.L.; Maquin, D.; Maillard, L. Blind source separation, wavelet denoising and discriminant analysis for EEG artefacts and noise cancelling. *Biomed. Signal Process. Control* **2012**, *7*, 389–400. [CrossRef]
17. Jung, T.P.; Makeig, S.; Westerfield, M.; Townsend, J.; Courchesne, E.; Sejnowski, T.J. Removal of eye activity artifacts from visual event-related potentials in normal and clinical subjects. *Clin. Neurophysiol.* **2000**, *111*, 1745–1758. [CrossRef]
18. Delorme, A.; Makeig, S. EEGLAB: An open source toolbox for analysis of single-trial EEG dynamics including independent component analysis. *J. Neurosci. Methods* **2004**, *134*, 9–21. [CrossRef]
19. Mur, A.; Dormido, R.; Duro, N. An Unsupervised Method for Artefact Removal in EEG Signals. *Sensors* **2019**, *19*, 2302. [CrossRef]
20. Wang, Y.; Wang, Y.T.; Jung, T.P. Translation of EEG spatial filters from resting to motor imagery using independent component analysis. *PLoS ONE* **2012**, *7*, e37665. [CrossRef]
21. Zhou, B.; Wu, X.; Ruan, J.; Zhao, L.; Zhang, L. How many channels are suitable for independent component analysis in motor imagery brain-computer interface. *Biomed. Signal Process. Control* **2019**, *50*, 103–120. [CrossRef]
22. Chiu, C.Y.; Chen, C.Y.; Lin, Y.Y.; Chen, S.A.; Lin, C.T. Using a novel LDA-ensemble framework to classification of motor imagery tasks for brain-computer interface applications. In Proceedings of the Intelligent Systems and Applications: Proceedings of the International Computer Symposium (ICS), Taichung, Taiwan, 12–14 December 2014.
23. Das, A.B.; Bhuiyan, M.I.H.; Alam, S.S. Classification of EEG signals using normal inverse Gaussian parameters in the dual-tree complex wavelet transform domain for seizure detection. *Signal Image Video Process.* **2016**, *10*, 259–266. [CrossRef]
24. Chatterjee, R.; Bandyopadhyay, T. EEG based Motor Imagery Classification using SVM and MLP. In Proceedings of the 2nd International Conference on Computational Intelligence and Networks (CINE), Bhubaneswar, India, 11 January 2016; pp. 84–89.
25. He, L.; Hu, D.; Wan, M.; Wen, Y.; Von Deneen, K.M.; Zhou, M. Common Bayesian network for classification of EEG-based multiclass motor imagery BCI. *IEEE Trans. Syst. Man Cybern. Syst.* **2015**, *46*, 843–854. [CrossRef]
26. Ma, X.; Dai, Z.; He, Z.; Ma, J.; Wang, Y.; Wang, Y. Learning traffic as images: A deep convolutional neural network for large-scale transportation network speed prediction. *Sensors* **2017**, *17*, 818. [CrossRef] [PubMed]
27. Dai, M.; Zheng, D.; Na, R.; Wang, S.; Zhang, S. EEG Classification of Motor Imagery Using a Novel Deep Learning Framework. *Sensors* **2019**, *19*, 551. [CrossRef] [PubMed]
28. Lu, N.; Li, T.; Ren, X.; Miao, H. A deep learning scheme for motor imagery classification based on restricted boltzmann machines. *IEEE Trans. Neural Syst. Rehabil. Eng.* **2016**, *25*, 566–576. [CrossRef] [PubMed]
29. Schirrmeister, R.T.; Springenberg, J.T.; Fiederer, L.D.J.; Glasstetter, M.; Eggensperger, K.; Tangermann, M.; Hutter, F.; Burgard, W.; Ball, T. Deep learning with convolutional neural networks for EEG decoding and visualization. *Hum. Brain Mapp.* **2017**, *38*, 5391–5420. [CrossRef]
30. Tang, Z.; Li, C.; Sun, S. Single-trial EEG classification of motor imagery using deep convolutional neural networks. *Opt. Int. J. Light Electron Opt.* **2017**, *130*, 11–18. [CrossRef]
31. Tayeb, Z.; Fedjaev, J.; Ghaboosi, N.; Richter, C.; Everding, L.; Qu, X.; Wu, Y.; Cheng, G.; Conradt, J. Validating deep neural networks for online decoding of motor imagery movements from EEG signals. *Sensors* **2019**, *19*, 210. [CrossRef]

32. Zhang, Z.; Duan, F.; Solé-Casals, J.; Dinarès-Ferran, J.; Cichocki, A.; Yang, Z.; Sun, Z. A novel deep learning approach with data augmentation to classify motor imagery signals. *IEEE Access* **2019**, *7*, 15945–15954. [CrossRef]
33. Zhang, X.; Yao, L.; Wang, X.; Monaghan, J.; Mcalpine, D. A Survey on Deep Learning based Brain Computer Interface: Recent Advances and New Frontiers. *arXiv* **2019**, arXiv:1905.04149.
34. Tabar, Y.R.; Halici, U. A novel deep learning approach for classification of EEG motor imagery signals. *J. Neural Eng.* **2016**, *14*, 016003. [CrossRef]
35. Roy, Y.; Banville, H.; Albuquerque, I.; Gramfort, A.; Falk, T.H.; Faubert, J. Deep learning-based electroencephalography analysis: A systematic review. *J. Neural Eng.* **2019**, *16*. [CrossRef] [PubMed]
36. Belouchrani, A.; Abed-Meraim, K.; Cardoso, J.; Moulines, E. Second-order blind separation of temporally correlated sources. *Proc. Int. Conf. Digit. Signal Process.* **1993**, 346–351.
37. Comon, P.; Jutten, C. *Handbook of Blind Source Separation: Independent Component Analysis and Applications*; Academic Press: Cambridge, MA, USA, 2010.
38. Hyvärinen, A.; Oja, E. Independent component analysis: Algorithms and applications. *Neural Netw.* **2000**, *13*, 411–430. [CrossRef]
39. Mallat, S. *A Wavelet Tour of Signal Processing*; Elsevier: Amsterdam, The Netherlands, 1999.
40. Zhang, W.; Itoh, K.; Tanida, J.; Ichioka, Y. Parallel distributed processing model with local space-invariant interconnections and its optical architecture. *Appl. Opt.* **1990**, *29*, 4790–4797. [CrossRef]
41. A Comprehensive Guide to Convolutional Neural Networks—The ELI5 Way. Available online: https://towardsdatascience.com/a-comprehensive-guide-to-convolutional-neural-networks-the-eli5-way-3bd2b1164a53 (accessed on 9 September 2019).
42. Data Set IVa. Available online: http://www.bbci.de/competition/iii/desc_IVa.html (accessed on 9 September 2019).
43. Blankertz, B.; Muller, K.R.; Krusienski, D.J.; Schalk, G.; Wolpaw, J.R.; Schlogl, A.; Pfurtscheller, G.; Millan, J.R.; Schroder, M.; Birbaumer, N. The BCI competition III: Validating alternative approaches to actual BCI problems. *IEEE Trans. Neural Syst. Rehabil. Eng.* **2006**, *14*, 153–159. [CrossRef]
44. Morse Wavelets. Available online: https://la.mathworks.com/help/wavelet/ug/morse-wavelets.html (accessed on 10 August 2019).
45. Oosugi, N.; Kitajo, K.; Hasegawa, N.; Nagasaka, Y.; Okanoya, K.; Fujii, N. A new method for quantifying the performance of EEG blind source separation algorithms by referencing a simultaneously recorded ECoG signal. *Neural Netw.* **2017**, *93*, 1–6. [CrossRef]
46. Klemm, M.; Haueisen, J.; Ivanova, G. Independent component analysis: Comparison of algorithms for the investigation of surface electrical brain activity. *Med. Biol. Eng. Comput.* **2009**, *47*, 413–423. [CrossRef]
47. Albera, L.; Kachenoura, A.; Comon, P.; Karfoul, A.; Wendling, F.; Senhadji, L.; Merlet, I. ICA-based EEG denoising: A comparative analysis of fifteen methods. *Bull. Pol. Acad. Sci. Tech. Sci.* **2012**, *60*, 407–418. [CrossRef]
48. Siuly, S.; Li, Y. Improving the separability of motor imagery EEG signals using a cross correlation-based least square support vector machine for brain–computer interface. *IEEE Trans. Neural Syst. Rehabil. Eng.* **2012**, *20*, 526–538. [CrossRef]
49. Wang, H.; Zhang, Y. Detection of motor imagery EEG signals employing Naïve Bayes based learning process. *Measurement* **2016**, *86*, 148–158.
50. Kevric, J.; Subasi, A. Comparison of signal decomposition methods in classification of EEG signals for motor-imagery BCI system. *Biomed. Signal Process. Control* **2017**, *31*, 398–406. [CrossRef]
51. Taran, S.; Bajaj, V. Motor imagery tasks-based EEG signals classification using tunable-Q wavelet transform. *Neural Comput. Appl.* **2018**, 1–8. [CrossRef]

Article

Most Relevant Spectral Bands Identification for Brain Cancer Detection Using Hyperspectral Imaging

Beatriz Martinez [1,*], Raquel Leon [1], Himar Fabelo [1], Samuel Ortega [1], Juan F. Piñeiro [2], Adam Szolna [2], Maria Hernandez [2], Carlos Espino [2], Aruma J. O'Shanahan [2], David Carrera [2], Sara Bisshopp [2], Coralia Sosa [2], Mariano Marquez [2], Rafael Camacho [3], Maria de la Luz Plaza [3], Jesus Morera [2] and Gustavo M. Callico [1]

1 Institute for Applied Microelectronics (IUMA), University of Las Palmas de Gran Canaria (ULPGC), 35017 Las Palmas de Gran Canaria, Spain; slmartin@iuma.ulpgc.es (R.L.); hfabelo@iuma.ulpgc.es (H.F.); sortega@iuma.ulpgc.es (S.O.); gustavo@iuma.ulpgc.es (G.M.C.)

2 Department of Neurosurgery, University Hospital Doctor Negrin of Gran Canaria, 35010 Barranco de la Ballena s/n, Las Palmas de Gran Canaria, Spain; pinerbrains1@yahoo.es (J.F.P.); adamszolna@wp.pl (A.S.); hhdez.maria@gmail.com (M.H.); carlosespinopostigo@gmail.com (C.E.); aruosha@gmail.com (A.J.O.); david__carrera@hotmail.com (D.C.); sarabisshop@hotmail.com (S.B.); coralia.sosa@gmail.com (C.S.); marquezrdguez@yahoo.es (M.M.); jmormol@gobiernodecanarias.org (J.M.)

3 Department of Pathological Anatomy, University Hospital Doctor Negrin of Gran Canaria, 35010 Barranco de la Ballena s/n, Las Palmas de Gran Canaria, Spain; rcamgal@gobiernodecanarias.org (R.C.); mplaperb@gobiernodecanarias.org (M.d.l.L.P.)

* Correspondence: bmartinez@iuma.ulpgc.es; Tel.: +34-928-451-220

Received: 25 October 2019; Accepted: 10 December 2019; Published: 12 December 2019

Abstract: Hyperspectral imaging (HSI) is a non-ionizing and non-contact imaging technique capable of obtaining more information than conventional RGB (red green blue) imaging. In the medical field, HSI has commonly been investigated due to its great potential for diagnostic and surgical guidance purposes. However, the large amount of information provided by HSI normally contains redundant or non-relevant information, and it is extremely important to identify the most relevant wavelengths for a certain application in order to improve the accuracy of the predictions and reduce the execution time of the classification algorithm. Additionally, some wavelengths can contain noise and removing such bands can improve the classification stage. The work presented in this paper aims to identify such relevant spectral ranges in the visual-and-near-infrared (VNIR) region for an accurate detection of brain cancer using in vivo hyperspectral images. A methodology based on optimization algorithms has been proposed for this task, identifying the relevant wavelengths to achieve the best accuracy in the classification results obtained by a supervised classifier (support vector machines), and employing the lowest possible number of spectral bands. The results demonstrate that the proposed methodology based on the genetic algorithm optimization slightly improves the accuracy of the tumor identification in ~5%, using only 48 bands, with respect to the reference results obtained with 128 bands, offering the possibility of developing customized acquisition sensors that could provide real-time HS imaging. The most relevant spectral ranges found comprise between 440.5–465.96 nm, 498.71–509.62 nm, 556.91–575.1 nm, 593.29–615.12 nm, 636.94–666.05 nm, 698.79–731.53 nm and 884.32–902.51 nm.

Keywords: brain cancer; hyperspectral imaging; intraoperative imaging; feature selection; image-guided surgery; genetic algorithm; particle swarm optimization; ant colony optimization; support vector machine; machine learning

---

1. Introduction

Globally, around 260,000 brain tumor cases are detected each year, with the main brain tumor type being detected the glioblastoma multiforme (GBM) that has the highest death rate (22%) [1]. This type of cancer leads to death in children under the age of 20, and also is one of the principal causes of death among 20- to 29-year-old males [2]. Surgery is one of the principal treatments alongside radiotherapy and chemotherapy [3]. During surgery, several image guidance tools, such as intra-operative neuro-navigation, intra-operative magnetic resonance imaging (iMRI) and fluorescent tumor markers, have been commonly used to assist in the identification of brain tumor boundaries. However, these technologies have several limitations, producing side effects in the patient or invalidating the patient-to-image mapping, reducing the effectiveness of using pre-operative images for intra-operative surgical guidance [4].

Hyperspectral imaging (HSI) is a technology that combines conventional imaging and spectroscopy to obtain simultaneously the spatial and the spectral information of an object [5]. Hyperspectral (HS) images provide abundant information that covers hundreds of spectral bands for each pixel of the image. Each pixel contains an almost continuous spectrum (radiance, reflectance or absorbance), acting as a fingerprint (the so called spectral signature) that can be used to characterize the chemical composition of that particular pixel [5]. One of the main advantages of this technique is that it uses non-ionizing light in a non-contact way, resulting in a non-invasive technology. For this reason, HSI is an emerging technique in the medical field and it has been researched in many different applications, such as oximetry of the retina [6–8], intestinal ischemia identification [9], histopathological tissue analysis [10–13], blood vessel visualization enhancement [14,15], estimation of the cholesterol levels [16], chronic cholecystitis detection [17], diabetic foot [18], etc. In particular, HSI has started to achieve promising results in the recent years with respect to cancer detection through the utilization of cutting-edge machine-learning algorithms [4,19–21]. Several types of cancer have been investigated using HSI including both in vivo and ex vivo tissue samples, such as gastric and colon cancer [22–25], breast cancer [26,27], head and neck cancer [28–33], and brain cancer [34–36], among others.

This imaging modality is mainly characterized by the curse of dimensionality, produced due to the high dimensionality of the data in contrast to the low number of available samples. This rich amount of data allows having more detailed information about the scene that is being captured. However, it also causes a large increase of the computing time required to process the data, containing normally redundant information [37]. For this reason, it is necessary to employ processing algorithms able to reduce the dimensionality of the HS data without losing the relevant information. This dimensional reduction process consists in the transformation of the data, characterized by their high dimensionality, into a significant representation of such data in a reduced dimension [38]. There are two main types of methods for dimensionality reduction: feature extraction [39] and feature selection [40]. Feature extraction algorithms are able to reduce, scale and rotate the original feature space of the HS data through a transformation matrix. This transformation optimizes a given criterion on the data so they can be formulated as a linear transformation that projects feature vectors on a transformed subspace defined by relevant directions. On the other hand, feature selection algorithms applied to HS images aim to find the optimal subset of bands in such images, performing several combinations of bands in a certain way until the best subset is found. This process reduces the dimensionality of the data by selecting the most discriminant bands of the dataset. Some of the most common algorithms for feature selection are optimization algorithms such as the Genetic Algorithm (GA) [41], Particle Swarm Optimization (PSO) [42], and Ant Colony Optimization (ACO) [43], among the most relevant.

The main goal of this work is to evaluate different band selection algorithms in order to identify the minimum number of wavelengths to sample in HS images using a supervised classifier that are necessary to process in vivo human brain HS data. This wavelength reduction will allow an accurate delineation of brain tumors during surgical procedures, obtaining similar results to the classification performed by using the original number of wavelengths. In this sense, a straightforward Support Vector Machine (SVM) classifier has been used instead of other more advanced ones to avoid hiding the band selection procedure effects. The use of feature selection algorithms was motivated by the goal of providing insights about the relevant spectral regions for this task, offering the possibility of reducing the number of spectral bands that the HS sensor has to capture. This will lead to a reduction of the acquisition system size and costs, as well as an acceleration of the execution time required by the processing algorithms. In this sense, the use of customized snapshot HS cameras coupled with a surgical microscope, could be considered to capture real-time HS data during brain surgery. This type of cameras can capture HS video imaging but with a reduced number of spectral channels. The surgical microscopic-based HS system will be the future replacement of the current macroscopic HS capturing system based on push-broom HS cameras, which requires at least 1 min to capture the entire HS cube, employed in the intraoperative HS brain cancer detection research [35]. Band reduction techniques will be crucial to allow real time acquisition using snapshot HS cameras.

2. Materials

This section describes the HSI instrumentation used to generate the in vivo HS brain cancer image database, as well as the proposed band selection-based methodologies. In addition, the evaluation methodology and the metrics employed to validate the results are presented.

2.1. *Intraoperative Hyperspectral (HS) Acqusition System*

In order to obtain the in vivo human brain cancer database, a customized intraoperative HS acquisition system was employed [35]. The acquisition system was composed of a VNIR (visible and near infra-red) push-broom camera (Hyperspec® VNIR A-Series, Headwall Photonics Inc., Fitchburg, MA, USA), providing HS images in the spectral range from 400 to 1000 nm. The HS cubes were formed by 826 spectral bands, having a spectral resolution of 2–3 nm and a spatial resolution of 128.7 μm. Due to the push-broom nature of the HS camera, the sensor is a 2-D detector with a dimension of 1004 × 826 pixels, capturing the complete spectral dimensions and one spatial dimension of the scene at once. For this reason, the system requires a scanning platform in order to shift the field of view of the camera relative to the scene that is going to be captured in order to obtain the second spatial dimension. Considering the employed scanning platform, the maximum size of the HS image is 1004 × 1787 pixels and 826 spectral bands, covering an area of 129 × 230 mm with a pixel size of 128.7 μm. Furthermore, a specific illumination system able to emit cold light in the spectral range between 400 and 2200 nm has been coupled to the system. A 150 W QTH (quartz tungsten halogen) source light is attached to a cold light emitter via fiber optic cable that avoids the high temperatures produced by the lamp in the exposed brain surface.

2.2. *In Vivo Human Brain Cancer Database*

The in vivo human brain cancer database employed in this study is composed by 26 HS images obtained from 16 adult patients. Patients underwent craniotomy for resection of intra-axial brain tumor or another type of brain surgery during clinical practice at the University Hospital Doctor Negrin at Las Palmas de Gran Canaria (Spain). Eleven HS images of exposed tumor tissue were captured from eight different patients diagnosed with grade IV glioblastoma (GBM) tumor. The remaining patients were affected by other types of tumors or underwent a craniotomy for stroke or epilepsy treatment. From these patients, only normal brain tissue samples were recorded and employed in this study. The tumor samples different from GBM were not included in this study. Moreover, in the GBM cases where the tumor area was not able to be captured in optimal conditions, mainly due to the presence of extravasated blood or surgical serum, these images were included in the database but no tumor samples were employed. Finally, only GBM tumor samples belonging to six HS images from four different patients were employed. Written informed consent was obtained from all participant subjects and the study protocol and consent procedures were approved by the Comité Ético de Investigación Clínica-Comité de Ética en la Investigación (CEIC/CEI) of the University Hospital Doctor Negrin.

The following protocol was performed to acquire the data during the surgical procedures. After craniotomy and resection of the dura, the operating surgeon initially identified the approximate location of the normal brain and tumor (if applicable). At that point, sterilized rubber ring markers were located on those places and the HS images were recorded with markers in situ. Next, the tissue inside the markers were resected, and histopathological examination was performed for the final diagnosis. Depending on the location of the tumor, images were acquired at various stages of the operation. In the cases with superficial tumors, some images were obtained immediately after the dura was removed. In the cases of deep-lying tumors, images were captured during the actual tumor resection. More details about this procedure can be found in [35].

From the obtained HS cubes, a specific set of pixels was labeled using four different classes: tumor tissue, normal tissue, hypervascularized tissue (mainly blood vessels), and background. The background class involves other materials or substances presented in the surgical scenario but not relevant for the tumor resection procedure, such as skull bone, dura, skin or surgical materials. The labeling of the images was performed using a combination of pathology assessment and neurosurgical criteria using a semi-automatic tool based on the Spectral Angle Mapper (SAM) algorithm [44]. In this procedure, the operating surgeon employed the semi-automatic labeling tool for a supervised selection of a reference pixel in the HS image where the neurosurgeon was very confident that it belonged to a certain class. Then, the SAM was computed in the entire image with respect to the reference pixel and a threshold was manually established to identify the pixels with the most similar spectral properties to the selected one. Tumor pixels were labeled according to the histopathological diagnosis obtained from the biopsies performed in a certain area (indicated by the rubber ring markers) during surgery. Normal, hypervascularized and background pixels were labeled according to the neurosurgeon experience and knowledge. During this supervised labeling procedure, special attention was paid to avoid the inclusion of pixels in more than one class. On average, 6% of the pixels where labeled from each HS cube available in the database.

It is worth noting that the non-uniformity of the brain tissue produced the presence of specular reflections in the HS image of the captured scene. As described in the previous section, the acquisition system was based on a push-broom HS camera, equipped with a high-power illumination device connected to a linear cold-light emitter, thus avoiding interference of the environment illumination in the capturing process. The incident light beam emitted over the brain surface only illuminates the line captured by the HS camera, and both the camera and the light beam were shifted to capture the entire HS cube. The use of the required powerful illumination together with the non-uniformity of the brain surface, the inherent movement of the exposed brain and the movement of the HS scanning platform make extremely difficult to avoid specular reflections in the HS image. This challenging problem has been investigated in many applications, especially in works related with the analysis of in vivo and ex vivo head and neck cancer samples through HSI [45,46]. In our case, we excluded the use of these glare pixels for the quantitative processing of the HS data. During the supervised labeling procedure, glare pixels were avoided, i.e., glare pixels are not included in the labelled dataset. Hence, both the training and the quantitative classification of the data were not affected by the specular reflections. However, in the qualitative results based on classification maps where the entire HS cube is classified, the glare pixels were classified and we realized that such pixels were mostly identified as background. More information about the acquisition of the HS data and the generation of the labeled dataset can be found in [47].

Table 1 shows the number of labeled samples per class employed in this work, while Table S1 from the supplementary material details the patients and the number of samples per class and per image that were involved in the experiments according to the dataset published in [47]. As previously mentioned, from the eight patients affected by GBM tumor, only tumor pixels from four patients were labeled. In total, six HS images were labeled with four classes and were employed as a test dataset. Figure 1 shows the synthetic RGB (red green blue) images of the HS cubes with the tumor areas surrounded in yellow and the ground truth maps of each HS image employed for the test evaluation of the algorithms throughout a leave-one-patient-out cross-validation methodology. This cross-validation methodology was selected in order to perform an inter-patient validation and to avoid overfitting in the supervised classification model generation. Due to the low number of HS images with tumor pixels labeled, it was not possible to perform another evaluation approach based on training, validation and test data partition. In the ground truth maps the green, red, blue, and black pixels represents the normal, tumor, hypervascularized, and background labeled samples, respectively. The white pixels represent the pixels that have no class assigned, so it is not possible to perform a quantitative evaluation of such pixels. The classification of the entire HS cube is only evaluated in a qualitative way. The identification numbers of the test HS cubes correspond with those presented in [47].

Table 1. Summary of the hyperspectral (HS)-labeled dataset employed in this study.

Class	#Labeled Pixels	#Images	#Patients
Normal Tissue	101,706	26	16
Tumor Tissue (Glioblastoma-GBM)	11,054	6	4
Hypervascularized Tissue	38,784	25	16
Background	118,132	24	15
Total	269,676	26	16

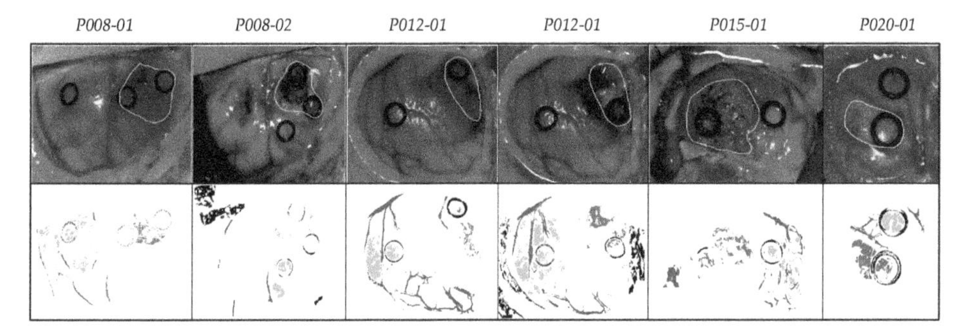

Figure 1. Synthetic RGB (Red, Green and Blue) images of HS test dataset with the tumor area surrounded in yellow (first row) and gold standard maps obtained with the semi-automatic labeling tool from the HS cube (second row). Normal, tumor, hypervascularized and background classes are represented in green, red, blue, and black color, respectively. White pixels correspond with non-labeled data.

Figure 2 shows the average and standard deviations of the spectral signatures available in both the original and the reduced training datasets. As can be seen, there are minimal differences in the average and standard deviation of the normal, tumor and hypervascularized classes while in the background class the differences are more noticeable. This is mainly produced due to this class involving several different materials that can be found in a neurosurgical scene, such as the skull bone, dura matter, gauzes with and without blood, plastic pins, etc. These materials have highly different spectral signatures which is evidenced by the high standard deviation obtained for this class.

From the point of view of the biological analysis, certain wavelength ranges have been associated to particular optical properties of cancer tissues [4]. The major spectral contribution of hemoglobin (Hb) is found in the range between 450 and 600 nm [48]. Particularly, deoxygenated Hb shows a single absorbance peak around 560 nm, while oxygenated Hb shows two equal absorbance peaks around 540 and 580 nm [49]. On the other hand, the region of the NIR spectrum from 700 to 900 nm corresponds with the scattering dominant optical properties of biological tissues, mainly composed of fat, lipids, collagen, and water [50]. Considering that the absorbance peaks are transformed to valleys in reflectance measurements, within the spectral signatures of the normal and tumor classes in Figure 2, these valleys in the range between 540 and 580 nm can be identified.

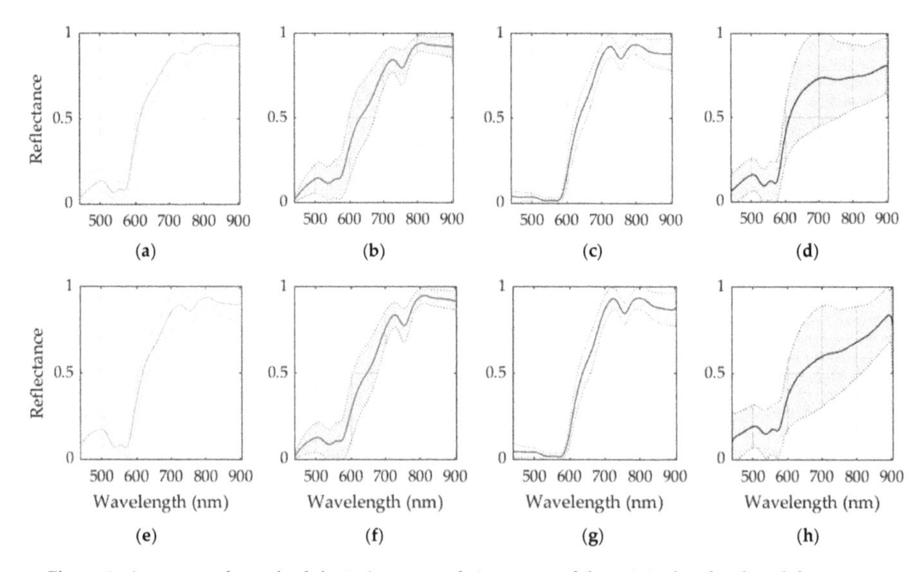

Figure 2. Average and standard deviation spectral signatures of the original and reduced dataset per classes. (**a–d**) Normal, tumor, hypervascularized and background spectral signatures respectively, extracted from the original training dataset. (**e–h**) Normal, tumor, hypervascularized and background spectral signatures respectively, extracted from the reduced training dataset.

3. Methodology

The general methodology employed in this work to evaluate the results obtained was based on a SVM classifier [51], following a data partition consisting on a leave-one-patient-out cross-validation. This method performs an inter-patient classification where the samples of an independent patient are used for the test, while the training group includes all the patients' samples except the ones to be tested. This process is repeated for each patient in the test database. The SVM classifier was selected in order to compare the results with previous published works [34,36,52]. However, although in the previous works deep learning approaches were evaluated, in this preliminary study only the SVM-based approaches were evaluated and compared, mainly because of the limited sample size. The SVM implementation was performed in the MATLAB® R2019a (The MathWorks, Inc., Natick, MA, USA) environment using the LIBSVM package developed by Chang et al. [53]. Following this general methodology, three different processing frameworks have been proposed.

The first processing framework (*PF1*) performs a sampling interval analysis of the HS data (composed by 826 bands) in order to evaluate the reduction of the number of bands in the HS images by modifying the sampling interval of the HS camera, i.e., decimating the bands to be employed in the classification process at certain steps. This procedure is intended to reduce the redundant information in the data due to the high dimensionality, allowing also a decrease in the execution time of the classification algorithm. In addition, in this processing framework, a training dataset reduction algorithm based on the K-means clustering algorithm is proposed with the goal of reducing the number of samples in the training dataset. By employing this method, the most relevant information is employed to train the SVM classifier, balancing the training samples for each class of the dataset and drastically reducing the training execution time. This time reduction obtained in the sampling interval analysis and the training dataset reduction will be crucial in the next processing frameworks, where the analysis of different optimization algorithms is performed. The block diagram of the *PF1* is shown in Figure 3a.

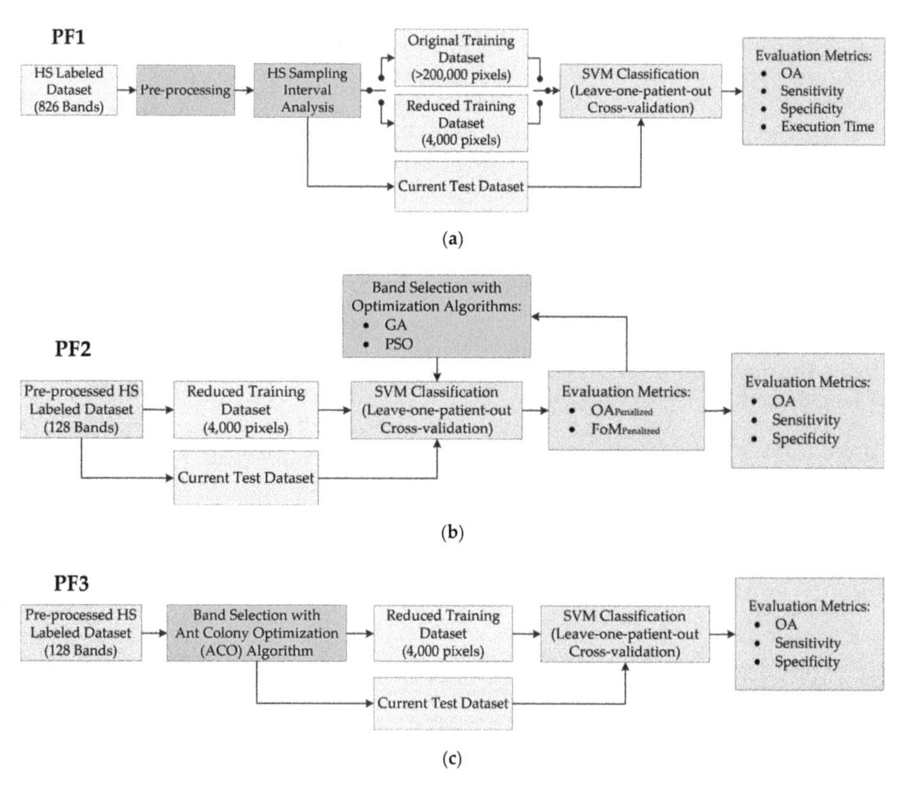

Figure 3. Different processing frameworks (PF) evaluated in this work. (**a**) *PF1* where the HS sampling interval analysis is performed with the original and the reduced training datasets independently. (**b**) *PF2* where the analysis of the *GA* and *PSO* optimization algorithms is performed using only the reduced training dataset and the optimal sampling interval selected in PF1. (**c**) *PF3* where the *ACO* algorithm is evaluated using the same training and input datasets as those employed in *PF2*. In these figures, green blocks represent the input data, orange blocks identify the main part of the proposed framework, blue blocks denote the train and test datasets employed for the supervised classification (pale brown blocks), and purple blocks represent the evaluation metrics employed.

The second processing framework (*PF2*) has the goal of evaluating the *GA* and *PSO* optimization algorithms as band selection methods. In this framework, the suitable solutions obtained in the *PF1* are used and specific evaluation metrics are employed to iteratively find the best solution. Finally, general evaluation metrics for classification tasks are employed to obtain the final results for each case. Figure 3b shows the block diagram of this processing framework.

The third processing framework (*PF3*) evaluates the use of the *ACO* algorithm to find the most relevant bands. This algorithm works in a different way than the *GA* and *PSO*. The *ACO* algorithm sorts the different bands according to their importance and correlation between them. Thus, the iterative procedure is not required. In Figure 3c, the block diagram of this process is shown.

In next sections, each one of these proposed processing frameworks are explained in detail. To improve the readability of the rest of the manuscript, a list of the acronyms of each proposed method that will be named in the results section is presented in Appendix A in Table A1.

3.1. Processing Framework 1 (PF1): Sampling Interval Analysis and Training Dataset Reduction

The *PF1* (Figure 3a) aims to perform a sampling interval analysis of the HS raw data in order to reduce as much as possible the number of bands required to obtain accurate classification results, removing the redundant information provided by the high spectral resolution of the HS camera. In addition, in this processing framework a training dataset reduction algorithm is proposed with the goal of reducing the high training times of the SVM classifier. The best pre-processing chain obtained in this analysis, involving also the sampling interval selection, together with the optimal training dataset selection, will be employed in the next proposed processing frameworks.

3.1.1. Data Pre-Processing

A pre-processing chain composed of three main steps was applied to the HS raw data in order to homogenize the spectral signatures of the dataset. The first stage in this chain is the radiometric calibration of the HS images, performed to avoid the interference of environmental illumination and the dark currents of the HS sensor. The raw data was calibrated using white and dark reference images following Equation (1), where C_i is the calibrated image, R_i is the raw image, and W_r and D_r are the white and dark reference images, respectively. The white reference image was captured from a material that reflects the 99% of the incoming radiation in the full spectral range considered in this work (Spectralon® tile). This tile was placed at the same location where the patient's head will be placed during the surgery, thus taking into account the real light condition. On the other hand, the dark reference image was acquired by keeping the camera shutter closed. This calibration procedure ensures the consistence of data and the reproducibility of the results independent of the operating room where the system is used.

$$C_i = 100 \cdot \frac{R_i - D_r}{W_r - D_r} \tag{1}$$

The second stage consists in the application of a noise filter in order to reduce the high spectral noise generated by the camera sensor using a smooth filter. Except for the first case of the sampling interval analysis, presented in the next section, the preprocessing applied involves an extreme band-removing step before the noise filter with the goal of eliminating the bands with high noise in the first and last bands of the HS data produced with the low performance of the sensor in these bands. After this procedure, the operating bandwidth of the HS data is between 440 and 902 nm. Finally, in the third stage, the spectral signatures are normalized between zero and one in order to homogenize reflectance levels in each pixel of the HS image produced by the non-uniform surface of the brain. Equation (2) shows this process, where the normalized pixel (P'_i) is computed by subtracting the minimum reflectance value in a certain pixel (P_{min}) by the reflectance value of such pixel in a certain wavelength (P_i) and dividing it by the difference between the maximum and minimum reflectance values ($P_{max} - P_{min}$).

$$P'_i = \frac{P_i - P_{min}}{P_{max} - P_{min}} \tag{2}$$

3.1.2. Sampling Interval Analysis

In order to simulate the use of different HS cameras where a different number of spectral bands are captured covering the same spectral rage, the following methodology was employed to vary the spectral sampling interval of the HS data. The main goal of this analysis is to reduce the number of bands employed for the classification in this particular application and, consequently, the HS camera size and cost, as well the computational effort.

The spectral sampling interval is the distance between adjacent sampling points in the spectrum or spectral bands. This spectral sampling interval is calculated by Equation (3), where $\lambda max - \lambda min$ is the difference between the maximum and minimum wavelength captured by the sensor, also named as the spectral range. The number of spectral bands have been reduced while the sampling interval increase in order to simulate diverse HS cameras that capture different number of bands.

$$Sampling\ Interval\ (nm) = \frac{\lambda_{max} - \lambda_{min}}{number\ of\ bands} \tag{3}$$

Table 2 shows the different sampling interval values obtained for each number of spectral bands chosen. The original raw HS image captured by the sensor is composed by 826 bands, having 2–3 nm of spectral resolution and 0.73 nm of sampling interval. This HS camera covers the range between 400 and 1000 nm. In order to avoid the noise produced by the CCD (charge-coupled device) sensor in the extreme bands, they were removed, obtaining a final spectral range from 440 to 902 nm with 645 spectral bands and the same sampling interval. Using this HS cube as a reference, several sampling intervals were applied, reducing the number of bands and, in consequence, the size of the HS image. The original raw image size was higher than 1 GB and by reducing spectral bands from 826 to 8 the image size obtained was ~12 MB. These data were obtained from an average value of the HS test cubes, since the spatial dimension of each HS cube were different.

Table 2. Summary of the HS dataset with different sampling intervals and number of spectral bands.

	#Spectral Bands										
	826	**645**	**320**	**214**	**160**	**128**	**80**	**64**	**32**	**16**	**8**
λmin (nm)	400	440	440	440	440	440	440	440	440	440	440
λmax (nm)	1000	902	902	902	902	902	902	902	902	902	902
Sampling Interval (nm)	0.73	0.73	1.44	2.16	2.89	3.61	5.78	7.22	14.44	28.88	57.75
Size (MB)	1328.3	1037.3	514.6	344.1	257.3	205.8	128.6	102.9	51.4	25.7	12.8

3.1.3. Training Dataset Reduction

Supervised classifiers rely on the quality and amount of the labeled data to perform the training and create a generalized model to produce accurate results. However, in some cases, the labeled data can be unbalanced between the different classes and may contain redundant information, increasing the execution time of the training process and even worsening the performance of the classification results. Taking into account that the optimization algorithms used in this work have to perform an iterative training of the classifier in order to find the most relevant wavelengths for obtaining an accurate classification, it is beneficial to accelerate the training process.

The methodology proposed in this section for optimizing the training dataset is based on the use of K-Means unsupervised clustering [54]. Figure 4 shows the block diagram of the proposed approach for reducing the training dataset. In this approach, the training dataset is separated in four groups that correspond with the different classes available in the labeled dataset: normal, tumor, hypervascularized and background. The total number of samples available in the entire dataset is 269,676 pixels, corresponding to 101,706 normal pixels, 11,054 tumor pixels, 38,784 hypervascularized pixels, and 118,132 background pixels (see Table 1). The K-Means clustering is applied independently to each group of labeled pixels in order to obtain 100 different clusters (K = 100) per group (400 clusters in total). Hence, 100 centroids that correspond to a certain class are obtained. In order to reduce the original training dataset, such centroids are employed to identify the most representative pixels of each class by using the SAM algorithm [44]. For each centroid, only the 10 most similar pixels are selected, having a total of 1000 pixels per class (100 centroids × 10 pixels). Thus, the reduced dataset is intended to avoid the inclusion of redundant information in the training of the supervised classifier. At the end, the reduced training dataset will be composed of 4000 pixels (after applying the K-Means four times independently) being the most representative pixels of the original dataset and obtaining a balanced dataset among the different classes. It is worth noticing that this procedure is executed within the leave-one-patient-out cross-validation methodology. Thus, the labeled pixels of the current test patient will not be included in the original dataset employed for the reduction process.

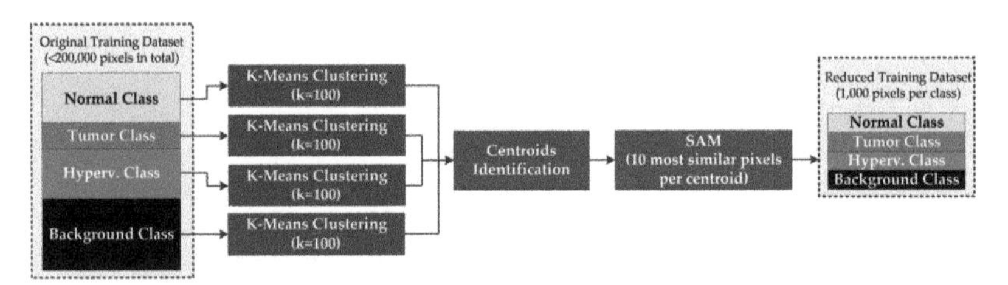

Figure 4. Block diagram of the training dataset reduction algorithm.

3.2. *PF2: Band Selection Using Genetic Algorithm (GA) and Particle Swarm Optimization (PSO)*

In this processing framework (Figure 3b), the use of the *GA* and *PSO* for the selection of the most representative wavelengths of the HS data for this particular application is proposed. The optimization algorithms are aimed to find the best combination of elements from an initial set of available elements. Normally, these algorithms are focused on reaching the global minimum of the function to be analyzed. The reduced training labeled dataset evaluated in *PF1* was employed to reduce the extremely high execution time of the SVM training procedure, due to the iterative nature of the optimization algorithms. The procedure is as follows. First, the training data are employed in the optimization algorithm, where the initial bands to be used for the classification are randomly selected. After this band selection procedure, a classification model is generated and evaluated with the test dataset, obtaining a classification result that is assessed using three custom metrics independently ($OA_{Penalized}$, FoM and $FoM_{Penalized}$). These metrics will be explained in detail later on in Section 3.5. After computing one of the custom metrics, its value is stored and then the procedure is iteratively executed using other bands selected by the optimization algorithm. The algorithm is executed until performs all possible combinations, returning the best metric value, or when after a high number of iterations, the metric value remains constant. Once the algorithm finishes its execution, it returns the identification of the optimal bands to obtain the best classification result.

3.2.1. Genetic Algorithm (GA)

The *GA* is an optimization algorithm that mimics the process of natural selection [41]. *GA* tries to find the optimal solution (usually the global minimum) of the function to be studied. The main advantage of this algorithm is its great ability to work with a large number of variables [55,56]. The objective of this algorithm is to optimize a series of parameters (called genes) that will then be concatenated with each other, when necessary, to provide the best results (called chromosomes). In order to find these most important parameters, it is necessary to generate populations in a random manner, whose size is chosen by the user. This population allows the performance of the algorithm to be improved. Once these parameters are defined, the *GA* must perform the following steps:

(1) **Initialization**: In this step, the selection of the population is performed in a random way.
(2) **Evaluation**: The goal is to study the results obtained from the initial population (*parents*) and each of the descendant generations (*children*).
(3) **Selection**: This point is responsible for keeping the best result obtained during the evaluation process.
(4) **Recombination**: In this step, the combination of the different initial contributions (*parents*) for the creation of better solutions (*children*) is performed. This crossing is performed by dividing the populations in two (or more) parts and exchanging part of those populations with each other.
(5) **Mutation**: This technique is performed in the same way as in the recombination step. However, instead of exchanging parts of the populations among themselves, a single value of each of the populations is modified.
(6) **Replacement**: After performing the recombination and mutation steps, these generations (*children*) replace the initial populations (*parents*).

Steps 2 to 6 are repeated as many times as necessary until the best solution is found [41].

3.2.2. Particle Swarm Optimization (PSO) Algorithm

The PSO algorithm is a stochastic technique based on the behavior of bee swarms [42]. This algorithm exhibits great effectiveness in multidimensional optimization problems. This methodology is based on the survival of some living beings (specifically the bees). As well as the genetic algorithm, PSO is initialized with a random initial population. However, in this case, each possible solution, known as a particle, has also been assigned a random velocity [42]. Each particle updates and stores the best position found so far (pbest) and also stores and updates the best position of the rest of the swarm (gbest). To represent the velocity update, the algorithm uses Equation (4), where v_i is the velocity vector, x_i is the position vector, α is the weight of the particle that controls the recognition of the place, c_1 and c_2 are the acceleration constants of the particles (usually take a value of 2 by default), and *rand* is a random number between 0 and 1.

$$v_i(t+1) = \alpha v_i + c_1 \cdot rand \cdot (pbest(t) - x_i(t)) + c_2 \cdot rand \cdot (gbest(t) - x_i(t)) \quad (4)$$

Once the parameters that conform the algorithm are obtained, the swarm is generated by means of the following steps:

(1) **Initialization**: This step initializes a random population with different positions and velocities.
(2) **Selection**: In this step, each particle evaluates the best location found and the best position found by the rest of the swarm.
(3) **Evaluation**: Here, a comparison of all the results and selection of the *pbest* is performed. The same process is applied to find the best *gbest*.
(4) **Replacement**: In this last step, the new results replace the initial population and the process is repeated up to a maximum number of generations established by the user or until the solution converges.

3.3. PF3: Band Selection Using Ant Colony Optimization (ACO)

The *PF3* (Figure 3c) has the goal of selecting the optimal bands employing directly the *ACO* algorithm to the train dataset. In this case, the algorithm restructures the spectral bands of the HS data in order to allocate the most relevant bands, based on endmember extraction [57] (selection of pure spectra signature of the different materials) in the first positions and the least important ones at the end. After this band reorganizing procedure, a classification model was generated with the SVM classifier and evaluated with the standard evaluation metrics (accuracy, sensitivity and specificity). This process was generated five times, evaluating the 20, 40, 60, 80 and 100 most relevant bands obtained with the *ACO* algorithm.

Ant Colony Optimization (ACO) Algorithm

The *ACO* algorithm is based on a metaheuristic procedure, which aims to obtain acceptable solutions in problems of combinatorial optimizations in a reasonable computational time [43]. As the name suggests, this algorithm is based on the composition of the ant colonies. The ants, when searching for food, separate and begin to make trips in a random way. Once an ant gathers food, while carrying the food to the nest, it expels pheromones along the way. Depending on the quality or quantity of the food found, the amount of pheromones will vary. On the other hand, the evaporation of the pheromones causes the pheromones to disappear, so that, if these routes are not reinforced, they end up disappearing. This process is repeated until the best possible route is found.

Taking into account this selection process, the algorithm is characterized by having a main component, the *pheromone* model. This model is a parameterized probabilistic model, which consists of a vector of parameters that indicates the trajectory followed by *pheromones*. These values are updated until the minimum value of the problem is reached.

3.4. Coincident Bands Evaluation Methodology

When the optimization algorithms employed in *PF2* and *PF3* selected the bands, different bands were identified for each test image. Thus, the six HS test images were evaluated employing the same selected bands, i.e., using the coincident and non-coincident selected bands in each test image obtained in the leave-one-patient-out cross-validation for the SVM training and classification. The procedure was as follows:

- **First level (*L1*):** the coincident and non-coincident bands from all the test images were used to generate and evaluate the results.
- **Second level (*L2*):** the coincident bands repeated in at least two test images were used to generate and evaluate the results.
- **Third level (*L3*)**: the coincident bands repeated in at least three test images were used to generate and evaluate the results.

This process was repeated until reaching the possible six coincidences. However, in our case, a maximum of three levels of coincidence were obtained.

3.5. Evaluation Metrics

The validation of the proposed algorithm was performed using inter-patient classification, following a leave-one-patient-out cross-validation. Overall accuracy (OA), sensitivity and specificity metrics were calculated to measure the performance of the different approaches. OA is defined by Equation (5), where TP is true positives, TN is true negatives, P is positives, and N is negatives. Sensitivity and specificity are defined in Equations (6) and (7), respectively, where FN is false negatives, and FP is false positives. In addition, the Matthews correlation coefficient (MCC) was employed to evaluate the different approaches (Equation (8)). This metric is mainly used to analyze classifiers that work with unbalanced data, which computes the correlation coefficient between the observed and the predicted values [58]. MCC has a value range between [−1, 1], where −1 represents a completely wrong prediction and 1 indicates a completely correct prediction. For comparison purposes with other metrics presented in this work, the MCC metric has been normalized within the [0, 1] range applying Equation (9).

$$OA = \frac{TP + TN}{P + N} \tag{5}$$

$$Sensitivity = \frac{TP}{TP + FN} \tag{6}$$

$$OA = \frac{TP + TN}{P + N} \tag{7}$$

$$MCC = \frac{TP \cdot TN - FP \cdot FN}{\sqrt{(TP + FP) \cdot (TP + FN) \cdot (TN + FP) \cdot (TN + FN)}} \tag{8}$$

$$MCC\prime = \frac{MCC + 1}{2} \tag{9}$$

Classification maps are another evaluation metric commonly used in HSI. This evaluation metric allows users to visually identify where each of the different classes are located. This metric is employed to visually evaluate the classification results obtained when the entire HS cube is processed, including labeled and non-labeled pixels. After performing the classification of the HS cube, a certain color is assigned to each class. This process allows mainly evaluating the results obtained in the prediction of non-labeled pixels. The colors that are represented in the classification map are the following: green was assigned to the first class (healthy tissue); red was assigned the second class (tumor tissue); blue was assigned to the third class (hypervascularized tissue); and black was assigned to the fourth class (background).

In addition to the standard OA, an additional metric has been proposed for the identification of the best results obtained with the optimization algorithms but taking into account also the number of selected bands. This $OA_{Penalized}$ is based on the OA presented in Equation (5) but including a penalty in the case that a high number of bands is used. Equation (10) presents the mathematical expression to compute this $OA_{Penalized}$, where λ is the number of bands selected by the algorithm and λ_{max} is the total number of bands in the original dataset.

The specific Figure of Merit (*FoM*) employed to obtain the most relevant bands with the optimization algorithms in the *PF2* has the goal of finding the most balanced accuracy results for each class, as observed in Equation (12), where n is the number of classes, i and j are the indexes of the classes that are being calculated. The mathematical expression of the $ACC_{perClass}$ in a multiclass classification is obtained by dividing the total number of successful results (TP) for a particular class by the total population of this class ($TP + FN$). This expression is equal to the sensitivity of a certain class in a multiclass classification problem. Equation (11) shows the mathematical expression of the $ACC_{perClass}$.

In addition to the previously presented *FoM*, another metric has been proposed for the identification of the best results obtained with the optimization algorithms but taking into account also the number of selected bands. This $FoM_{Penalized}$ is based on the *FoM* presented in Equation (12) but including a penalty in the case that a high number of bands is used. Equation (13) presents the mathematical expression to compute this $FoM_{Penalization}$, where λ is the number of bands selected by the algorithm and λ_{max} is the total number of bands in the original dataset.

$$OA_{Penalized} = 1 - \frac{OA}{1 + \frac{\lambda}{\lambda_{max}}} \tag{10}$$

$$ACC_{perClass} = \frac{TP}{TP + FN} \tag{11}$$

$$FoM = \frac{1}{2} \cdot \left(\sum_{\substack{i,j \\ i<j}}^{n} \frac{ACC_i + ACC_j}{|ACC_i - ACC_j| + 1} \right) \cdot \binom{n}{2}^{-1} \tag{12}$$

$$FoM_{Penalized} = 1 - \frac{FoM}{1 + \frac{\lambda}{\lambda_{max}}} \tag{13}$$

4. Experimental Results and Discussion

This section will present the results obtained in the three proposed processing frameworks, as well as the overall discussion of the results. Table A1 in the Appendix A shows the acronym list of each proposed method named in the next sections in order to help the reader to follow the experimental results explanation.

4.1. Sampling Interval Analysis (PF1)

The *PF1* has the goal of performing a comparison between the use of different numbers of spectral bands in the HS database, modifying the sampling interval of the spectral data in order to simulate the use of different HS cameras and reducing the size of the database. This will lead to a reduction of the execution time of the processing algorithm. In addition, as shown in Figure 3a, the *PF1* was evaluated using two different training datasets for the SVM model generation: the original and the reduced training dataset.

Figure 5a shows the classification results obtained for each sampling interval (different number of bands), training the SVM algorithm with the original dataset, while Figure 5b shows the results using the reduced training dataset. It can be observed in both figures that the overall accuracy of the classifier decreases as the number of bands is reduced. However, the sensitivity values for each class are quite similar from 826 to 64 bands. When the number of bands is lower than 64, the sensitivity values drop drastically. Respect to the standard deviation, the behavior obtained in both datasets are similar. It can be observed that for the OA, normal sensitivity and hypervascularized sensitivity, the results remain stable as the number of bands are reduced. In the case of tumor sensitivity, the standard deviation is higher than in the other cases. This behavior is caused by one of the HS test images (*P020-01*) presenting problems in the classification and not being able to correctly identify any of the tumor pixels. In the background class, as the number of bands decrease, the standard deviation increases. On the other hand, as it can be seen in Figure S1 of the supplementary material, the specificity results are very similar in both cases being higher than 80% in general. In addition, Figure S2 of the supplementary material shows the results of the normalized MCC metric for the original and reduced training datasets, which takes into account the unbalance of the test labeled dataset. As can be seen, the obtained results in all the classes are similar except for the tumor tissue class, which improve an average of ~5% when the reduced training dataset is employed.

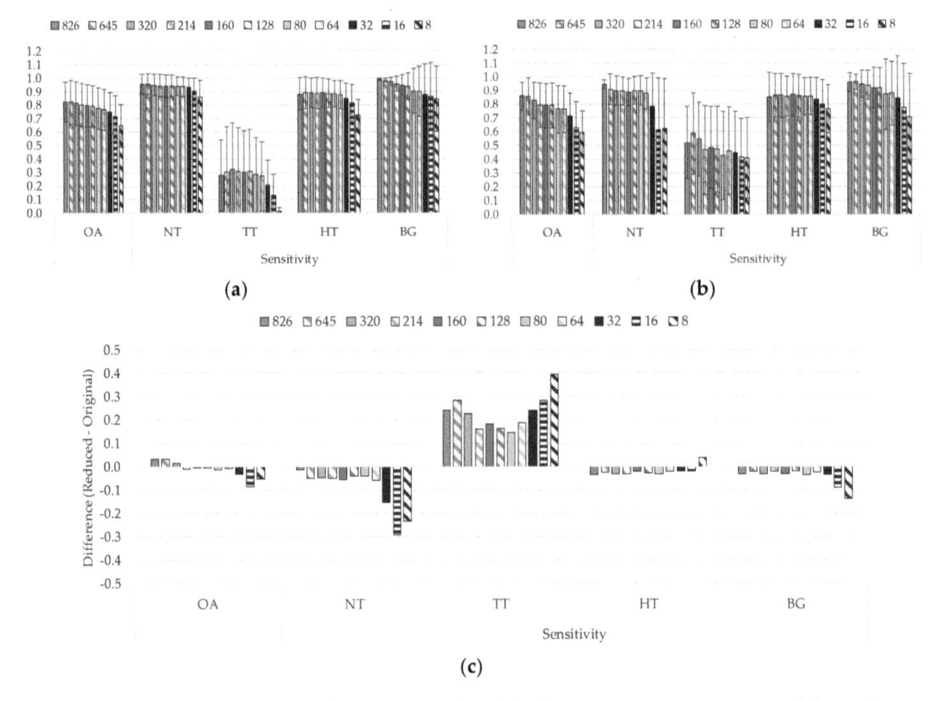

Figure 5. Average and standard deviation results of the leave-one-patient-out cross-validation for each band reduction. (**a**) Using the original training dataset. (**b**) Using the reduced training dataset. (**c**) Difference of the results computed using the reduced dataset respect to the original dataset.

Figure 5c shows the differences in the results between the reduced and original training dataset. As it can be seen, the reduced dataset provides better accuracy results in the tumor class respect to the original dataset, reaching an average increment of ~20%. Since the main goal of this work is to accurately identify the tumor pixels, this increment provides a significant improvement on this goal. It is worth noticing that the test dataset was not reduced in the number of samples, only the training dataset was reduced.

On the other hand, the image size is directly related to the execution time of the processing algorithm. Figure 6a shows the execution time of the SVM training and classification processes computed by using the MATLAB® programming environment together with the LIBSVM package. This figure presents the execution time results (expressed in minutes) for both training schemes (original training dataset, and reduced training dataset). The times depicted in such a table include both the time required to train the model for one leave-one-patient-out cross-validation iteration, and the time needed to classify the correspondent patient data. In order to compare the results, a logarithmic scale was used. On one side, as the number of bands decreases, the execution time also decreases, being practically stable from 128 to 8 bands in both cases. On the other side, it is clear that the use of the reduced training dataset offers a significant execution time reduction. For example, using 826 bands, the execution time for the original training dataset is ~778 min, while for the reduced dataset it is ~16 min, obtaining a speedup factor of ~48×. Taking this into account, the reduced training dataset was selected for the next experiments.

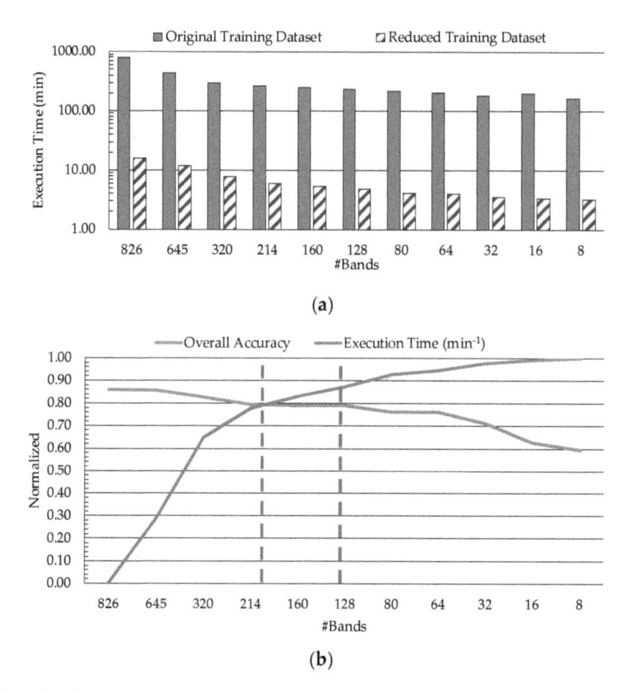

Figure 6. (**a**) Band reduction execution time for original and reduced training datasets (representation in logarithmic scale). (**b**) Overall accuracy and inverse normalized execution time achieved using the reduced training dataset with respect to the number of bands.

In order to select a good trade-off sampling interval, which provides a reduction on the execution time of the algorithm while keeping high discrimination, a relation between these two metrics was performed. Figure 6b shows the relation between the accuracy and the inverse normalized execution time depending on the number of bands employed in the HS dataset, ranging from 826 to 8 bands. The analysis of the overall accuracy shows that when all the bands are used the overall accuracy is high, but the execution time is also very high. However, when only a few bands are used the overall accuracy decreases more than 20%, but the execution time is quite low. In this sense, the suitable range that provides a good compromise between the execution time and the overall accuracy is found between the 214 and 128 bands (dashed red lines in Figure 6b). In this range, the accuracy results are stabilized in the value of 80% and the execution times are acceptable. For this reason, the number of bands selected to conform the HS dataset in the next experiments was 128 (lowest number of bands with the same overall accuracy), involving a sampling interval of 3.61 nm.

The use of the reduced training dataset together with the selection of 128 spectral bands will ensure that we reduce the execution time (mainly in the SVM training process), allowing us to perform the band selection using the optimization algorithms in the next processing framework (PF2). This will avoid large processing times during the huge number of iterations required by the optimization algorithms until reaching the optimal solution.

4.2. Band Selection Using Optimization Algorithms (PF2 and PF3)

The *PF2* and *PF3* aim to use optimization algorithms in order to find the most relevant bands able to perform an accurate classification of the brain tumors, using the lower possible number of features. The evaluation of both processing frameworks was performed using the reduced training dataset, and a data partition scheme following a leave-one-patient-out cross-validation to create the SVM model and evaluate the results (see Figure 3b,c). In these processing frameworks, the six test HS images were evaluated with different optimization algorithms to find out which offers the best results. In addition, in case of *GA* and *PSO* algorithms (PF2), two different metrics were employed to evaluate the selected bands: $OA_{Penalized}$ (OA_P) and $FoM_{Penalized}$ (FoM_P). In summary, the band selection techniques evaluated were: *GA* using $OA_{Penalized}$ (PF2-GA-OA_P); *PSO* using $OA_{Penalized}$ (PF2-PSO-OA_P); *GA* using $FoM_{Penalized}$ (PF2-GA-FoM_P); *PSO* using $FoM_{Penalized}$ (PF2-PSO-FoM_P) and *ACO* using 60 bands (PF3-ACO-60). Furthermore, all the results were compared with the reference values obtained with the *PF1* using 128 bands (PF1-128).

Figure 7 shows the boxplot results of the OA and the normal, tumor and hypervascularized tissue sensibilities obtained after the evaluation of the *GA*, *PSO* and *ACO* algorithms. The *ACO* algorithm was evaluated selecting different numbers of bands (20, 40, 60, 80 and 100), but only the results obtained with 60 bands have been reported because they were found to be the most competitive ones. Figures S3 and S4 in the supplementary material present the detailed results obtained with the *ACO* algorithm. Figure 7a shows the OA results of all the techniques, where it should be noted that the median values are around 80%, offering the PF2-GA-OA_P the best result. However, the results obtained in the tumor sensitivity boxplot (Figure 7b) shows that the technique that uses the *GA* algorithm with the $FoM_{Penalized}$ metric (PF2-GA-FoM_P) provides the best results, achieving a tumor sensitivity median of ~79%. This represents an increment of ~21% with respect to the PF2-GA-OA_P method. On the other hand, in Figure 7c, it can be seen that the results of the normal tissue sensitivity are similar in both cases (PF2-GA-OA_P and PF2-GA-FoM_P), with a median value of 89% and 90%, respectively. The same behavior can be observed in the hypervascularized tissue sensitivity results (Figure 7d). As can be observed in Figure 8, the MCC metric, which takes into account the unbalance of the test labeled dataset, shows the same behavior as the other metrics. Thus, the PF2-GA-FoM_P is the method that provides on average the best results. This is especially highlighted in the tumor class results.

Figure 9 illustrates the qualitative results represented in the classification maps obtained for each method. These maps allow visualization of the identification of the different structures for each class found in the complete HS cube, i.e., it is possible to visually evaluate the results obtained in the non-labeled pixels of the HS test cubes. In addition, this figure indicates the number of bands selected with each method for each HS test cube. Figure 9a shows the synthetic RGB images for each HS test cube, indicating the location of the tumor area surrounded by a yellow line, while Figure 9b shows the classification results obtained with the reference method (PF1-128).

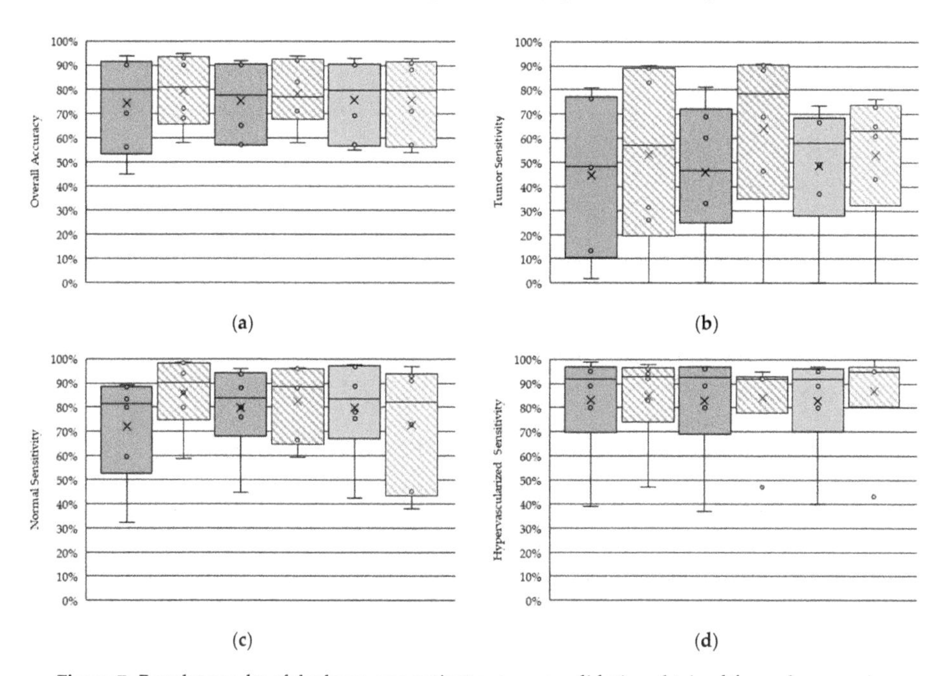

Figure 7. Boxplot results of the leave-one-patient-out cross-validation obtained for each processing framework. (**a**) Overall accuracy. (**b**) Tumor tissue sensitivity. (**c**) Normal tissue sensitivity. (**d**) Hypervascularized tissue sensitivity.

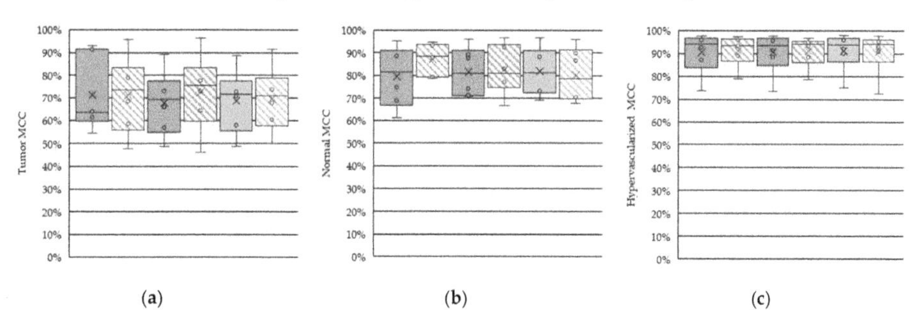

Figure 8. Boxplot results of the Matthews correlation coefficient (MCC) metric using the leave-one-patient-out cross-validation obtained for each processing framework. (**a**) Tumor tissue. (**b**) Normal tissue. (**c**) Hypervascularized tissue.

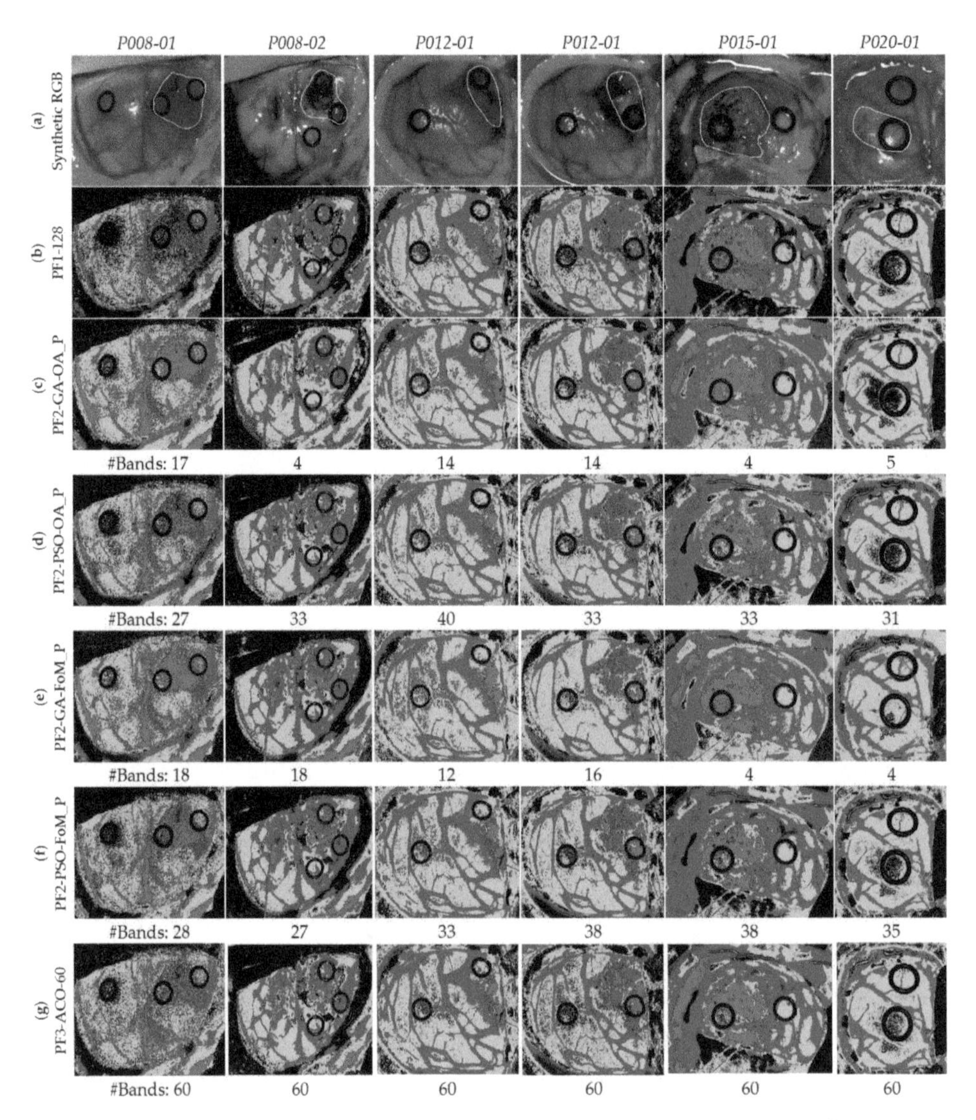

Figure 9. Classifications maps of the test database. (**a**) Synthetic RGB images with a yellow line identifying the tumor area. (**b**) Reference results using 128 bands. (**c**) Results of the *GA* algorithm using $OA_{Penalized}$. (**d**) Results of the *PSO* algorithm using $OA_{Penalized}$. (**e**) Results of the *GA* algorithm using $FoM_{Penalized}$. (**f**) Results of the *PSO* algorithm using $FoM_{Penalized}$. (**g**) Results of the *ACO* algorithm using the 60 bands per image.

Considering these images as reference, it is observed that the results in all cases are very similar. Nonetheless, when analyzing the images one-by-one, in the case of the first image, *P008-01*, it is observed that using either PF2-PSO-OA_P (Figure 9d) or PF3-ACO-60 (Figure 9g) the results are more accurate with fewer false positives. The best case that visualizes the *P008-02* image is the PF2-GA-FoM_P (Figure 9e), since it is observed a higher number of tumor pixels in the area of the tumor. As for the *P012-01*, all the techniques visualize the tumor area correctly, but the PF1-128 (Figure 9b) is the one that shows fewer false positives pixels. In the case of *P012-02*, the best technique is the PF2-GA-FoM_P (Figure 9e) due to it shows less false positive pixels. With respect to the *P015-01* image, it is observed that using PF2-GA-OA_P (Figure 9d) and PF2-GA-FoM_P (Figure 9e) the tumor area is clearly identified, although they have a small group of false positives in the upper left image due to some extravasated blood out of the parenchymal area. Finally, image *P020-01* offers practically the same result in all cases without a successful identification of the tumor area, even in the reference results. Regarding the number of bands selected to perform the classification, the PF2-GA-OA_P and PF2-GA-FoM_P are the methods that achieved the less number of bands for each HS test image, being lower than 18 bands in all the cases.

After conducting a thorough analysis of the results obtained, it was observed that the best technique, which provided the best balance between qualitative and quantitative results, is the PF2-GA-FoM_P. Quantitatively, the GA-FoM_P provided the best average OA value of 78% (improving ~4% with respect to the reference results with 128 bands) and the best median tumor sensitivity value of 79%, which represents an increment of ~21% with respect to the best solution provided with the other optimization approaches. Moreover, the GA-FoM_P increases the tumor sensitivity value in 30% with respect to the reference results.

4.3. Coincident Bands Evaluation of the GA Algorithm with Figure of Merit ($FoM_{Penalized}$)

Taking into account the decision made in the previous section, the next step is the evaluation of the HS test images using the coincident bands in order to generate a general SVM model that can provide accurate results for all the HS test images. In this sense, the *PF2* using the *GA* algorithm and the $FoM_{Penalized}$ metric was evaluated by using the coincident and non-coincident bands selected during the cross-validation method over the six HS test images.

The procedure followed for this evaluation consists of three levels. The index *i*, in *Li*, indicates the number of HS test images where the bands are common between all the test set. Figure 10 identifies the bands that were selected by the PF2-GA-FoM_P for each HS test image and the coincident bands between them. The number of bands for *L1*, *L2* and *L3* are 48, 22 and 2, respectively.

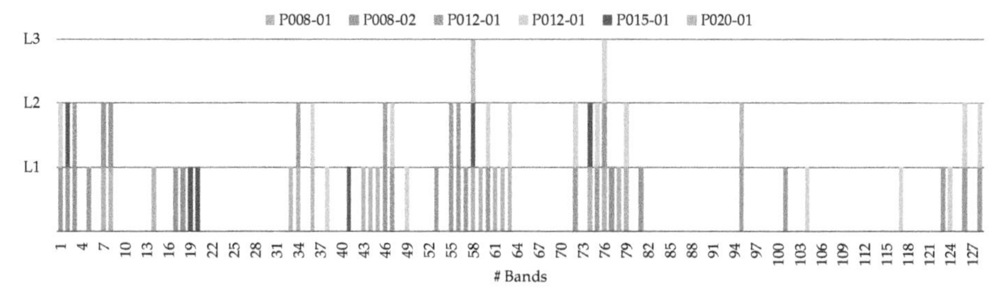

Figure 10. Graphical representation of the coincident and non-coincident bands obtained with the PF2-GA-FoM_P method (Processing Framework 2 using Genetic Algorithm and the Figure of Merit Penalized evaluation metric) for the three different levels.

Table 3 shows the quantitative results obtained for the evaluation of the different levels. These results are the average and standard deviation of the six HS test images. In terms of OA, it is observed that the best result was obtained in *L1*, with 77.9%, followed closely by *L2* with 77.0%. However, *L3* worsens notably the results, achieving only 54%, mainly because of the low number of bands employed for the generation of the SVM model. With respect to the sensitivity results, *L1* and *L2* remain practically the same for all classes, being *L2* more accurate in the tumor tissue class. Nevertheless, *L3* worsens, especially in the normal and hypervascularized tissue classes. Regarding the specificity, it follows the same trend as in sensitivity, having *L1* and *L2* similar results and *L3* bears off from these results in the normal and tumor tissue classes.

Table 3. Average and standard deviation (STD) of overall accuracy (OA), Matthews correlation coefficient (MCC), sensitivity and specificity of all images.

Level (#bands)	OA (%) (STD)	MCC (%) (STD)	Sensitivity (%) - (STD)				Specificity (%) - (STD)			
			NT	TT	HT	BG	NT	TT	HT	BG
L1 (48)	77.9 (17.0)	83.6 (9.1)	85.1 (17.6)	52.7 (29.8)	83.5 (20.9)	92.5 (14.2)	87.3 (12.2)	94.6 (8.3)	96.7 (5.1)	85.3 (18.0)
L2 (22)	77.0 (16.8)	83.3 (8.6)	83.7 (19.9)	57.0 (32.6)	81.9 (23.0)	90.1 (20.1)	85.2 (13.4)	91.2 (14.4)	97.1 (4.9)	87.7 (17.6)
L3 (2)	53.8 (21.2)	68.9 (11.4)	52.8 (42.6)	57.6 (36.5)	48.8 (26.4)	84.8 (27.1)	72.9 (13.2)	70.3 (30.8)	93.1 (8.0)	80.0 (21.1)

NT: Normal Tissue; TT: Tumor Tissue; HT: Hypervascularized Tissue; BG: Background.

On the other hand, Figure 11 shows the qualitative results of each HS test image for the different levels, indicating below each classification map the number of bands employed to generate the classification model. Figure 11a shows the synthetic RGB images of each HS test image where the tumor area has been surrounded by a yellow line. Figure 11b shows the reference results obtained without applying the optimization methodology, so the 128 bands were employed. Figure 11c presents the classification results generated using the best methodology (PF2-GA-FoM_P) selected in the previous section, and employing the specific wavelengths obtained for each HS test image independently. Figure 11d–f show the classification results obtained using the *L1*, *L2*, and *L3* levels, respectively. In these results it is observed that *L3* (Figure 11f) provides several false positives in all the classes. For example, in the *P012-01* and *P012-02* images, a large number of tumor pixels (left side of the image) are presented in the normal tissue parenchymal area that are out of the surrounded yellow line presented in Figure 11a. Regarding *L1* (Figure 11d) and *L2* (Figure 11e), the results are very similar between them, with the only difference in the *P012-01* image which shows more false positive pixels in the tumor class in *L2* than in *L1*. Both quantitative and qualitative results obtained in *L3* show that the two selected wavelengths are not representative enough to generalize a classification model that offers accurate results for all the HS test images compared to *L1* and *L2*. By contrast, it is worth noticing that the results obtained in *L1* using only 48 bands are very similar and even better in some cases with respect to the results obtained with the reference method employing 128 bands (Figure 11b).

Taking into account the quantitative and qualitative results obtained in these experiments, it has been concluded that the *L1* method provides the best accuracy results using only 48 bands. These selected bands represent the following spectral ranges: 440.5–465.96 nm, 498.71–509.62 nm, 556.91–575.1 nm, 593.29–615.12 nm, 636.94–666.05 nm, 698.79–731.53 nm and 884.32–902.51 nm. Figure 12 graphically represents the identification of the selected bands over an example of tumor, normal and hypervascularized spectral signatures. In addition, Table S2 of the supplementary material details the specific 48 wavelengths selected.

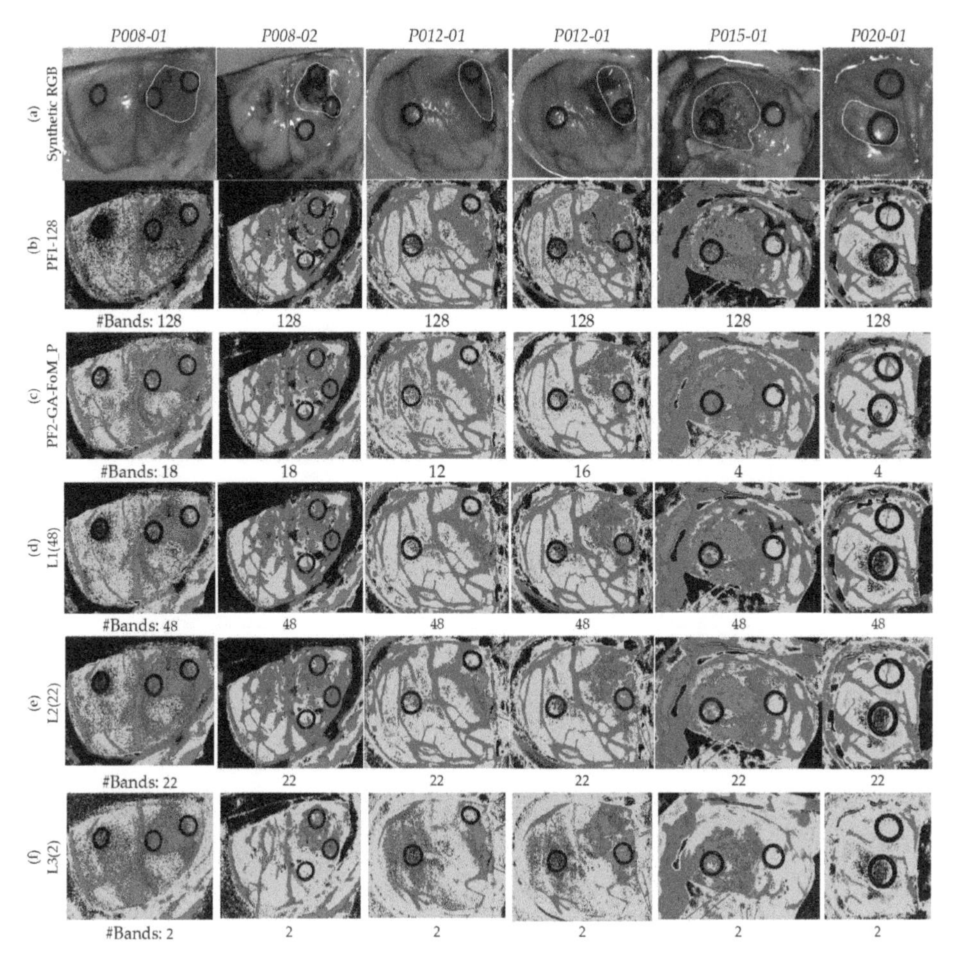

Figure 11. Maps of the test database using the different coincident levels. (**a**) Synthetic RGB images with a yellow line identifying the tumor area. (**b**) Reference results using 128 bands. (**c**) Results of the PF2-GA-FoM_P using the specific wavelengths identified for each HS test image. (**d**–**f**) Results of the PF2-GA-FoM_P in *L1*, *L2* and *L3*, respectively.

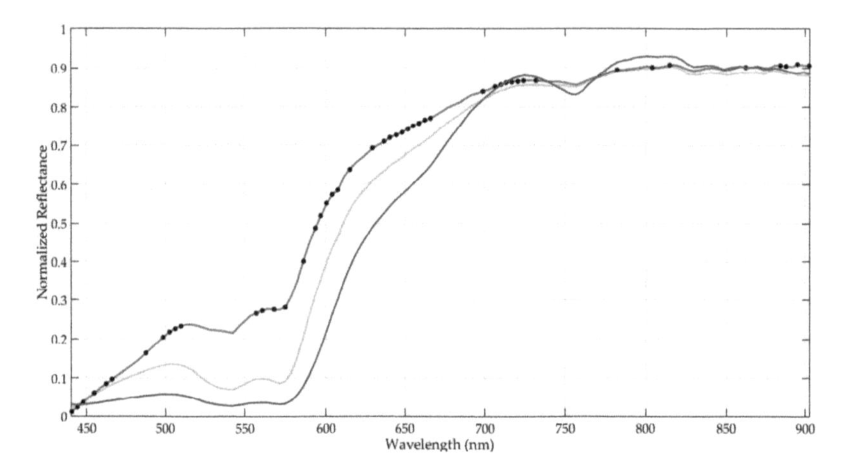

Figure 12. Example of the normalized average spectral signatures of the tumor (red), normal (green) and hypervascularized (blue) tissue classes. The black spots over the tumor spectral signature represent the 48 selected bands using the *GA* algorithm with $FoM_{Penalized}$ and the *L1* coincidence level.

5. Conclusions

Hyperspectral images are able to capture a large number of spectral bands per pixel, conforming the so-called spectral signature. This type of images has high amount of information acquired by the sensors. Depending on the HS camera type, the HS images can be composed of thousands of spectral channels (involving large sizes in the range of gigabytes) and their processing requires high-performance computing in order to reduce the processing time as much as possible. In addition, the number of bands captured by the sensor normally implies different camera sizes and different acquisition methodologies, which in some cases are difficult to employ in certain applications. Thus, the large amount of data is one of the main challenges of HSI.

The work presented in this paper had the goal of analyzing the use of different sampling intervals to reduce the number of bands employed in the HS data. This led to accurate classification results with a reduced processing time being obtained and a possible future use for a reduced-size HS camera. Furthermore, a methodology to optimize the training dataset, employed to generate the SVM model, was proposed. This methodology offered a reduced training processing time and even achieved more accurate classification results due to the redundant information elimination and noise reduction. The reduced processing time for training is extremely important in the next steps of the work, aiming to evaluate different optimization algorithms (*GA*, *PSO* and *ACO*) for the selection of the most relevant bands in the delineation and identification of brain tumors.

The employed VNIR HS database was composed of 26 HS images of the in vivo human brain obtained during neurosurgical procedures. For each image, a certain number of pixels were labeled by the experts in four different classes (normal tissue, tumor tissue, hypervascularized tissue and background) in order to create a labeled dataset that was employed to generate and evaluate a SVM classification model. A leave-one-patient-out cross-validation methodology was followed using 6 HS test images of exposed brain from four different patients affected by GBM tumors pathologically confirmed.

Different processing frameworks were defined during the development of this work. The *PF1* demonstrated that the use of a sampling interval of 3.61 nm (128 bands, ~200 MB) instead of 0.73 nm (826 bands, ~1300 MB), together with the employment of the reduced training dataset (4000 pixels vs. ~200,000 pixels), provides quite an excellent compromise between the execution time and the accuracy of the results. Specifically, the speedup factor achieved in the execution time employing the reduced training dataset was ~48× with respect to the reference results (826 bands). Taking into account these results, the next step had the goal of identifying the most representative bands for each HS test image with different optimization algorithms (*PF2* and *PF3*). The results obtained showed that the GA provided the most accurate and balanced results in terms of sensitivity between all the classes using the proposed evaluation metric ($FoM_{Penalized}$), increasing the median tumor sensitivity by ~21% with respect to the second best approach and by ~30% with respect to the reference results obtained with 128 bands. After identifying the most relevant bands for each HS test image, the coincident and non-coincident bands were evaluated. The quantitative and qualitative results showed that the selected bands (48 in total that involved the coincident and non-coincident bands) offered similar and even better results in some cases than the reference results obtained with 128 bands. Therefore, for this particular brain cancer detection application, the most relevant spectral ranges identified were: 440.5–465.96 nm, 498.71–509.62 nm, 556.91–575.1 nm, 593.29–615.12 nm, 636.94–666.05 nm, 698.79–731.53 nm and 884.32–902.51 nm.

Although this preliminary study achieved a reduction of the information captured by the HS sensor within the spectral range between 400 to 1000 nm, the rationale of this experiment was to try to identify the most relevant wavelengths and provide an accurate differentiation of the tissue types and materials presented in a neurosurgical scene. As has been demonstrated in this work, the use of the entire number of spectral bands captured by the HS sensor (826) is unnecessary to achieve an accurate classification. Furthermore, the proposed reduction in the number of bands can remove spurious spectral information that produces misclassifications between the different classes, especially between the tumor class and the other classes. As has been presented in the quantitative and qualitative results, no loss of detail is obtained within the proposed methodology.

Due to the challenges of obtaining good-quality HS data during surgical procedures, especially in brain surgery, using the intraoperative macroscopic HS acquisition system based on push-broom HS cameras, the number of patients currently included in the HS brain cancer database is not high enough to state that the proposed method is robust and general. The work presented in this paper is a preliminary study where we demonstrate, as a proof-of-concept, that using a reduced number of wavelengths the accuracy of the results remains constant with respect to the employment of the original number of wavelengths. Future works will be focused in the inclusion of more patients in the training and test datasets in order to validate the spectral ranges preliminarily identified in this study as the most relevant for this application. The use of animal studies with a large number of subjects could be considered as a complement to strongly validate the proposed methodology. In addition, the use of an improved methodology to select the final coincident and non-coincident bands with the goal of reducing as much as possible the number of bands, preserving the accuracy of the results, will be explored.

Further experiments will be carried out to improve the classification results achieved with the reduced number of bands by including the spatial information in conjunction with the spectral information. The inclusion of the spatial relationship among pixels could lead to a reduction of the false positives/negatives in the classification results, achieving better delineation of the tumor areas. Nevertheless, the inclusion of the spatial information is out of the scope of this research, where we are targeting identification of the most relevant spectral features, which will allow a cost reduction in the instrumentation and in the time required to train a classifier. Moreover, future works will explore the use of deep learning approaches to improve the classification results using more data but with a reduced number of bands to evaluate if deep-learning methods outperform traditional SVM-based approaches when the number of spectral bands is extremely reduced.

In this preliminary study, the evaluation of the tumor margin delineation provided by the proposed algorithm was performed through visual inspection of the classification results by the operating surgeon, due to the impossibility of performing a pathological assessment of the entire brain tissue sample. This limitation should be addressed in future studies in order to confirm the validity of the results. A possible approach for this validation could be performing several biopsies, during the surgical procedure, in different points of the tumor area (especially in the margins) that will be intraoperatively identified by the system. Then, the histopathological analysis of such samples will be carried out in order to confirm the accuracy of the results obtained by the classification algorithm. In addition, further experiments should be accomplished in controlled environments with the goal of establishing a correlation between the selected wavelengths and the biological properties of the tissue, especially the contributions of hemoglobin and water for the tissue-type differentiation. Moreover, a preliminary segmentation of the parenchymal area and an accurate identification of the blood vessels' map of the exposed brain, performed before classification, could improve the results in the discrimination of the normal and tumor tissues using a binary classifier specifically trained to identify the relevant biological differences between these two tissue types.

The methodologies proposed in this preliminary study could be extrapolated to intraoperatively analyze other types of cancers in other organs using HSI. Several studies have been performed in the literature to analyze the use of HSI for cancer detection in different tissue types [4]. In this sense, the use of the proposed methodology could be applied to these databases in order to find the most relevant wavelengths for each particular application.

Finally, the results obtained with the proposed methodology based on the GA optimization (PF2-GA-FoM_P in *L1*) demonstrated that using only 48 bands, the quantitative classification results for the tumor class identification are slightly improved in ~5% with respect to the reference results employing the 128 bands. Although this result is not highly significant, especially taking into account the high standard deviations, it is worth noticing that the use of a reduced number of bands for the acquisition will accelerate both the acquisition time (customized HS sensors could be developed to provide real-time HS imaging) and the processing time of the data. The results of our experiments motivate the use of simpler HS cameras for the acquisition of intraoperative brain images, reducing the complexity of the instrumentation and enabling the possibility of its integration in surgical microscopes. In addition, this will motivate the use of snapshot HS cameras with an optimized spectral range tuned to this application, which would make possible the acquisition of intraoperative HS video during surgical procedures. In this sense, the number of data available for training machine-learning models would increase, and thus, the classification would be further improved. The enhancement of the HS database will be mandatory to fulfill our long-term goal that focuses on providing hospitals all over the world with a new generation of aid-visualization systems based on HSI technology that could help surgeons in routine clinical practice.

Supplementary Materials: The following are available online at http://www.mdpi.com/1424-8220/19/24/5481/s1: Table S1. Summary of the labeled dataset employed in the experiments. The numbering of the data corresponds with the dataset publically available in [47]; Figure S1. Average and standard deviation results of the leave-one-patient-out cross-validation for each band reduction with different sampling interval values. (a) Using the original training dataset. (b) Using the reduced training dataset; Figure S2. Average and standard deviation results of the normalized MCC metric using leave-one-patient-out cross-validation for each band reduction with different sampling interval values. (a) Using the original training dataset. (b) Using the reduced training dataset; Figure S3. Boxplot results of the leave-one-patient-out cross-validation obtained for the *PF3* employing the *ACO* algorithm with different numbers of bands (20, 40, 60, 80 and 100). (a) Overall accuracy. (b) Tumor tissue sensitivity. (c) Normal tissue sensitivity. (d) Hypervascularized tissue sensitivity; Figure S4. Boxplot results of the normalized MCC metric using leave-one-patient-out cross-validation obtained for the *PF3* employing the *ACO* algorithm with different numbers of bands (20, 40, 60, 80 and 100). (a) Tumor tissue. (b) Normal tissue. (c) Hypervascularized tissue; Table S2. Identification of the specific 48 wavelengths selected by the *GA* in the *PF2* using the $FoM_{Penalized}$ evaluation metric and the *L1* band combination.

Author Contributions: Conceptualization, B.M., R.L., H.F., S.O. and G.M.C.; software, B.M. and R.L.; validation, B.M. and R.L.; investigation, B.M., R.L., H.F. and S.O.; resources, J.M. and G.M.C.; data curation, A.S., J.F.P., M.H., C.S., A.J.O., S.B., C.E., D.C., M.M., R.C., M.d.l.L.P. and J.M.; writing—original draft preparation, B.M., R.L. and H.F.; writing—review and editing, S.O., and G.M.C.; supervision, J.M. and G.M.C.; project administration, G.M.C.; funding acquisition, J.M, and G.M.C.

Funding: This research was supported in part by the Canary Islands Government through the ACIISI (Canarian Agency for Research, Innovation and the Information Society), ITHACA project "Hyperspectral Identification of Brain Tumors" under Grant Agreement ProID2017010164 and it has been partially supported also by the Spanish Government and European Union (FEDER funds) as part of support program in the context of Distributed HW/SW Platform for Intelligent Processing of Heterogeneous Sensor Data in Large Open Areas Surveillance Applications (PLATINO) project, under contract TEC2017-86722-C4-1-R. Additionally, this work was completed while Samuel

Ortega was beneficiary of a pre-doctoral grant given by the "*Agencia Canaria de Investigacion, Innovacion y Sociedad de la Información (ACIISI)*" of the "*Conserjería de Economía, Industria, Comercio y Conocimiento*" of the "*Gobierno de Canarias*", which is part-financed by the European Social Fund (FSE) (POC 2014-2020, *Eje 3 Tema Prioritario 74* (85%)).

Conflicts of Interest: The authors declare no conflict of interest. The founding sponsors had no role in the design of the study; in the collection, analyses, or interpretation of data; in the writing of the manuscript; and in the decision to publish the results.

Appendix A

Table A1. Acronym list of the proposed methods employed in this study.

PF1-128	Processing Framework 1 using 128 bands
PF2-GA-OA_P	Processing Framework 2 using Genetic Algorithm and the Overall Accuracy Penalized evaluation metric
PF2-GA-FoM_P	Processing Framework 2 using Genetic Algorithm and the Figure of Merit Penalized evaluation metric
PF2-PSO-OA_P	Processing Framework 2 using Particle Swarm Optimization algorithm and the Overall Accuracy Penalized evaluation metric
PF2-PSO-FoM_P	Processing Framework 2 using Particle Swarm Optimization algorithm and the Figure of Merit Penalized evaluation metric
PF3-ACO-60	Processing Framework 3 using Ant Colony Optimization algorithm taking into account 60 bands

References

1. Hammill, K.; Stewart, C.G.; Kosic, N.; Bellamy, L.; Irvine, H.; Hutley, D.; Arblaster, K. Exploring the impact of brain cancer on people and their participation. *Br. J. Occup. Ther.* **2019**, *82*, 162–169. [CrossRef]
2. Joshi, D.M.; Rana, N.K.; Misra, V.M. Classification of Brain Cancer using Artificial Neural Network. In Proceedings of the 2010 2nd International Conference on Electronic Computer Technology, Kuala Lumpur, Malaysia, 7–10 May 2010; IEEE: Piscataway, NJ, USA, 2010; pp. 112–116.
3. Perkins, A.; Liu, G. Primary Brain Tumors in Adults: Diagnosis and Treatment—American Family Physician. *Am. Fam. Physician* **2016**, *93*, 211–218. [PubMed]
4. Halicek, M.; Fabelo, H.; Ortega, S.; Callico, G.M.; Fei, B. In-Vivo and Ex-Vivo Tissue Analysis through Hyperspectral Imaging Techniques: Revealing the Invisible Features of Cancer. *Cancers* **2019**, *11*, 756. [CrossRef] [PubMed]
5. Kamruzzaman, M.; Sun, D.W. Introduction to Hyperspectral Imaging Technology. *Comput. Vis. Technol. Food Qual. Eval.* **2016**, 111–139. [CrossRef]
6. Mordant, D.J.; Al-Abboud, I.; Muyo, G.; Gorman, A.; Sallam, A.; Ritchie, P.; Harvey, A.R.; McNaught, A.I. Spectral imaging of the retina. *Eye* **2011**, *25*, 309–320. [CrossRef]
7. Johnson, W.R.; Wilson, D.W.; Fink, W.; Humayun, M.; Bearman, G. Snapshot hyperspectral imaging in ophthalmology. *J. Biomed. Opt.* **2007**. [CrossRef]
8. Gao, L.; Smith, R.T.; Tkaczyk, T.S. Snapshot hyperspectral retinal camera with the Image Mapping Spectrometer (IMS). *Biomed. Opt. Express* **2012**, *3*, 48–54. [CrossRef]
9. Akbari, H.; Kosugi, Y.; Kojima, K.; Tanaka, N. Detection and Analysis of the Intestinal Ischemia Using Visible and Invisible Hyperspectral Imaging. *IEEE Trans. Biomed. Eng.* **2010**, *57*, 2011–2017. [CrossRef]
10. Ortega, S.; Fabelo, H.; Camacho, R.; Plaza, M.L.; Callicó, G.M.; Sarmiento, R. Detecting brain tumor in pathological slides using hyperspectral imaging. *Biomed. Opt. Express* **2018**, *9*, 818–831. [CrossRef]
11. Zhu, S.; Su, K.; Liu, Y.; Yin, H.; Li, Z.; Huang, F.; Chen, Z.; Chen, W.; Zhang, G.; Chen, Y. Identification of cancerous gastric cells based on common features extracted from hyperspectral microscopic images. *Biomed. Opt. Express* **2015**, *6*, 1135–1145. [CrossRef]

12. Lu, C.; Mandal, M. Toward automatic mitotic cell detection and segmentation in multispectral histopathological images. *IEEE J. Biomed. Health Inform.* **2014**, *18*, 594–605. [CrossRef] [PubMed]
13. Khouj, Y.; Dawson, J.; Coad, J.; Vona-Davis, L. Hyperspectral Imaging and K-Means Classification for Histologic Evaluation of Ductal Carcinoma In Situ. *Front. Oncol.* **2018**, *8*, 17. [CrossRef] [PubMed]
14. Bjorgan, A.; Denstedt, M.; Milanič, M.; Paluchowski, L.A.; Randeberg, L.L. Vessel contrast enhancement in hyperspectral images. In *Optical Biopsy XIII: Toward Real-Time Spectroscopic Imaging and Diagnosis*; Alfano, R.R., Demos, S.G., Eds.; SPIE—International Society For Optics and Photonics: Bellingham, WA, USA, 2015.
15. Akbari, H.; Kosugi, Y.; Kojima, K.; Tanaka, N. Blood vessel detection and artery-vein differentiation using hyperspectral imaging. In Proceedings of the 31st Annual International Conference of the IEEE Engineering in Medicine and Biology Society: Engineering the Future of Biomedicine, EMBC 2009, Minneapolis, MN, USA, 3–6 September 2009; pp. 1461–1464.
16. Milanic, M.; Bjorgan, A.; Larsson, M.; Strömberg, T.; Randeberg, L.L. Detection of hypercholesterolemia using hyperspectral imaging of human skin. In Proceedings of the SPIE—European Conference on Biomedical Optics, Munich, German, 21–25 June 2015; Brown, J.Q., Deckert, V., Eds.; SPIE—The International Society for Optical Engineering: Bellingham, WA, USA, 2015; p. 95370C.
17. Zhi, L.; Zhang, D.; Yan, J.; Li, Q.L.; Tang, Q. Classification of hyperspectral medical tongue images for tongue diagnosis. *Comput. Med. Imaging Graph.* **2007**, *31*, 672–678. [CrossRef] [PubMed]
18. Yudovsky, D.; Nouvong, A.; Pilon, L. Hyperspectral Imaging in Diabetic Foot Wound Care. *J. Diabetes Sci. Technol.* **2010**, *4*, 1099–1113. [CrossRef] [PubMed]
19. Lu, G.; Fei, B. Medical hyperspectral imaging: A review. *J. Biomed. Opt.* **2014**, *19*, 10901. [CrossRef] [PubMed]
20. Calin, M.A.; Parasca, S.V.; Savastru, D.; Manea, D. Hyperspectral imaging in the medical field: Present and future. *Appl. Spectrosc. Rev.* **2014**, *49*, 435–447. [CrossRef]
21. Ortega, S.; Fabelo, H.; Iakovidis, D.; Koulaouzidis, A.; Callico, G.; Ortega, S.; Fabelo, H.; Iakovidis, D.K.; Koulaouzidis, A.; Callico, G.M. Use of Hyperspectral/Multispectral Imaging in Gastroenterology. Shedding Some–Different–Light into the Dark. *J. Clin. Med.* **2019**, *8*, 36. [CrossRef]
22. Akbari, H.; Uto, K.; Kosugi, Y.; Kojima, K.; Tanaka, N. Cancer detection using infrared hyperspectral imaging. *Cancer Sci.* **2011**, *102*, 852–857. [CrossRef]
23. Kiyotoki, S.; Nishikawa, J.; Okamoto, T.; Hamabe, K.; Saito, M.; Goto, A.; Fujita, Y.; Hamamoto, Y.; Takeuchi, Y.; Satori, S.; et al. New method for detection of gastric cancer by hyperspectral imaging: A pilot study. *J. Biomed. Opt.* **2013**, *18*, 026010. [CrossRef]
24. Baltussen, E.J.M.; Kok, E.N.D.; Brouwer de Koning, S.G.; Sanders, J.; Aalbers, A.G.J.; Kok, N.F.M.; Beets, G.L.; Flohil, C.C.; Bruin, S.C.; Kuhlmann, K.F.D.; et al. Hyperspectral imaging for tissue classification, a way toward smart laparoscopic colorectal surgery. *J. Biomed. Opt.* **2019**, *24*, 016002. [CrossRef]
25. Han, Z.; Zhang, A.; Wang, X.; Sun, Z.; Wang, M.D.; Xie, T. In vivo use of hyperspectral imaging to develop a noncontact endoscopic diagnosis support system for malignant colorectal tumors. *J. Biomed. Opt.* **2016**, *21*, 016001. [CrossRef] [PubMed]
26. Panasyuk, S.V.; Yang, S.; Faller, D.V.; Ngo, D.; Lew, R.A.; Freeman, J.E.; Rogers, A.E. Medical hyperspectral imaging to facilitate residual tumor identification during surgery. *Cancer Biol. Ther.* **2007**, *6*, 439–446. [CrossRef] [PubMed]
27. Pourreza-Shahri, R.; Saki, F.; Kehtarnavaz, N.; Leboulluec, P.; Liu, H. Classification of ex-vivo breast cancer positive margins measured by hyperspectral imaging. In Proceedings of the 2013 IEEE International Conference on Image Processing, ICIP 2013, Melbourne, Australia, 15–18 September 2013; pp. 1408–1412.
28. Lu, G.; Halig, L.; Wang, D.; Chen, Z.G.; Fei, B. Hyperspectral imaging for cancer surgical margin delineation: Registration of hyperspectral and histological images. In *SPIE Medical Imaging 2014: Image-Guided Procedures, Robotic Interventions, and Modeling*; Yaniv, Z.R., Holmes, D.R., Eds.; International Society for Optics and Photonics: Bellingham, WA, USA, 2014; Volume 9036, p. 90360S.
29. Pike, R.; Lu, G.; Wang, D.; Chen, Z.G.; Fei, B. A Minimum Spanning Forest-Based Method for Noninvasive Cancer Detection With Hyperspectral Imaging. *IEEE Trans. Biomed. Eng.* **2016**, *63*, 653–663. [CrossRef] [PubMed]
30. Fei, B.; Lu, G.; Wang, X.; Zhang, H.; Little, J.V.; Patel, M.R.; Griffith, C.C.; El-Diery, M.W.; Chen, A.Y. Label-free reflectance hyperspectral imaging for tumor margin assessment: A pilot study on surgical specimens of cancer patients. *J. Biomed. Opt.* **2017**, *22*, 1–7. [CrossRef] [PubMed]

31. Halicek, M.; Little, J.V.; Wang, X.; Chen, A.Y.; Fei, B. Optical biopsy of head and neck cancer using hyperspectral imaging and convolutional neural networks. *J. Biomed. Opt.* **2019**, *24*, 1–9. [CrossRef]
32. Regeling, B.; Thies, B.; Gerstner, A.O.H.; Westermann, S.; Müller, N.A.; Bendix, J.; Laffers, W. Hyperspectral Imaging Using Flexible Endoscopy for Laryngeal Cancer Detection. *Sensors* **2016**, *16*, 1288. [CrossRef]
33. Halicek, M.; Dormer, J.D.; Little, J.V.; Chen, A.Y.; Myers, L.; Sumer, B.D.; Fei, B. Hyperspectral Imaging of Head and Neck Squamous Cell Carcinoma for Cancer Margin Detection in Surgical Specimens from 102 Patients Using Deep Learning. *Cancers* **2019**, *11*, 1367. [CrossRef]
34. Fabelo, H.; Ortega, S.; Ravi, D.; Kiran, B.R.; Sosa, C.; Bulters, D.; Callicó, G.M.; Bulstrode, H.; Szolna, A.; Piñeiro, J.F.; et al. Spatio-spectral classification of hyperspectral images for brain cancer detection during surgical operations. *PLoS ONE* **2018**, *13*, e0193721. [CrossRef]
35. Fabelo, H.; Ortega, S.; Lazcano, R.; Madroñal, D.; M Callicó, G.; Juárez, E.; Salvador, R.; Bulters, D.; Bulstrode, H.; Szolna, A.; et al. An Intraoperative Visualization System Using Hyperspectral Imaging to Aid in Brain Tumor Delineation. *Sensors* **2018**, *18*, 430. [CrossRef]
36. Fabelo, H.; Halicek, M.; Ortega, S.; Shahedi, M.; Szolna, A.; Piñeiro, J.F.; Sosa, C.; O'Shanahan, A.J.; Bisshopp, S.; Espino, C.; et al. Deep Learning-Based Framework for In Vivo Identification of Glioblastoma Tumor using Hyperspectral Images of Human Brain. *Sensors* **2019**, *19*, 920. [CrossRef]
37. Ghamisi, P.; Plaza, J.; Chen, Y.; Li, J.; Plaza, A.J. Advanced Spectral Classifiers for Hyperspectral Images: A review. *IEEE Geosci. Remote Sens. Mag.* **2017**, *5*, 8–32. [CrossRef]
38. Van Der Maaten, L.J.P.; Postma, E.O.; Van Den Herik, H.J. Dimensionality Reduction: A Comparative Review. *J. Mach. Learn. Res.* **2009**, *10*, 1–41. [CrossRef]
39. Lunga, D.; Prasad, S.; Crawford, M.M.; Ersoy, O. Manifold-Learning-Based Feature Extraction for Classification of Hyperspectral Data: A Review of Advances in Manifold Learning. *IEEE Signal Process. Mag.* **2014**, *31*, 55–66. [CrossRef]
40. Dai, Q.; Cheng, J.H.; Sun, D.W.; Zeng, X.A. Advances in Feature Selection Methods for Hyperspectral Image Processing in Food Industry Applications: A Review. *Crit. Rev. Food Sci. Nutr.* **2015**, *55*, 1368–1382. [CrossRef]
41. Sastry, K.; Goldberg, D.E.; Kendall, G. Genetic Algorithms. In *Search Methodologies*; Springer: Boston, MA, USA, 2014; pp. 93–117.
42. Perez, R.E.; Behdinan, K. Particle Swarm Optimization in Structural Design. *Swarm Intell. Focus Ant Part. Swarm Optim.* **2012**, 1–24. [CrossRef]
43. Sharma, S.; Buddhiraju, K.M. Spatial-spectral ant colony optimization for hyperspectral image classification. *Int. J. Remote Sens.* **2018**, *39*, 2702–2717. [CrossRef]
44. Rashmi, S.; Addamani, S.; Ravikiran, A. Spectral Angle Mapper algorithm for remote sensing image classification. *IJISET Int. J. Innov. Sci. Eng. Technol.* **2014**, *1*, 201–205. [CrossRef]
45. Halicek, M.; Fabelo, H.; Ortega, S.; Little, J.V.; Wang, X.; Chen, A.Y.; Callicó, G.M.; Myers, L.; Sumer, B.; Fei, B. Cancer detection using hyperspectral imaging and evaluation of the superficial tumor margin variance with depth. In *Medical Imaging 2019: Image-Guided Procedures, Robotic Interventions, and Modeling*; Fei, B., Linte, C.A., Eds.; SPIE: Bellingham, WA, USA, 2019; Volume 10951, p. 45.
46. Lu, G.; Qin, X.; Wang, D.; Chen, Z.G.; Fei, B. Quantitative wavelength analysis and image classification for intraoperative cancer diagnosis with hyperspectral imaging. In Proceedings of the Progress in Biomedical Optics and Imaging—Proceedings of SPIE, Orlando, FL, USA, 21–26 February 2015; Volume 9415.
47. Fabelo, H.; Ortega, S.; Szolna, A.; Bulters, D.; Pineiro, J.F.; Kabwama, S.; J-O'Shanahan, A.; Bulstrode, H.; Bisshopp, S.; Kiran, B.R.; et al. In-Vivo Hyperspectral Human Brain Image Database for Brain Cancer Detection. *IEEE Access* **2019**, *7*, 39098–39116. [CrossRef]
48. Chen, P.C.; Lin, W.C. Spectral-profile-based algorithm for hemoglobin oxygen saturation determination from diffuse reflectance spectra. *Biomed. Opt. Express* **2011**, *2*, 1082–1096. [CrossRef]
49. Eaton, W.A.; Hanson, L.K.; Stephens, P.J.; Sutherland, J.C.; Dunn, J.B.R. Optical spectra of oxy- and deoxyhemoglobin. *J. Am. Chem. Soc.* **1978**, *100*, 4991–5003. [CrossRef]
50. Sekar, S.K.V.; Bargigia, I.; Mora, A.D.; Taroni, P.; Ruggeri, A.; Tosi, A.; Pifferi, A.; Farina, A. Diffuse optical characterization of collagen absorption from 500 to 1700 nm. *J. Biomed. Opt.* **2017**, *22*, 015006. [CrossRef] [PubMed]
51. Camps-Valls, G.; Bruzzone, L.; Electr, E.; Escola, T.; Val, U. De Kernel-based methods for hyperspectral image classification. *IEEE Trans. Geosci. Remote Sens.* **2005**, *43*, 1351–1362. [CrossRef]

52. Fabelo, H.; Halicek, M.; Ortega, S.; Szolna, A.; Morera, J.; Sarmiento, R.; Callicó, G.M.; Fei, B. Surgical aid visualization system for glioblastoma tumor identification based on deep learning and in-vivo hyperspectral images of human patients. In *Medical Imaging 2019: Image-Guided Procedures, Robotic Interventions, and Modeling*; SPIE: Bellingham, WA, USA, 2019; Volume 10951, p. 35.
53. Chang, C.C.; Lin, C.J. LIBSVM: A library for support vector machines. *ACM Trans. Intell. Syst. Technol.* **2011**, *2*, 1–27. [CrossRef]
54. Moore, A. K-means and Hierarchical Clustering. Available online: http://www.cs.cmu.edu/afs/cs/user/awm/web/tutorials/kmeans11.pdf (accessed on 24 November 2019).
55. Akhter, N.; Dabhade, S.; Bansod, N.; Kale, K. Feature Selection for Heart Rate Variability Based Biometric Recognition Using Genetic Algorithm. In *Intelligent Systems Technologies and Applications*; Springer: Cham, Switzerland, 2016; pp. 91–101.
56. Haupt, S.E.; Haupt, R.L. Genetic algorithms and their applications in environmental sciences. In *Advanced Methods for Decision Making and Risk Management in Sustainability Science*; Nova Science Publishers: Hauppauge, NY, USA, 2007; pp. 183–196.
57. Zortea, M.; Plaza, A. Spatial Preprocessing for Endmember Extraction. *IEEE Trans. Geosci. Remote Sens.* **2009**, *47*, 2679–2693. [CrossRef]
58. Boughorbel, S.; Jarray, F.; El-Anbari, M. Optimal classifier for imbalanced data using Matthews Correlation Coefficient metric. *PLoS ONE* **2017**, *12*, e0177678. [CrossRef] [PubMed]

Article

Adaptive Sampling of the Electrocardiogram Based on Generalized Perceptual Features

Piotr Augustyniak

AGH University of Science and Technology, 30-059 Krakow, Poland; august@agh.edu.pl; Tel.: +48-69703-2858

Received: 19 November 2019; Accepted: 7 January 2020; Published: 9 January 2020

Abstract: A non-uniform distribution of diagnostic information in the electrocardiogram (ECG) has been commonly accepted and is the background to several compression, denoising and watermarking methods. Gaze tracking is a widely recognized method for identification of an observer's preferences and interest areas. The statistics of experts' scanpaths were found to be a convenient quantitative estimate of medical information density for each particular component (i.e., wave) of the ECG record. In this paper we propose the application of generalized perceptual features to control the adaptive sampling of a digital ECG. Firstly, based on temporal distribution of the information density, local ECG bandwidth is estimated and projected to the actual positions of components in heartbeat representation. Next, the local sampling frequency is calculated pointwise and the ECG is adaptively low-pass filtered in all simultaneous channels. Finally, sample values are interpolated at new time positions forming a non-uniform time series. In evaluation of perceptual sampling, an inverse transform was used for the reconstruction of regularly sampled ECG with a percent root-mean-square difference (PRD) error of 3–5% (for compression ratios 3.0–4.7, respectively). Nevertheless, tests performed with the use of the CSE Database show good reproducibility of ECG diagnostic features, within the IEC 60601-2-25:2015 requirements, thanks to the occurrence of distortions in less relevant parts of the cardiac cycle.

Keywords: visual perception; electrocardiogram (ECG) coder; non-uniform sampling; telecardiology; compressed sensing

---

1. Introduction

Huge amounts of ECG data are nowadays collected worldwide due to achievements made in the storage of media technology during the last decade. Data compression, as it has been identified in the 1990s, is no longer a necessary condition for the operation of long-term Holter recorders or wireless sensors. Nevertheless, the scientific problem of intelligent adaptive coding remains valid [1–5] and, in the context of cardiac-based home care and surveillance, smart solutions have considerable impact on performance and costs. As such systems commonly use a wireless link, millions of recording hours are difficult to manage and cause high expenses for data transmission [6–8].

The ECG is sampled with a constant frequency, mainly for the reason of commodity, despite the full bandwidth of the data stream being used only for a very limited period within the QRS complex (of a typical duration of 100 ms). In the remaining part of the heartbeat, the discrete time series is significantly oversampled causing high correlation of neighboring samples unless blurred with noise. A family of short-time decorrelation techniques uses this feature for signal compression (e.g., [9,10]). The oversampled sections are used as a reference for local measurement of noise level (e.g., [11]) or as a host of watermark data (e.g., [12]). A class of algorithms perform compressed sensing based on local statistics (e.g., [13–15]) or adaptive sampling based on the mutual data dependence (e.g., [16,17]) but with no regard for distribution of medical relevance in the record.

In previous research projects we have studied the irregular distribution of medical information in the ECG record with use of various methods: local bandwidth of ECG waves [18], susceptibility of diagnostic result to signal distortion caused by a local data loss, and local conspicuity of the ECG trace [19,20]. Results reported in the latter work include a generalized quantitative estimate of local temporal distribution of the electrocardiogram medical relevance, here referred to as generalized medical relevance function (gMRF). This function is briefly recalled hereafter and proposed as a background for the adaptive ECG sampling technique presented in this paper.

Due to non-uniform distribution of diagnostic information in the ECG, simple metrics for time series comparisons such as signal to noise ratio (SNR) or percentage root-mean-square difference (PRD) do not adequately represent the degradation of quality of diagnostic results. For the same reasons the lossy compression of the ECG is distrusted [3,8,10], and currently not allowed in clinical applications. At the same time, various techniques for ECG recording with constant sampling frequency ranging from 125 to 1000 Hz are used depending on the medical goals and provide either more concise or more detailed background for diagnostic analyses. This justifies the seamless adaptation of the sampling frequency in relation to the patient status and—like in case of the proposed method—with progression of the cardiac cycle.

Different phases of the cardiac cycle can be distinguished in the surface-recorded ECG as a representation of sequence of cell action potentials in the heart conduction system and myocardium. Due to different electric properties of the conducting tissue, waves representing the progression of the cardiac cycle show different bandwidths as the stimulus travels through the heart. They also show significant variability in duration on a beat-to-beat basis not directly related to the RR interval even in normal rhythms. In case of abnormal rhythms, the order of waves in the sequence may also be altered: in ectopic beats P-wave is absent and in atrial fibrillation a continuous wave is observed instead of P-wave. Therefore, the progression of the cardiac cycle is adequately represented by the time relative to wave borders, that in turn become adequate reference points for estimation of medical information density in each individual heartbeat. Several algorithms were developed for automatic recognition and delineation of ECG waves and proven to perform with precision and accuracy acceptable for medical use. They can roughly be classified as signal derivative-, geometric template- or semantic sequence-based. The methods by Almeida et al. [21], by Martinez et al. [22], and by Dumont et al. [23] are based on the discrete wavelet transform (DWT). Other delineation methods reported use continuous wavelet transform (CWT) [24].

In this paper we propose a method for adaptive sampling of the electrocardiogram. The local sampling interval is driven by the expected conspicuity of the trace (i.e., medical relevance) calculated for a given progression of the cardiac cycle relative to its beginning. Therefore, a procedure for automated wave borders detection is necessary to control the piecewise projection of the gMRF to each particular heartbeat. The rest of this paper is organized as follows. In Section 2 the idea and processing scheme are presented together with a method for transforming the ECG trace conspicuity to an estimate of local information density, details on piecewise adaptation of the local relevance function, and ECG signal resampling. In Section 3 the evaluation of the method is reported with details on test signal set, error metrics and experiment results. In Section 4 a discussion is presented together with concluding remarks.

2. Materials and Methods

2.1. The Idea and Processing Scheme

The key idea of proposed adaptive ECG sampling consists of allocating space in the output data stream accordingly to the information density in the input series (i.e., uniformly sampled ECG). Unlike in many signal compression methods or compressed sensing based on statistical description of information density, in our method the local information distribution is estimated on the background of perceptual studies [19,20]. Although an inverse transform does not bit-accurately reconstruct the regularly sampled ECG, the adaptive sampling preserves the most important sections of the heartbeat and maintains all diagnostic features of the original signal. Moreover, the method accepts various (e.g., application-specific) information distribution models arbitrarily specifying parts of the heart cycle as more or less important. Both scenarios have been experimentally proven and results are described in the later part of this paper.

The adaptive sampling scheme (Figure 1) uses two procedures typical for the ECG interpretation-oriented processing chain: detection of heart beats (QRS complex) [25] and detection of ECG wave borders [21,22]. Both algorithms comply with the performance requirements for interpretive electrocardiographs [26], which guarantees high precision of wave borders delineation. Since both algorithms are designed to work with regular ECG signals, the proposed adaptive coding assumes uniformly sampled discrete time series at the input. Direct non-uniform sampling of analog signals (such as [17] or [27]) falls beyond the scope of this paper.

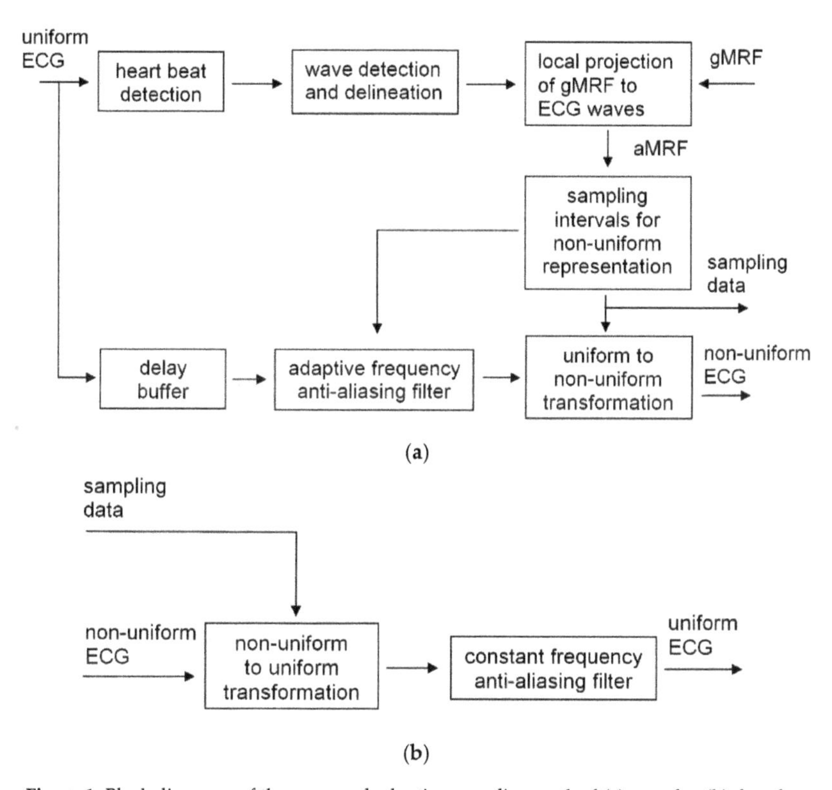

Figure 1. Block diagrams of the proposed adaptive sampling method (**a**) encoder, (**b**) decoder.

The second step of the coding scheme is the local projection of gMRF to the ECG waves i.e., adaptation to the actual duration of consecutive electrocardiogram sections. The heart cycle rhythmicity is influenced by many factors of physiological origin: escape and premature discharges, conduction abnormalities, P-wave absence or multitude, simultaneous multi-center discharges, etc. [28]. Consequently, even if the wave order within a beat is predictable, the length of individual sections may vary independently in a broad range. As the description of expected data density is based on wave-related progression of the heart cycle, the gMRF has to be projected individually on each heartbeat. This operation described in detail in Section 2.4 yields the adapted medical relevance function (aMRF).

The third step of the adaptive sampling scheme is the individual aMRF-based calculation of consecutive sampling intervals in the target irregular ECG. We used linear relationship (see Equation (4)), however other functions can also be employed. The sequence of sampling intervals determines the time points in which the ECG values are actually sampled (i.e., interpolated from the original discrete signal). Usually the number of intervals in the irregularly sampled representation is significantly lower than the number of original samples in the equivalent recording time. The sampling interval can be expressed by any real number, but for practical reasons we had to quantize the interval value. A high quantization step yields a concise representation of sampling rate at the price of spectral discontinuities causing signal distortions. A low quantization step guarantees good signal reproduction, but the sampling description consumes extra space. In the case of a multilead ECG record, the wave borders are determined jointly for all leads and stored in a single interval sequence.

2.2. The ECG Trace Conspicuity as an Estimate of Local Information Density

The proposed adaptive sampling scheme uses the gMRF as a quantitative estimate of local wave-related density of medical information. As it was proven with different scenes, visual interpretation of their content by a human observer involves alternating perception-cognition processes. Based on the relationship between the gaze time and the information amount, visual tasks are widely applied for objective assessment of preferences in graphic interfaces, advertising and many others [29,30]. Following these examples we studied ECG trace conspicuity and perceptual strategies of human experts for visual interpretation of records on display. The research presented in detail in [19] was focused on following the human expert reasoning and revealed different strategies of visual ECG interpretation for people with different degrees of expertise. Recorded eyeball trajectories were analyzed in the context of medical features represented in the displayed ECG traces (i.e., the wave borders were known but not displayed, Figure 2a).

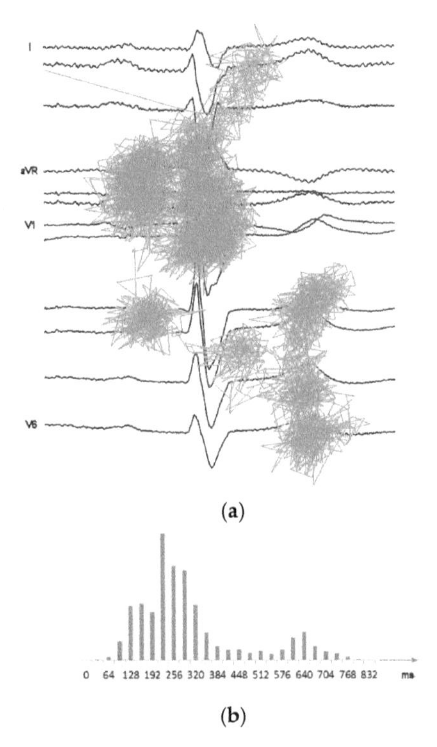

Figure 2. (**a**) The example of expert's eyeglobe trajectory over a 12-lead ECG plot (CSE-Mo-001); (**b**) Corresponding bar graph of the attention density (each bar collects gaze time in a 32 ms interval in ECG record)—the background for gMRF (modified from [20]).

First, the ECG traces were section-wise normalized so as to equalize the length of corresponding waves from different records. Respective scanpaths were contracted or dilated accordingly. Next, gaze points were detected in the scanpaths and their center point horizontal coordinates were discretized with a step of 0.78 mm on display, corresponding to 32 ms of the ECG trace (at 25 mm/s). Finally, the time of gazing was collected in each ECG time slot separately and normalized to reveal interesting information about diagnostic data distribution (Figure 2b). As the histogram of the visual attention density was collected with respect to wave-dependent progression of the heart cycle, reference wave borders played a role of landmarks in the resulting gMRF.

The scanpath-based gMRF benefits from oculomotoric habits gathered by experts in their everyday practice, unconscious mutual perception-recognition interactions not affected by human intention, memory or verbalization limits. For these reasons the perceptual quantitative description of medical data distribution provides more objective assessment than any other method willingly controlled by the human. The perceptual model of the ECG proposed here as gMRF is universal since it was built from scanpaths of experts interpreting a wide range of normal and pathological records. Nevertheless, disease-specific models (i.e., specific medical relevance functions, sMRF) may also be built and applied accordingly to the status of particular patient.

2.3. ECG Waves Delineation

The positions of wave borders are a background for individual calculations of the Medical Relevance Function for each heartbeat. As particular segments of a heartbeat independently vary in time, a piecewise projection of the gMRF does the necessary local stretching or contracting of the gMRF in order to fit its landmarks to the actual duration of consecutive cardiac events.

Moreover, calculation of diagnostic parameters and comparing them between the original ECG and its counterpart reconstructed from the adaptive sampling, as well as the use of weighted diagnostic distortion (WDD) [5] (see Section 3.2) as a measure of signal reconstruction error, requires calculation of 18 diagnostic features by an ECG interpretive software.

To this point we used a development version of Ascard 6 (® by Aspel S.A., Zabierzow, Poland) software that meets the performance requirements of international IEC standard [26] and on the other hand allows for access to interpretation of metadata. The wave delimitation procedure embedded in Ascard 6 software uses the first and second derivatives of various versions of filtered signal in order to determine the point where the wave energy emerges from the background noise [31]. The algorithm also uses wave-specific features individual for each wave's onset and endpoint.

The Common Standard for Quantitative Electrocardiography (CSE) database, used in the experimental part for local adaptation of gMRF, provides reference wave border points of a representative heartbeat of each record [32] (see Section 3.1). The opportunity was taken to check whether substituting the values calculated with Ascard 6 software by their reference counterparts influences the adaptive ECG sampling result.

2.4. Piecewise Adaptation of the Local Relevance Function

Assigning $k = \{1 \ldots 5\}$ to the consecutive borders of waves: P-onset, P-end, QRS-onset, QRS-end and T-end (T-onset is not considered as standard fiducial point), the projection of gMRF to aMRF consists in calculating the values of the latter,

$$\forall_{b \in aMRF} \exists_{a \in gMRF} : aMRF(b) = gMRF(a) \tag{1}$$

where a and b express integer sample numbers in adapted and generalized medical relevance functions, respectively, and

$$a = \frac{(a_k - a_{k-1}) \times (b - b_{k-1})}{(b_k - b_{k-1})} + c \tag{2}$$

where c is a complement transferring fraction of border sampling interval between ECG waves, i.e.,

$$c = \begin{cases} \frac{a_{k+1}-a_k}{b_{k+1}-b_k} & for\, k = 2 \\ 0 & for\, k \in \{3,4\} \\ \frac{a_{k-2}-a_{k-1}}{b_{k-2}-b_{k-1}} & for\, k = 5 \end{cases} \tag{3}$$

We applied the piecewise linear projection of gMRF to aMRF (Figure 3) for its computational simplicity and without noticing any consequences of singularities in aMRF caused by stepwise changes of gMRF sampling. Otherwise, either digital filtering of the resulting aMRF or projection with the use of cubic splines with nodes at the landmarks are possible alternatives. All calculations use real number representation in time and value domains of these functions.

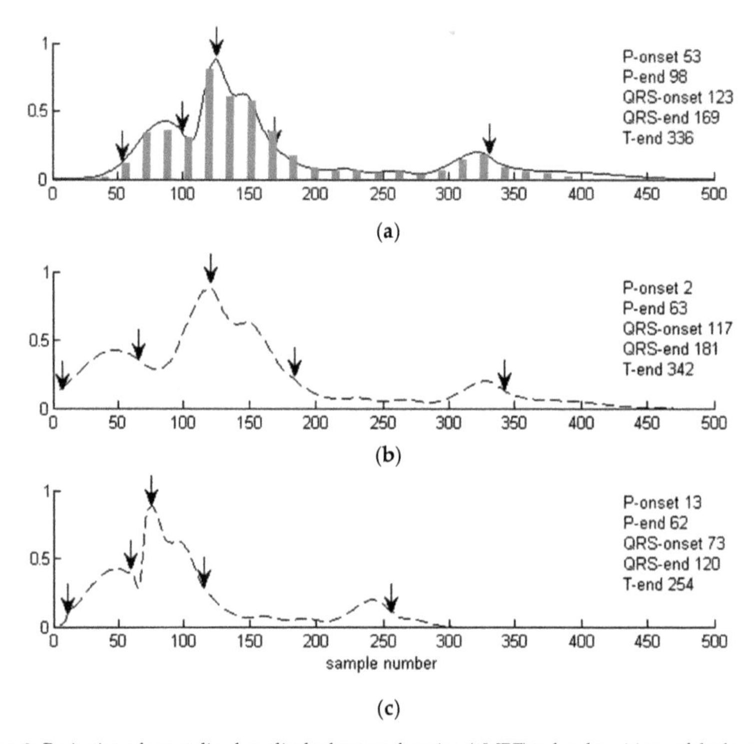

Figure 3. Projection of generalized medical relevance function (gMRF) to local positions of the heartbeat sections: (**a**) gMRF (with the bar graph it stems from), (**b**) aMRF (adapted medical relevance function) calculated for CSE-Mo001 record, (**c**) aMRF calculated for CSE-Mo003 record.

Finally, for each heartbeat the values of the aMRF are used to control the local sampling interval $ls(t)$ within the range corresponding to frequency limits (f_m, f_s) accordingly to the linear relationship:

$$ls(t) = T_m + (T_s - T_m) \times aMRF(t) \tag{4a}$$

or conversely,

$$ls(t) = \frac{1}{f_m} + \frac{f_m - f_s}{f_m \times f_s} \cdot aMRF(t) \tag{4b}$$

In the proposed implementation, the adaptive algorithm is dedicated to the ECG signal sampled at f_s = 500 Hz and the minimum usable value of local sampling frequency tested in two experiments was set to f_{m1} = 100 Hz (Figure 4) and f_{m2} = 50 Hz respectively.

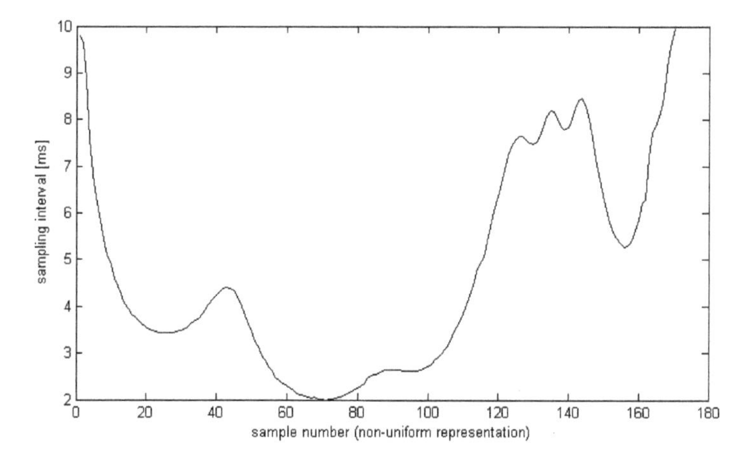

Figure 4. Interval of consecutive samples in a non-uniform representation (f_{m1} = 100 Hz) calculated for the reference beat from file CSE001.

2.5. ECG Signal Resampling

The objective of the sampling problem is to recover a function f on $\Re^d$ from its samples $\{f(x_j): j \in J.\}$, where $J.$ is a countable indexing set, and f satisfies some a priori constraints [33]. Extension of the classical Shannon theory to the non-uniform sampling of bandlimited functions specifies that for the exact and stable reconstruction of such function f from its samples $\{f(x_j): x_j \in X\}$, it is sufficient that the Beurling density,

$$D(X) = \lim_{r\to\infty} \inf_{y \in R} \frac{\#X \cap (y + [0, r])}{r} \tag{5}$$

(where r is radius of sampling grid and y—sample position), satisfies $D(X) > 1$ [34,35]. Conversely, if f is uniquely and stably determined by its samples on $X \subset \Re$, then $D(X) \geq 1$.

Solution of the sampling problem f in non-uniform shift-invariant bases $V_v^p(\phi)$ (where p is space dimension, v is the weight function and ϕ is the space generator) consists of two parts.

- Given a generator φ, conditions on X have to be defined, usually in the form of a density, such that the norm equivalence (6) holds.

$$c_p \|f\|_{L_v^p} \leq \left(\sum_{x_j \in X} |f(x_j)|^p |v(x_j)|^p \right)^{\frac{1}{p}} \leq C_p \|f\|_{L_v^p} \tag{6}$$

  Then, at least in principle, $f \in V_v^p(\phi)$ is uniquely and stably determined by $f|_X$.
- Reconstruction procedures useful and efficient in practical applications have to be designed as fast numerical algorithms which recover f from its samples $f|_X$, when (6) is satisfied.

Since the iterative frame algorithm is often slow to converge and its convergence is not even guaranteed beyond $V^2(\varphi)$, alternative reconstruction procedures based on Neumann series have been designed [34]. In [33] Aldroubi presented the example iterative algorithm with the proof of the convergence of results. The reconstruction of the uniform biosignal from an incomplete time series was also developed by Candes et al. [36] and Needell and Tropp [37].

In the proposed algorithm for adaptive ECG sampling we used the cubic splines interpolation to transform the ECG from its native uniform representation to the adaptively sampled representation and vice-versa. Considering the uniform representation as a particular case of non-uniform time series, the approximation first projects the input time series $N_j(\{n, v(n)\})$ to the continuous space with the use of 3rd order polynomial function,

$$S_n(t) = a_n + b_n(t - t_n) + c_n(t - t_n)^2 + d_n(t - t_n)^3 \tag{7}$$

where $t \in [t_n, t_{n+1}]$, $n \in \{0, 1, \ldots N - 1\}$ is best fitted to the time series N_j. Next, the output signal representation is obtained by sampling the $S_n(t)$ at desired time points m:

$$N'_j(m) = \sum_m S_n(t) \times \delta(t - m \times ls(t)) \tag{8}$$

In the case of forward transformation, the positions of input sampling points n are equispaced whereas the positions of output sampling points m are determined by the local sampling interval $ls(t)$ (see Equation (4), Figure 5). In the case of inverse transformation, the input time series comes as non-uniformly sampled, and considering the information about local distances between samples, the cubic splines interpolation yields the uniform ECG representation.

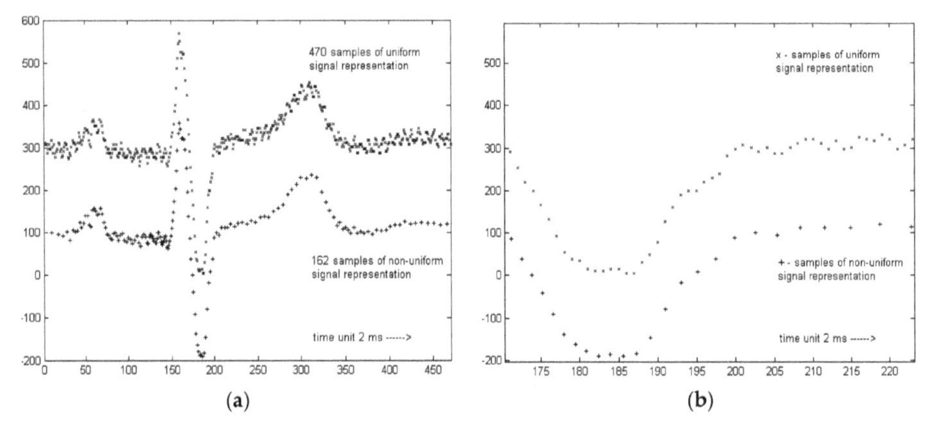

Figure 5. Regular and irregular representations of the same heartbeat (**a**) global scale, (**b**) local scale (terminal section of QRS). Vertical axes represent the ECG voltage, approximately 2.44 μV per unit.

2.6. Implementation Details

For the purpose of experimental assessment of our method, the adaptive ECG sampling algorithm was implemented in C++ on a Windows-based PC platform, however, due to moderate complexity, it may be easily transferred to a portable or wearable device with floating-point arithmetic. Heartbeat detection and wave delineation procedures were used without modification from the firmware of Ascard 6 bedside interpretive electrocardiograph. The accuracy of wave delineation is not critical due to a smooth shape of aMRF. Apart from the cardiology-oriented procedures, second contribution to the computational complexity comes from the translation of uniform to non-uniform representation. The use of cubic splines-based interpolation instead Aldroubi iterative algorithm reduces the calculation costs.

The finite impulse response (FIR) filter with 6 dB/octave (i.e., 20 dB/decade) slope has been used for low-pass anti-alias filtering of discrete uniform ECG. Unlike a regular design where the number of taps weighting the sequence of delayed input samples is fixed, we used a tunable fractional-delay filter. Respective theory and literature review are presented in [38] or [39], and such filters are also available in a recent Matlab toolbox and available as FPGA implementations [40]. As the result of exploring different design variants, we selected the filter order $N = 10$ which was the best compromise between cut-off frequency tuning range, phase flateness degree, and computational cost. Anti-alias filtering is directly followed by signal resampling. This procedure uses regular cubic splines interpolation, which is known as optimally matching two differently sampled signals and avoids discontinuities in the sampling rate function.

The sampling interval information (see Equation (4a), Figure 4) was collectively used for all simultaneous leads. Its value was quantized to 6 bits, yielding $P = 64$ possible values linearly assigned to durations of sampling interval so as $Is = T_s + \frac{p}{64}(T_m - T_s)$ i.e., between 2 ms and 10 ms with a 0.125 ms step (or between 2 ms and 20 ms with a 0.281 ms step).

3. Evaluation of the Method

The proposed method has been implemented and validated in an experimental environment Matlab (MathWorks, Natick, MA, USA) with two external executable procedures compiled from C++ source. The evaluation procedure follows a common scheme and includes:

1. Selecting a set of test signals complying with international standards (the CSE Database);
2. Selecting tools (the ECG interpretive software) and error measures (PRD, local PRD and WDD);
3. Selecting the range of method parameters (gMRF, sMRF, sampling interval);
4. Modifying signals from the test set with the proposed adaptive sampling method in all combinations of parameters by encoding the original (see Figure 1a) and decoding the encoded (see Figure 1b) records;
5. Comparing differences between original and processed records with error measures (PRD compares discrete values of samples, whereas the WDD compares values of diagnostic results);
6. Statistical processing of error values estimated for each file with each combination of parameters.

Steps 1–3 are described in detail in Sections 3.1–3.3.

3.1. The Test Signal Set

The proposed adaptive ECG sampling algorithm was tested with the use of CSE Multilead Database [32] recommended for electrocardiographs performance tests by the IEC [26]. It is worth a remark that the CSE Database is used for industrial validation of ECG wave delineation accuracy, and the perceptual model is adapted to the information on wave start and endpoint positions. Consequently, tests with the CSE files are sufficient for a complete evaluation of the method and the algorithm performance in the case of ectopic or missing beats, arrhythmias, etc., can be inferred from the reliability of wave border detection in those cases.

We used 125 automatically annotated 12-lead signals from CSE Dataset 3. The Dataset 3 consists of proportionally represented examples of normal ECGs, myocardial infarction, bundle branch blocks, premature ventricular contractions, ischemic ST changes, atrial fibrillation and many others. It is sampled with a 12-bit resolution at 500 samples per second to a data stream of 72,000 bits per second (bps). The average additional data stream carrying the local sampling rate is 1034 bps (i.e., 1.44% of the original record's data volume).

We first compared adaptive ECG sampling results for representative heartbeats of each CSE record with use of either their reference or the Ascard 6-calculated wave border values. Although the Ascard 6 software (see Section 2.3) is certified, the respective fiducial points differ by up to 2–4 ms depending on the point type and the presence of noise. Substituting these values was possible with just writing a few lines of I/O instructions to the development version of the Ascard 6 software. With any of the 123 CSE files (pacemaker-originated records 67 and 70 were discarded for the lack of annotation) we did not notice any difference of sampling results. Consequently, we stated that the inaccuracy of wave delineation procedure within Ascard 6 software is not significant and can be neglected. In all remaining tests, additionally taking into account beat-to-beat variability in each 12 s ECG sequence, individual wave borderlines were determined with the Ascard 6 software.

3.2. The Error Metrics

In order to make our results comparable with existing methods, we use the PRD (Equation (9)) and data compression ratio CR as a first-step approach to the assessment of transparency of proposed adaptive ECG sampling:

$$PRD = \sqrt{\frac{\sum_{i=1}^{n}[x_1(i) - x_2(i)]^2}{\sum_{i=1}^{n}[x_1(i)]^2}} \cdot 100\% \tag{9}$$

However, due to intrinsic inadequacy of PRD that averages all errors in time without consideration of their position with regard to medical information, we also calculated its values separately for each ECG wave, as suggested in [1]. This method is reported as a local distortion measure that depends on estimated ECG sections and possible weighting of their significance [41]. The distortion level was also provided in microvolts as an average peak-to-peak error value, for the reason of compatibility with performance requirements for interpretive electrocardiographs [26].

To avoid the abovementioned flaw of the PRD, we applied a more comprehensive error metric based on diagnostic features to reveal differences of primary diagnostic outcomes derived from the original and adaptively sampled ECG record. A widely discussed, but yet recently applied global quality estimate based on comparing the PQRST complex features of the two ECG signals is the weighted diagnostic distortion (WDD) introduced in [5]. It was defined as

$$WDD(\beta, \hat{\beta}) = \Delta\beta^T \frac{\Lambda}{tr(\Lambda)} \Delta\beta \cdot 100 \tag{10}$$

where $\Delta\beta$ is the normalized difference vector between original and processed PQRST features, where β and $\hat{\beta}$ represent two vectors of 18 diagnostic features ($RR_{int.}$, $QRS_{dur.}$, $QT_{int.}$, $QTp_{int.}$, $P_{dur.}$, $PR_{int.}$, $QRS_{peaks\_no.}$, $Q_{wave\_exsist.}$, $\Delta_{wave\_exsist.}$, T_{shape}, P_{shape}, ST_{shape}, $QRS(+)_{amp.}$, $QRS(-)_{amp.}$, $P_{amp.}$, $T_{amp.}$, $ST_{elevation}$, ST_{slope}) of compared beats and Λ is a diagonal matrix of weights heuristically set to [42],

$$\Lambda = \text{diag}\,[2.5\ 2.5\ 1\ 1\ 2\ 2\ 1\ 0.5\ 0.1\ 1.5\ 1\ 3\ 1.5\ 1.5\ 1\ 1\ 3\ 3] \tag{11}$$

This ECG-specific error metric, although requiring interpretive calculation of diagnostic outcomes, yields results related to medical findings equivalence rather than to signal representation accuracy. Therefore it is more adequate to evaluate the medical content preservation in signals with irregular distribution of information. Since the definition of WDD roughly reflects the local density of medical data in the ECG, expressed by diagnostic parameters, we consider it as the principal quality estimator of proposed adaptive ECG sampling. Moreover, the use of WDD shifts the evaluation of our adaptive sampling method from the signal domain to the parameter domain and thus reliably reflects the possible alteration of medical content. A serious cost of this is the necessity of using an interpretive software, of which none shows a 100% accuracy.

3.3. Performance Assessment

With the aim of exploring the possible flexibility of the adaptive ECG sampling, we applied two different ranges of sampling frequency adaptation: (1) 100 Hz ... 500 Hz and (2) 50 Hz ... 500 Hz. We also used two different medical relevance functions. Besides the original gMRF, obtained directly from scanpath studies (see Figure 2b), we simulated a specific medical relevance function (sMRF) with a purpose-oriented region of interest (ROI) focused on the end of QRS complex (e.g., for the investigation of an infarct or a conduction defect [29]). This variant of sMRF emphasizes the relevance and increases the accuracy of QT section at the price of saving samples mostly in P-wave vicinity (Figure 6).

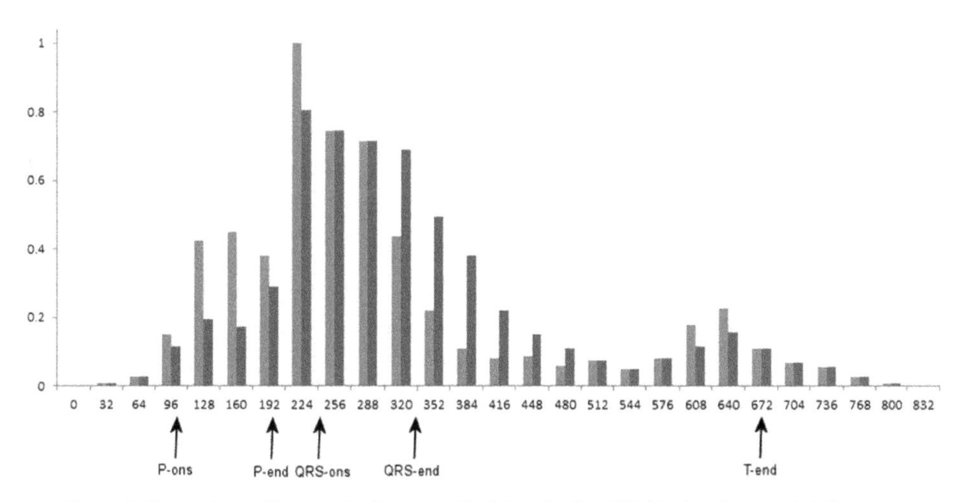

Figure 6. Comparison of bar graphs for scanpath-determined gMRF (blue) and example infarct or a conduction defect-oriented sMRF (red). Wave borders are: P-onset: 106, P-end: 196, QRS-onset 246, QRS-end: 338 and T-end 672.

Results of experiments for compression, distortions and medical parameter differences are presented in Table 1.

Table 1. Data compression, distortions and medical parameter differences for adaptive ECG sampling.

Parameter		gMRF (Perceptual)		sMRF (with ROI)	
f_s Range (Hz)		100 ... 500	50 ... 500	100 ... 500	50 ... 500
Compression ratio		3.01	3.61	3.83	4.72
PRD [%] ([μV] *)	global	3.11 (46.6)	3.73 (55.9)	3.96 (59.3)	4.88 (73.1)
	within P-wave	0.16 (2.4)	0.18 (2.7)	0.35 (5.3)	0.41 (6.2)
	within QRS complex	0.22 (3.3)	0.22 (3.3)	0.22 (3.3)	0.24 (3.6)
	within T-wave	0.37 (5.6)	0.41 (6.2)	0.44 (6.7)	0.47 (7.1)
	out of waves	1.11 (16.6)	1.77 (26.5)	2.70 (40.4)	3.93 (58.8)
WDD [%]		0.21	0.23	0.37	0.41
RR interval std [ms]		1.5	1.5	1.5	1.5
P-wave duration std [ms] (15) **		10.3	10.7	12.4	14.1
PQ interval length std [ms] (10)		7.1	7.2	8.6	9.7
QRS duration std [ms] (10)		7.6	7.6	7.6	7.6
QT interval length std [ms] (30)		14.7	16.5	16,1	18.2
P axis std [deg]		7.5	8.8	7.8	9.7
QRS axis std [deg]		2.1	2.7	2.4	3.0
T axis std [deg]		3.1	3.5	3.3	3.5

* acceptable value specified in IEC 60601-2-51 is 25 μV or 5% for amplitudes above 500 μV. ** acceptable value of standard deviation with reference to CSE Database results specified in IEC 60601-2-51.

4. Discussion

The presented adaptive ECG sampling algorithm, although not bit-accurate, shows interesting compression efficiency, making it worth considering in clinical applications. It may be classified as an alternative to bit-accurate methods yielding the compression ratio in the order of 3 at the price of high computational complexity [10] to the quality-on-demand (lossy-to-lossless) algorithms [3,43] and to recently proposed compressed sensing [13,15] or adaptive sensing [16,17] methods. The main advantage of the newly proposed algorithm is that unlike methods in the last two categories, the temporal distribution of distortions is based on medical rather than on statistical features of the signal (Table 1). For this reason, the quality estimate based on primary ECG diagnostic parameters (WDD) shows only little difference caused by the transformation.

We do not follow the example of several authors [3,7,42,44–48] using the MIT-BIH database [49] for tests of adaptive ECG sampling technique. The reason is threefold:

The sampling frequency of the MIT-BIH database is too low; following the paradigm typical for long-term recording, sampling at 360 Hz avoids oversampling in low-frequency components, limits the bandwidth of the high-frequency sections, and reduces the interval range (f_s–f_m) available for adaptive sampling.

The CSE Database provides reference positions of wave border points that enable evaluation of whether the proposed algorithm is robust to possible inaccuracy of wave delimitation.

The proposed algorithm uses the aMRF commonly calculated and stored for all leads in a multidimensional signal, therefore its efficiency decreases with the leads number.

A broad discussion was held with collaborators and experts in the field whether to apply the proposed method to the MIT-BIH standards, namely arrhythmia and compression databases. This would allow us to compare the proposed method to the values reported by Sayadi et al. [42] on that dataset or to compare the compression efficiency to the variety of algorithms like the ones proposed by Fira et al. [45] or Kim et al. [50] who reported the quality score (QS) above 15 for the MIT-BIH database. For three reasons we decided not to follow this thinking.

The proposed method is not yet another data compression algorithm—its novelty consists in the use of a priori knowledge about the ECG content derived from the human perception instead of local statistics of the signal.

The use of reference wave border points provided by the CSE database makes it possible to prove the robustness of the coding method to the inexact performance of the delineation software.

The QS proposed as a ratio of CR to PRD [45] inherits the principal drawback of the PRD that is the negligence of local variations of ECG signal relevance.

Additionally, as we are skeptical of the PRD as a medically-justified measure of distortion and propose using the WDD instead, we need a diagnostic software in order to calculate all 18 heartbeat features given in Equations (10) and (11). Following [51] we also did preliminary tests with the European QT Database providing reference wave borders available from Physionet at no charge. The database consists of 2-lead ECG signals, and we tried to simulate the missing leads in order to feed them to Ascard 6 12-lead diagnostic software. At this point we had to give up since the results we got for wave delineation were significantly different from the values given in the database.

Parkale and Nalbalwar give a complete survey of the CS techniques in [52]. The main performance results of the most significant methods are summarized in Table 2 for two scenarios: Scenario 1, distortion-optimized (with reasonable compression ratio), and Scenario 2, compression-optimized (with acceptable distortion level).

Table 2. Performance of the proposed adaptive sampling method compared to recent landmark systems for ECG compressed sensing; NB authors use different data sets for testing.

Work (Test Set)	Method	Scenario 1		Scenario 2	
		CR	PRD [%]	CR	PRD [%]
Mamaghanian [14] (MIT-BIH)	wavelet db10	3.70	2.00	10.00	9.00
	CS	2.04	2.00	3.45	9.00
Mishra [53] (10 ECGs custom set)	db	2.00	1.31	6.00	17.37
	rbio3.9	2.00	0.32	6.00	10.84
Craven [13] (MIT-BIH)	SPIHT	6.10	1.95	12.00	4.00
	AD-Q6	6.75	3.20	11.10	4.50
Polania [15] (subset of MIT–BIH)	MMB–IHT	6.40	3.76		
	MMB–CoSaMP	6.40	3.96		
Polania [51] (European QT DB)	RBM-OMP-like	2.00	1.20	5.00	6.30
	BPON	2.00	1.30	5.00	8.90
Chen [54] (MIT-BIH)	rbio5.5	2.00	10.03	5.00	46.63
	rbio5.5-JBHI	2.00	3.85	5.00	9.10
this work (CSE)	**perceptual**	**3.01**	**3.11**	**4.72**	**4.88**

Comparing our work to the CS algorithms one should note different approaches to testing the quality of the output ECG. Mamaghanian et al. [14] and other followers refer to the paper by Zigel et al. [5] in specifying the 'quality class' as 'very good' for PRD < 2% or 'good' for 2% < PRD < 9%. For Luo et al. [55] the ECGs recovered with PRD < 6% are 'essentially undistorted' and Craven et al. [13] allow for PRD > 5% for 'clinically relevant metrics' but, despite the title, they focus on heart rate variability (HRV) parameters, more tolerant to amplitude distortion.

In works by Rieger et al. [16], Kim et al. [56], and Yazicioglu et al. [17], the bi-frequency sampling is controlled by the algorithm recognizing high activity and low activity sections. In [16] the ECG trace has to meet given peak and curvature conditions (i.e., first and second derivatives are calculated) to switch basic sampling frequency 50 Hz to the fast rate of 400 Hz. The threshold was set to effectively recognize the QRS beginning, maximum and end points in normal heartbeats. As a result, the data compression ratio of 1.6 was achieved. In [56], following the wavelet-based detection of QRS the basic sampling frequency of 64 Hz is stepwise raised to 512 Hz. As a result the data compression ratio of 4.5 was achieved. In [17] the activity detector circuit senses the rate of change of the input signal by using a switching capacitance differentiator. As the activity passes a given threshold, the basic sampling frequency of 64 Hz is stepwise raised to 1024 Hz. The data compression ratio depends on the duty cycle controlled by the value of activity threshold. For duty cycles in the range of 4–12%, the compression ratios of 9.94–5.72 are expected respectively. In all these works authors focus on the circuitry design and power efficiency rather than on ECG signal diagnostability, and thus do not provide information on the ECG distortion level. Despite a quantitative comparison of these methods with our work not being possible, they share a common concept of non-uniform distribution of medical information in the ECG. Therefore, the gMRF proposed in this paper can be seen as an advanced version of 'activity detector' controlling the actual ECG sampling in 64 steps instead of 2, which allows us to preserve best signal quality within the ECG waves.

A bi-frequency compression scheme based on local sub-sampling (decimation) of the signal is also recommended by the SCP-ECG communication protocol (ENV 1064) [57]. In our approach, the sampling interval is adjusted in a nearly continuous way without precisely distinguishing bordered zones. This provides a fair tolerance margin for the accuracy of waves delineation.

Aiming at future industrial implementation, we thoroughly checked the temporal distribution of distortions and their possible influence to the diagnostic result. The global PRD value (Table 1) seems to be high, but it is noteworthy that within the waves the distortion level in terms of wideband noise meets the requirements of industrial IEC standards [26] (25 μV, accordingly to 51.106.4) and is very close to quantization error requirements (5 μV, accordingly to 51.107.4). The duration of intervals is also little affected by the adaptive sampling, and extending the sampling frequency to as low as 50 Hz together with application of a QRS-focused region of interest still yields results acceptable by the IEC standard (marked by a double star in Table 1).

CS methods are often presented in the context of computational efficiency, particularly stressed in low-power wireless networks of sensing nodes. Although our primary goal is the quality of medical content, we estimated the computational complexity of the method by implementing the detection and wave delineation parts of the Ascard 6 software to a mobile platform with PXA-270 CPU running at 624 MHz with 64 MB of operation random access memory and 32 MB of flash memory under Linux OS. Additionally, tunable low-pass filters and cubic splines-based procedures for translation of uniform to non-uniform ECG representation were also implemented in the platform. All the evaluation procedures (e.g., WDD) were kept on the PC and used offline. Since the adaptive sampling is conceived as a continuously running procedure, the results on processing time are expressed as a percentage of real ECG duration as follows:

- On a PC platform i7 3770 (® Intel, Santa Clara, CA, USA), 3400 MHz, 8 GB RAM—0.943% (i.e., 106 times faster than the ECG acquisition);
- on a mobile platform PXA-270 (® Toradex AG., Horw, Switzerland), 624 MHz, 64 MB RAM—7.518% (i.e., 13.3 times faster than the ECG acquisition).

The gMRF is derived as a result of the pursuit for local conspicuity of the ECG trace and represents common perceptual habits of cardiology experts participating in our visual experiment [19]. This function generalizes the knowledge from cardiology expert perception of the ECG trace and reflects the local relevance of the signal that would be difficult to express in another way. This relationship can easily be modified in order to create various application-oriented or user-tailored profiles, differing by temporal allocation of regions of interest and the remaining zones where distortions are tolerable. An example of such approach was also tested (sMRF) and yielded a promising compromise of coding efficiency to distortion ratio (CR = 4.72, PRD = 4.88, WDD = 0.41).

The gMRF, being an experimental perception-derived relevance curve, has been applied to modulating the ECG sampling. For this task alternative sMRFs may be developed and used accordingly to a specific medical purpose. Since the shape of the sMRF is a principal factor imposing the compression efficiency, a question arose as to whether one could predict the CR from the sum of MRF bins (ΣMRF). In the particular case of CSE Database records, accompanied by reference specifications of the length of each particular heartbeat, thanks to the performed temporal normalization such a strict relationship can be proven. In this case the output data stream d may be approximated as

$$d = [f_m + (f_s - f_m) \times bl \times \sum_b MRF(b)] \times sr \tag{12}$$

where bl stands for bin length [s] (in our work being equal to 0.032 s), b is the bin number, and sr is the sample resolution (equal to 12 bits per sample).

In a general case, such a precise estimate cannot be calculated due to the presence of additional factors. Specifically, the CR will be higher than estimated by ΣMRF in case of slow rhythm (long T-P interval) due to prolonged use of f_s between the adjacent heart beats. Otherwise, the CR will be lower than estimated by ΣMRF in the case of fast rhythms because main shortenings in the ECG pattern take place out of the waves (i.e., the accelerating heart first reduces its inactivity periods). A separate approach should be studied in cases of abnormal ECG when wave borders cannot be determined.

The proposed adaptive sampling technique makes sole use of local signal oversampling and therefore does not pretend to compete with existing ECG compression methods. In irregularly sampled ECG series, a significant short time (i.e., sample-to-sample), long time (i.e., heartbeat-to-heartbeat) and spatial (i.e., lead-to-lead) correlation is preserved. Therefore, further improvement of sampling efficiency is expected as a result of combining the adaptive sampling with long-term prediction technique (the use of beat-to-beat similarity of the ECG) and/or with ECG leads decorrelation (reduction of signal dimensionality) considered in future versions. In our opinion, this kind of sampling may have a similar impact on the telemedicine of tomorrow as the perceptual coding had on the audio and video broadcasting techniques of today.

Funding: This research was funded by AGH University of Science and Technology in 2019 as research project No. 16.16.120.773.

Acknowledgments: The author expresses their gratitude to the President of Aspel S.A. Zabierzów, Poland for his consent to use selected procedures of Ascard 6 interpretive software free of charge.

Conflicts of Interest: The author declares no conflict of interest. The funders had no role in the design of the study; in the collection, analyses, or interpretation of data; in the writing of the manuscript, and in the decision to publish the results.

References

1. Chen, J.; Itoh, S. A wavelet transform-based ECG compression method guaranteeing desired signal quality. *IEEE Trans. Biomed. Eng.* **1998**, *45*, 1414–1419. [CrossRef] [PubMed]
2. Lu, Z.; Kim, D.Y.; Pearlman, W.A. Wavelet compression of ECG signals by the set partitioning in hierarchical trees (SPIHT) algorithm. *IEEE Trans. Biomed. Eng.* **2000**, *47*, 849–856. [PubMed]
3. Miaou, S.G.; Chao, S.N. Wavelet-based lossy-to-lossless ECG compression in a unified vector quantization framework. *IEEE Trans. Biomed. Eng.* **2005**, *52*, 539–543. [CrossRef] [PubMed]
4. Rajoub, B.A. An efficient coding algorithm for the compression of ECG signals using the wavelet transform. *IEEE Trans. Biomed. Eng.* **2002**, *49*, 355–362. [CrossRef] [PubMed]
5. Zigel, Y.; Cohen, A.; Katz, A. The weighted diagnostic distortion (WDD) measure for ECG signal compression. *IEEE Trans. Biomed. Eng.* **2000**, *47*, 1422–1430.
6. Alesanco, S.; Olmos, R.S.H.; Garcıa, I.J. Enhanced Real-Time ECG Coder for Packetized Telecardiology Applications. *IEEE Trans. Inf. Tech. Biomed.* **2006**, *10*, 229–236. [CrossRef]
7. Huang, C.Y.; Miaou, S.G. Transmitting SPIHT compressed ECG data over a next-generation mobile telecardiology testbed. In Proceedings of the 23rd IEEE EMBS Annual International Conference, Istanbul, Turkey, 25–28 October 2001; pp. 3525–3528.
8. Kannan, R.; Eswaran, C. Lossless compression schemes for ECG signals using neural network predictors. *EURASIP J. Adv. Signal Process.* **2007**, *2007*, 035641. [CrossRef]
9. Bradie, B. Wavelet Packet-Based Compression of Single Lead ECG. *IEEE Trans. Biomed. Eng.* **1996**, *43*, 493–501. [CrossRef]
10. Duda, K.; Turcza, P.; Zielinski, T.P. Lossless ECG compression with lifting wavelet transform. In Proceedings of the IEEE Instrumentation and Measurement Technology Conference, Budapest, Hungary, 21–23 May 2001; pp. 640–644.
11. Augustyniak, P. Time-frequency modelling and discrimination of noise in the electrocardiogram. *Physiol. Meas.* **2003**, *24*, 753–767. [CrossRef]
12. Liji, C.A.; Indiradevi, K.P.; Babu, K.A. Integer-to-integer wavelet transform based ECG steganography for securing patient confidential information. *Procedia Technol.* **2016**, *24*, 1039–1047. [CrossRef]
13. Craven, D.; McGinley, B.; Kilmartin, L.; Glavin, M.; Jones, E. Adaptive Dictionary Reconstruction for Compressed Sensing of ECG Signals. *IEEE J. Biomed. Health Inform.* **2017**, *21*, 645–654. [CrossRef] [PubMed]
14. Mamaghanian, H.; Khaled, N.; Atienza, D.; Vandergheynst, P. Compressed Sensing for Real-Time Energy-Efficient ECG Compression on Wireless Body Sensor Nodes. *IEEE Trans. Biomed. Eng.* **2011**, *58*, 2456–2466. [CrossRef]
15. Polanía, L.F.; Carrillo, R.E.; Blanco-Velasco, M.; Barner, K.E. Exploiting Prior Knowledge in Compressed Sensing Wireless ECG Systems. *IEEE J. Biomed. Health Inform.* **2015**, *19*, 508–519. [CrossRef] [PubMed]

16. Rieger, R.; Taylor, J.T. An adaptive sampling system for sensor nodes in body area networks. *IEEE Trans. Neural Syst. Rehabil. Eng.* **2009**, *17*, 183–189. [CrossRef] [PubMed]
17. Yazicioglu, R.F.; Kim, S.; Torfs, T.; Kim, H.; van Hoof, C. A 30 μW analog signal processor ASIC for portable biopotential signal monitoring. *IEEE J. Solid-State Circuits* **2011**, *46*, 209–223. [CrossRef]
18. Augustyniak, P. Moving window signal concatenation for spectral analysis of ECG waves. *Comput. Cardiol.* **2010**, *37*, 665–668.
19. Augustyniak, P. How a Human Perceives the Electrocardiogram. *Comput. Cardiol.* **2003**, *30*, 601–604.
20. Augustyniak, P.; Tadeusiewicz, R. Assessment of electrocardiogram visual interpretation strategy based on scanpath analysis. *Physiol. Meas.* **2006**, *27*, 597–608. [CrossRef]
21. Almeida, R.; Martınez, J.P.; Rocha, A.P.; Laguna, P. Multilead ECG delineation using spatially projected leads from wavelet transform loops. *IEEE Trans. Biomed. Eng.* **2009**, *56*, 1996–2005. [CrossRef]
22. Martınez, J.P.; Almeida, R.; Olmos, S.; Rocha, A.P.; Laguna, P. A wavelet-based ECG delineator: Evaluation on standard databases. *IEEE Trans. Biomed. Eng.* **2004**, *51*, 570–581. [CrossRef]
23. Dumont, J.; Hernandez, A.I.; Carrault, G. ECG beat delineation with an evolutionary optimization process. *IEEE Trans. Biomed. Eng.* **2010**, *57*, 607–615. [CrossRef] [PubMed]
24. Yochum, M.; Renaud, C.; Jacquir, S. Automatic detection of P, QRS and T patterns in 12 leads ECG signal based on CWT. *Biomed. Signal Process. Control* **2016**. [CrossRef]
25. Kohler, B.; Hennig, C.; Orglmeister, R. The principles of software QRS detection. *IEEE Eng. Med. Biol. Mag.* **2002**, *21*, 42–57. [CrossRef] [PubMed]
26. IEC 60601-2-51. *Medical Electrical Equipment Part 2-51: Particular Requirements for Safety, Including Essential Performance, of Recording and Analyzing Single Channel and Multichannel Electrocardiographs*; IEC: Geneva, Switzerland, 2003; ISBN 2-8318-6880-7.
27. Chen, F.; Chandrakasan, A.P.; Stojanovic, V.M. Design and analysis of a hardware-efficient compressed sensing architecture for data compression in wireless sensors. *IEEE J. Solid-State Circuits* **2012**, *47*, 744–756. [CrossRef]
28. Malik, M.; Camm, A.J. *Dynamic Electrocardiography*; Futura; Blackwell: Oxford, UK, 2004; pp. 112–177.
29. Boccignone, G. An information-theoretic approach to active vision. In Proceedings of the 11th International Conference on Image Analysis and Processing (ICIAP '01), Palermo, Italy, 26–28 September 2001.
30. Pelz, J.B.; Canosa, R. Oculomotor behavior and perceptual strategies in complex tasks. *Vis. Res.* **2001**, *41*, 3587–3596. [CrossRef]
31. Morlet, D. Contribution a L'analyse Automatique des Electrocardiogrammes—Algorithmes de Localisation, Classification et Delimitation Precise des Ondes dans le Systeme de Lyon. Ph.D. Thesis, INSA-Lyon, Lyon, France, 1986. (In French).
32. Willems, J.L. *Common Standards for Quantitative Electrocardiography 10-th CSE Progress Report*; ACCO Publication: Leuven, Belgium, 1990.
33. Aldroubi, A.; Groechenig, K. Non-Uniform Sampling and Reconstruction in Shift-Invariant Spaces. *SIAM Rev.* **2001**, *43*, 585–620. [CrossRef]
34. Aldroubi, A.; Feichtinger, H. Exact iterative reconstruction algorithm for multivariate irregularly sampled functions in spline-like spaces: The Lp theory. *Proc. Am. Math. Soc.* **1998**, *126*, 2677–2686. [CrossRef]
35. Landau, H. Necessary density conditions for sampling and interpolation of certain entire functions. *Acta Math.* **1967**, *117*, 37–52. [CrossRef]
36. Candés, E.; Romberg, J.; Tao, T. Stable signal recovery from incomplete and inaccurate measurements. *Commun. Pure Appl. Math.* **2006**, *59*, 1207–1223. [CrossRef]
37. Needell, D.; Tropp, J.A. COSAMP: Iterative signal recovery from incomplete and inaccurate samples. *Appl. Comput. Harmon. Anal.* **2008**, *26*, 301–321. [CrossRef]
38. Pei, S.-C.; Lin, H.-S. Tunable FIR and IIR Fractional-Delay Filter Design and Structure Based on Complex Cepstrum. *IEEE Trans. Circuits Syst. I Regul. Pap.* **2009**, *56*, 2195–2206.
39. Wei, X.; Anyu, L.; Boya, S.; Jiaxiang, Z. A Novel Design of Sparse FIR Multiple Notch Filters with Tunable Notch Frequencies. *Math. Probl. Eng.* **2018**, *2018*, 3490830. [CrossRef]
40. Senthilkumar, E.; Manikandan, J.; Agrawa, V.K. FPGA Implementation of Dynamically Tunable Filters. In Proceedings of the International Conference on Advances in Computing, Communications and Informatics (ICACCI), Greater Noida, India, 24–27 September 2014; pp. 1852–1857.

41. Al-Fahoum, S. Quality Assessment of ECG Compression Techniques Using a Wavelet-Based Diagnostic Measure. *IEEE Trans. Inf. Technol Biomed.* **2006**, *10*, 182–191. [CrossRef] [PubMed]
42. Sayadi, O.; Shamsollahi, M.-B. ECG Denoising and Compression Using a Modified Extended Kalman Filter Structure. *IEEE Trans. Biomed. Eng.* **2008**, *55*, 2240–2248. [CrossRef] [PubMed]
43. Miaou, S.G.; Yen, H.L.; Lin, C.L. Wavelet-based ECG compression using dynamic vector quantization with tree codevectors in single codebook. *IEEE Trans. Biomed. Eng.* **2002**, *49*, 671–680. [CrossRef] [PubMed]
44. Filho, E.B.L.; Rodrigues, N.M.M.; da Silva, E.A.B.; de Faria, S.M.M.; da Silva, V.M.M.; de Carvalho, M.B. ECG signal compression based on DC equalization and complexity sorting. *IEEE Trans. Biomed. Eng.* **2008**, *55*, 1923–1926. [CrossRef]
45. Fira, C.M.; Goras, L. An ECG signals compression method and its validation using NNs. *IEEE Trans. Biomed. Eng.* **2008**, *55*, 1319–1326. [CrossRef]
46. Ku, C.-T.; Wang, H.-S.; Hung, K.-C.; Hung, Y.-S. A novel ECG data compression method based on nonrecursive discrete periodized wavelet transform. *IEEE Trans. Biomed. Eng.* **2006**, *53*, 2577–2583.
47. Sun, C.-C.; Tai, S.-C. Beat-based ECG compression using gain-shape vector quantization. *IEEE Trans. Biomed. Eng.* **2005**, *52*, 1882–1888. [CrossRef] [PubMed]
48. Tai, S.-C.; Sun, C.-C.; Yan, W.-C. A 2-D ECG compression method based on wavelet transform and modified SPIHT. *IEEE Trans. Biomed. Eng.* **2005**, *52*, 999–1008. [CrossRef] [PubMed]
49. MIT-BIH Arrhythmia Database. Available online: https://www.physionet.org/content/mitdb/1.0.0/ (accessed on 21 December 2019).
50. Kim, H.; Yazicioglu, R.F.; Merken, P.; Van Hoof, C.; Yoo, H.J. ECG signal compression and classification algorithm with quad level vector for ECG Holter system. *IEEE Trans. Inf. Technol. Biomed.* **2010**, *14*, 93–100. [PubMed]
51. Polania, L.F.; Plaza, R.I. Compressed Sensing ECG using Restricted Boltzmann Machines. *Biomed. Signal Process. Control* **2018**, *45*, 237–245. [CrossRef]
52. Parkale, Y.V.; Nalbalwar, S.L. Application of compressed sensing (CS) for ECG signal compression: A Review. In *Advances in Intelligent Systems and Computing, Proceedings of the International Conference on Data Engineering and Communication Technology, Pune, India, 15–16 December 2017*; Satapathy, S., Bhateja, V., Joshi, A., Eds.; Springer: Singapore, 2017; Volume 469.
53. Mishra, A.; Thakkar, F.; Modi, C.; Kher, R. ECG Signal Compression using Compressive Sensing and Wavelet Transform. In Proceedings of the 34th Annual International Conference of the IEEE EMBS, San Diego, CA, USA, 28 August–1 September 2012; pp. 3402–3407.
54. Chen, J.; Xing, J.; Zhang, L.Y.; Qi, L. Compressed sensing for electrocardiogram acquisition in wireless body sensor network: A comparative analysis. *Int. J. Distrib. Sens. Netw.* **2019**, *15*, 1550147719864884. [CrossRef]
55. Luo, K.; Cai, Z.; Du, K.; Zou, F.; Zhang, X.; Li., J. A digital compressed sensing-based energy-efficient single-spot Bluetooth ECG node. *J. Healthc. Eng.* **2018**. [CrossRef]
56. Kim, H.; Van Hoof, C.; Yazicioglu, R.F. A mixed signal ECG processing platform with an adaptive sampling ADC for portable monitoring applications. In Proceedings of the 33rd Annual International Conference of the IEEE EMBS, Boston, MA, USA, 30 August–3 September 2011; pp. 2196–2199.
57. Zywietz, C.; Fischer, R. Integrated content and format checking for processing of SCP ECG records. In Proceedings of the Computers in Cardiology, Chicago, IL, USA, 19–22 September 2004; pp. 37–40.

Article

The Rehapiano—Detecting, Measuring, and Analyzing Action Tremor Using Strain Gauges

Norbert Ferenčík [1], Miroslav Jaščur [1], Marek Bundzel [1,*] and Filippo Cavallo [2]

[1] Department of Cybernetics and Artificial Intelligence, Faculty of Electrical Engineering and Informatics, Technical University of Košice, Letná 9, 07602 Košice, Slovakia; norbert.ferencik@tuke.sk (N.F.); miroslav.jascur@tuke.sk (M.J.)

[2] The Biorobotics Institute, Scuola Superiore Sant'Anna, 560 25 Pisa, Italy; filippo.cavallo@santannapisa.it

* Correspondence: marek.bundzel@tuke.sk; Tel.: +421-55-602-2564

Received: 18 November 2019 ; Accepted: 23 January 2020; Published: 24 January 2020

Abstract: We have developed a device, the Rehapiano, for the fast and quantitative assessment of action tremor. It uses strain gauges to measure force exerted by individual fingers. This article verifies the device's capability to measure and monitor the development of upper limb tremor. The Rehapiano uses a precision, 24-bit, analog-to-digital converter and an Arduino microcomputer to transfer raw data via a USB interface to a computer for processing, database storage, and evaluation. First, our experiments validated the device by measuring simulated tremors with known frequencies. Second, we created a measurement protocol, which we used to measure and compare healthy patients and patients with Parkinson's disease. Finally, we evaluated the repeatability of a quantitative assessment. We verified our hypothesis that the Rehapiano is able to detect force changes, and our experimental results confirmed that our system is capable of measuring action tremor. The Rehapiano is also sensitive enough to enable the quantification of Parkinsonian tremors.

Keywords: strain gauge; tremor quantification; Parkinson's disease; action tremors

1. Introduction

Parkinson's disease (PD) is among the most common neurological diseases. In 2016, approximately 6.1 million people worldwide were affected. This is a substantial increase compared to the 2.5 million PD patients reported in 1990. Furthermore, between four and 20 new cases per 100,000 people are reported annually [1]. Improved longevity and more precise diagnostic techniques for PD have both contributed to this increase [2].

PD is a chronic, progressive disease involving a gradual loss of motor and non-motor functions [3]. Non-motor symptoms include apathy, anhedonia, and depression. The loss of motor functions manifests as tremor, bradykinesia, rigidity, and a loss of postural reflexes [4]. Parkinsonian tremor (PT) is an approximately sinusoidal oscillatory motion that varies based on the severity of the disease, the activity performed by the affected individual, and their level of stress. Tremor may have a varying amplitude with a fixed frequency, which complicates its detection [5]. Because of the periodic nature of PT, Fourier transform (FT) is commonly used to analyze sensor data in the frequency domain [6]. Therefore, wearable sensors, for instance, inertial measurement units (IMUs) and electromyography (EMG), with subsequent spectral analysis are the standard approach in the evaluation of tremors [7–10].

The Rehapiano has strain gauges placed on ergonomically designed handles for both hands. These gauges measure the forces—both voluntary and involuntary—that the fingers apply to the gauges, with an appropriate sampling frequency and high precision. The Rehapiano is an alternative to tremor quantification using accelerometers, gyroscopes, and EMG. It does not require a sensor to be attached to the subject, which is certainly an advantage over wearable sensors. As the Rehapiano measures action tremor under specific conditions, wearable sensors have a significant advantage as they can monitor patients in different situations. The Rehapiano is applicable for evaluation/diagnosis and for fine motor function rehabilitation in clinical use. A quantitative assessment of a subject's tremor serves as decision support for the physician who determines the dosage and the type of medication to be administered. Several studies suggest that there is a link between the severity of the PD tremor and the stage of PD [11].

The authors consider the contribution presented here to be as follows,

- an introduction of the Rehapiano device for fast detection and quantification of action tremor using strain gauges,
- validation of the system by the comparison of measurements made by the Rehapiano to those made using optical encoders,
- an experimental analysis of the Rehapiano on healthy subjects and patients with PD, and
- an adaptation of an algorithm [9] that was previously developed for use with accelerometers and gyroscopes to asses tremor severity.

Section 4 describes the novelty of the device hardware. The methodology of measuring tremor with strain gauges is presented in Sections 5 and 6. The results of our experiments support this unique approach towards evaluation of PT. Therefore, this framework could provide an alternative to standard approaches to PT assessment.

2. Problem Statement

Tremor is characterized in medicine as an involuntary rhythmic and periodic movement of body parts. All body parts may be affected, including the head, chin, and soft palate [12]. Muscle contractions during a tremor have a regular frequency [13]. However, identification of the frequency may be complicated by the signal's amplitude changes. The amplitude changes may occur spontaneously, but are often correlated with change in the limb's position, fatigue, or emotional stress [5]. Frequency is the tremor's basic descriptive criterion, being categorized as low (<4 Hz), medium (4–7 Hz), or high (>7 Hz). The frequencies of tremors of different etiology have differential diagnostic value: cerebellar, Holmes, or palatal tremor have slow frequency, whereas orthopedic tremor is very rapid [14].

Tremor, according to its etiology, is categorized as rest tremor or action tremor, the latter being further subdivided into action postural and action kinetic tremor (see Figure 1) [15]. A rest tremor appears on relaxed muscles and should be measured on a lying subject. Action tremor appears on muscles that have been voluntarily engaged. A specific isometric tremor appears also in healthy subjects during strong isometric muscle contractions but can be superimposed with a tremor of another type [16].

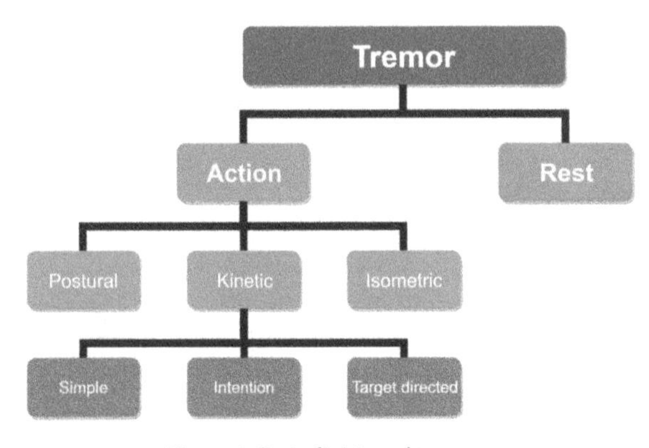

Figure 1. Basic division of tremors.

From a clinical point of view, we recognize physiological, essential, Parkinsonian, and orthostatic tremor, tremor associated with another neurological disease, and psychogenic tremor [17].

Physiological tremor occurs in fine motor activities and is normal. It can be accentuated by anxiety, emotional stress, and some medications or by underlying conditions such as hypothermia, hypoglycemia, or hyperthyroidism. The frequency of a physiological tremor ranges from 4 Hz to 8 Hz (usually more than 7 Hz) [18]. Essential tremor is a sign of a health problem. It is characterized by the presence of bilateral and predominantly symmetrical, action postural or action kinetic, permanent and visible tremor, especially of the upper limbs. The essential tremor frequency is between 4 Hz and 10 Hz. More than 50% of affected subjects report a tremor reduction after consuming alcohol [14].

PT manifests itself in three types: resting, action postural, and action kinetic tremor [19]. Action tremor often has been observed in Parkinson's disease (PD). The prevalence may be as high as 92% [20,21]. PT is often reduced during movement, although sometimes it is not, and then usually has the same frequency as the rest component. Typically, there is a pause in the tremor during a change from rest to posture [22]. PT frequency is usually above 5 Hz [23], although the upper frequency limit has not yet been established. Especially in the early stages of PD, the PT frequency ranges up to 8 Hz. Tremor is the most common initial symptom of PD, occurring in approximately 70% of Parkinson patients [24]. Its onset is usually one-sided. Some patients with clinical manifestations describe a strange feeling of internal shivering in the limb(s) or inside the body. Initially, the tremor occurs only intermittently from fatigue, agitation, or excessive concentration, and it disappears during sleep [25]. PT may resemble coin-counting or pill-rolling. The thumb moves with simultaneous flexion and extension of the metacarpophalangeal and interphalangeal joints [26]. During the course of the disease, the tremor may expand, and forearm and shoulder movements appear.

The primary orthostatic tremor manifests itself as a perceived sense of instability that reduces when walking. Tremor and the feeling of instability worsen during prolonged standing. Fine, high-frequency tremor (from 14 Hz to 16 Hz) may be felt by palpating the lower limbs rather than detected visually [27,28]. Tremor can be associated with a multitude of neurological diseases. Cerebellar tremor, Holmes tremor, and dystonic or neuropathic tremor have specific features, as described in [17]. Psychogenic tremor (a synonym for functional tremor) is the most common form (55%) of psychogenic motor disorders. Seventy-five percent of the affected people are female. A psychogenic tremor's frequency is less than 7 Hz [29].

Tremor is a significant symptom and its quantification can aid in diagnosing the related problem, determining the right dosage and the right type of medication, and evaluating the development of the symptom over time. When measuring Parkinsonian, psychological, or essential tremor with existing devices, it is often difficult to repeat the exercise and the related measurement under the same conditions.

3. Background and Related Works

Determining subjects' activity recognition and monitoring is important for understanding their condition and the development of the disease [8]. In Elble and McNames [30], a practical overview of the use of portable motion transducers in the quantification of tremor is provided. Rather than a comprehensive review of the transducers available for the assessment of tremor, it is a practical guide to the selection and use of portable transducers in tremor analysis. Elble and McNames determined what sensitivity, amplitude, and frequency ranges that transducers should use for high fidelity tremor detection. Tremor measuring devices can be divided into wearable sensors and fixed devices. The wearable sensors must be small, light, and securely affixed to the related body part.

In Haubenberger et al. [7], various types of devices for quantification and characterization of tremor are compared. The authors cover electromyography, accelerometry, gyroscopy, activity monitoring, digitizing tablets, and acoustic analysis of voice tremor. Availability on the market, ability to use, acceptability, reliability, and responsiveness were reviewed for each measurement method. The following criteria were adopted to evaluate each measurement method; (1) The use in the assessment of tremor, (2) use in published studies by people other than the developers, and (3) adequate clinical testing. Based on the criteria set out in this review, accelerometry, gyroscopy, electromyography, and tablet digitization met all three criteria for use in quantifying and detecting abnormal tremor. Some studies have indicated that accelerometer and gyroscope measurements correlate strongly with the unified PD rating scale (UPDRS) [31]. The United States Food and Drug Administration (FDA) approved the Kinesia™system (Great Lakes NeuroTechnologies Inc., Cleveland, OH, USA), which is used to assess Parkinsonian symptoms with an inertial measurement unit (IMU), which embeds a three-axis gyroscope and a three-axis accelerometer in a single chip on top of a finger [32]. In Niazmand et al. [33], a study about using a wireless wearable sensor system for evaluation of the severity of motor dysfunction in PD is presented. The system was integrated into a smart glove equipped with two touch sensors, two 3D-accelerometers, and a force sensor to assess the cardinal motor symptoms of PD (bradykinesia, tremor, and rigidity of the hand and arm). The study focused on the hardware, which includes a glove with a control and transmit unit, a receiver unit, and a computer for storage and analysis of the data. In Khan et al. [34], BioMotion Suite's wearable system kit, equipped with a triaxial accelerometer, was used. The sensor samples at a rate of 32 Hz and has a range of ± 3 g. Data are processed using proprietary BioMotion Suite software implemented in Matlab. Khan et al. processed their experimental data with six classification algorithms to classify PD data. Accelerometer data from measurements often produce noisy data, which complicates their processing.

4. The Rehapiano

The Rehapiano (see Figure 2) is a system that is comprised of dedicated hardware and analytic software that provides repeatable results, measures and quantifies the subjects' tremor, and can monitor progression of their disease. The primary use of our system is to measure and analyze tremor based on measuring force applied by the fingers. The Rehapiano can also be used for the rehabilitation of fine motor skills. We plan a clinical trial that will include measuring the tremor of healthy subjects over a period of time and the tremor of subjects with PT after each cycle of medication. These measurements serve to determine the individual's progress (improvement, deterioration, or stagnation), and in the case of the affected subjects, constitute decision support for further treatment.

Figure 2. The Rehapiano device.

The base of the device consists of an aluminum alloy frame 600 mm wide and 360 mm long. Aluminum beams have a square cross section of 30×30 mm^2. The device weighs 16 kg and is stable when subjects work with it. Subjects place their forearms into low-temperature thermoplastic splints designed specially for the Rehapiano. The splints are equipped with three Velcro fasteners, each of which has a soft lining on the inside. The hands are in the plate from the wrist to the elbow in a fixed position.

The padded splints are hygienically harmless, non-allergic, and washable. They can be positioned and adjusted. Spacing is adjustable from 10 cm to 50 cm, tilt aside angle ∠120°, and ahead angle ∠40°. The ergonomic handles provide support for the palms and fingers and feature 5 strain gauges for each hand. The areas where subjects must put their fingertips are marked pink. The strain gauges measure the force applied by the subject's fingertip in one axis. The splints that hold the forearm isolate the hand during the measurement. The strain gauges are powered by 5 V and measure a maximum force of 50 N (5 kg). The strain gauges measuring the thumbs are placed on the vertical surface for increased ergonomy. PT frequency is usually above 5 Hz, [17,35]; therefore, we used an Hx711 (24 bit) analog-to-digital converter used in industrial control applications and sampled the measurements at 40 Hz [36].

The experimental data was acquired by an Arduino Mega and sent to a computer via USB interface (see Figure 3).

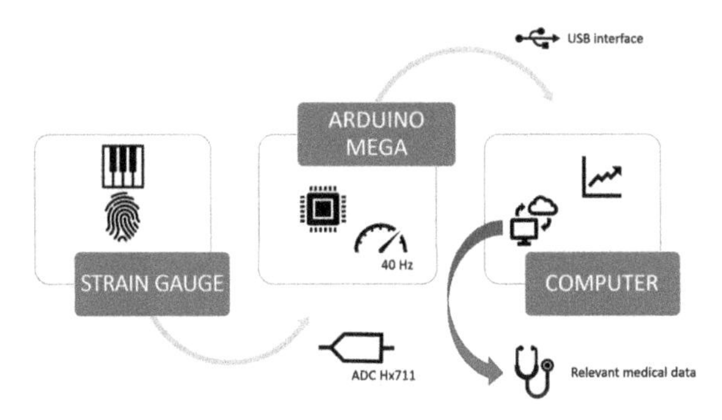

Figure 3. Rehapiano device data flow.

The measured values were sent together with the timestamps. The computer application was developed in C#. For easier transportation of the device, we used a minicomputer connected to the monitor. However, any monitor or projector can be used. The keyboard (see Figure 2) was used to enter the subject's initials, date of birth, gender, and current health problems. This information was used for knowledge retrieval from the experimental data and was anonymized.

5. Methods

Three hypotheses were formulated:

1. The Rehapiano is able to detect force changes with frequencies between 1 and 20 Hz.
2. Rehapiano measurements can be used to detect tremors of Parkinson patients.
3. The Rehapiano is sensitive enough to enable quantification of PT.

The first hypothesis was conducted for the evaluation of tremor simulation; the other two hypotheses involved human subjects. We conducted the experiments in laboratory conditions, on healthy and Parkinson subjects. Physicians of Parkinson patients reported to us that the PD patients also manifest action tremor. The facilities that enabled us to conduct these experiments were involved in the development of the Rehapiano. We have complied with all legal and legislative conditions, including GDPR patient protection. Each patient was personally acquainted with the use of their data from the device and signed an agreement permitting processing the medical data.

5.1. Verification of Hypothesis 1

To verify Hypothesis 1, a device (see Figure 4) that simulates force oscillations of varying frequency was constructed. L298N H bridge connected to an Arduino Uno and an RDO-37KE50G9A 12V DC motor (a 7.4 W output, a 0.17 mN torque, and a 400 rpm rotation speed, with a gearbox and gear ratio of 1:9) and an encoder. We used a pulley with an eccentrically mounted elastic rubber band. The other side of the elastic rubber was placed on the strain gauge. Because we could control the DC motor's revolutions per second precisely, we were able to generate a regularly varying force at the strain gauge. We validated the Rehapiano at three frequencies between 1.5 and 7 Hz. This range was chosen with regard to the usual PT frequencies.

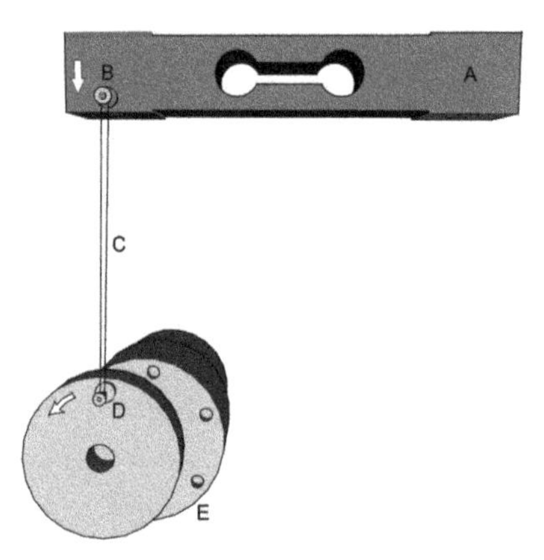

Figure 4. Experimental setup for Rehapiano validation. The direction of rotation of the DC motor and the force applied to the strain gauge are indicated by arrows. (A: fixed, B: force applied, C: elastic band, D: pulley, E: DC motor).

5.2. Verification of Hypothesis 2

We created a subject measurement protocol. The Rehapiano is suited for measuring action tremor produced during voluntary muscle contraction [37,38]. Each measured subject was prompted to apply force with a given finger. The amplitude of this force is displayed using vertical bars. By inducing targeted muscle contraction, the tremor may manifest and the device measures the force at 40 Hz frequency.

Thirty-six healthy volunteers (average age: 41.72 y, 21 females, 15 males) and seven PD patients (average age: 76.10 y, 6 females, 1 male) participated in this study. We also had a patients with leg tremors and dyskinesia, without any tremor in their hands, and were therefore not recruited. The PD subjects were evaluated also by the physician according to the Fahn–Tolosa–Marin Tremor Rating Scale (FTMTRS) (see Table 1), which quantifies rest, postural, and action/intention tremor. The Sessions column indicates the number of measurements for all fingers. Patient 3 had a higher tremor when he tried to write by hand, so he was rated 4 (unable to hold a pencil) by a physician. The FTMTRS is a widely used clinical rating scale quantifying severity of tremor from 0 (none) to 4 (severe) for the given body part [39,40].

Table 1. Fahn–Tolosa–Marin tremor rating scale values of the participants.

PD Patient	Tremor	Handwriting	Sessions
Patient 1.	1	2	2
Patient 2.	2	3	1
Patient 3.	3	3	1
Patient 4.	3	4	1
Patient 5.	3	2	7
Patient 6.	3	2	1
Patient 7.	4	3	1

The application shows which finger is presently being measured and the force value to be achieved. During the entire measurement, the patient is guided by a virtual nurse that is giving the subject instructions on what to do. When the subject reaches the desired value, the virtual nurse prompts the subject to maintain the force for a given time. If the measurement is valid, the application continues and the measurement is performed with the next finger. If not, the measurement is repeated. The measurement protocol can be modified by changing the required sequence of fingers, time, and force to be maintained. However, we used the same protocol parameters for all subjects.

- The target value—set to 300 g—is the force produced by pressing on the strain gauge. All PD subjects could exert a finger force of 450 g on average. The experimental target value was set to 2/3 of 450 g—300 g for all fingers.
- The hold time—set to 3 s—is the period during which a patient should keep the force around the target value. This value was selected based on two aspects: we assumed that a longer exercise than 3 s for all fingers would lead to fatigue and that a shorter exercise would not contain enough data to evaluate the tremor.
- The sequence of fingers represents the sequence of fingers without repetition from the left little finger to the right little finger. In these experiments, we chose the most simple sequence to iterate through the fingers from left to right to make the exercise as simple as possible.

The example in Figure 5 shows the measurement of the subject patient's left hand with the middle finger. The subject does not see the target value.

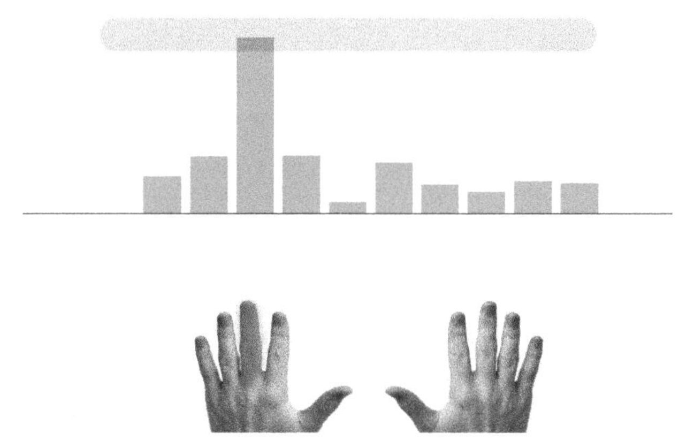

Figure 5. Example of finger force measurement. The required value is in the red area. The measured finger is highlighted and corresponds to the third vertical green bar from the left, and the y-axis represents the applied force.

5.3. *Verification of Hypothesis 3*

To determine whether the Rehapiano is sensitive enough to quantify the frequency of tremor on both hands of a patient, we measured a PD patient with both sides of the body affected. The whole exercise was performed under the supervision of a physician. For health reasons, the subject was not under the influence of medication. The subject also suffered from pain in the right shoulder and lowered fine motor skills in the right arm, as compared to the left arm. We repeated the aforementioned measurement protocol seven times for both hands with the hold time extended from 3 to 5 s for each finger, with the target value and sequence of fingers remaining the same. We acquired 28 measurements for each hand, excluding the thumbs. Based on the methodology presented in [9], we evaluated the validity of the measurements based on the peak power proportion. The ratio of the area below the power spectral density (PSD) curve around the detected dominant frequency (± 0.3 Hz) and the whole area below the curve had to be higher than a set threshold to consider the measurement valid; see Figure 6, Equation (2).

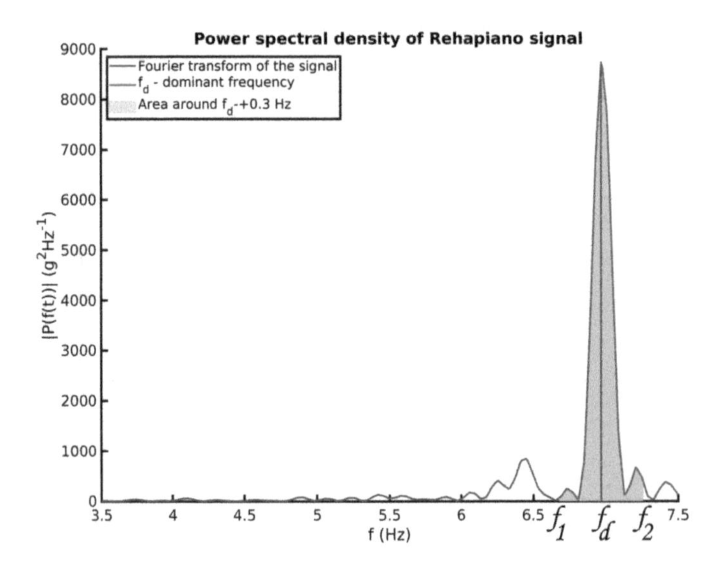

Figure 6. Power spectral density (PSD) of a PD patient measurement of a single finger. f_d represents the dominant frequency of the signal.

Equation (1) shows how the peak power from Figure 6 is calculated:

$$P_{\text{peak}} = \int_{f_1}^{f_2} \frac{FT^*(\text{ signals }) \times \text{FT (signals)}}{N^2} df, \tag{1}$$

where FT* is the FT conjugate, and N is the number of received samples of the signal.

The peak power proportion V_f is calculated as

$$V_f = \frac{P_{peak}}{\sum P_i}, \tag{2}$$

where P_i is the power estimation of the specific frequency, and $\sum P_i$ is the sum over all powers in a frequency domain.

6. Experimental Results

6.1. Validation of the Rehapiano

Hypothesis:

- $H_0 : \mu_{oe} = \mu_r$: Mean frequency measured by the optical encoder and by the Rehapiano is equal.
- $H_a : \mu_{oe} \neq \mu_r$: Mean frequency measured by optical encoder and by the Rehapiano is not equal.

We simulated tremors at three frequencies: low (≈1.5 Hz), medium (≈3 Hz), and high (≈7 Hz). For each frequency, we conducted 40 measurements, each 3 s long. Both the optical encoder data and strain gauge signal were recorded, and we calculated the frequency from optical encoder measurements for each 10 s measurement. We then used Fourier transform on the signal recorded by the Rehapiano. The resulting frequency is the frequency at maximal value of the single side power spectrum. The results of these measurements for high frequency are shown in Figure 7.

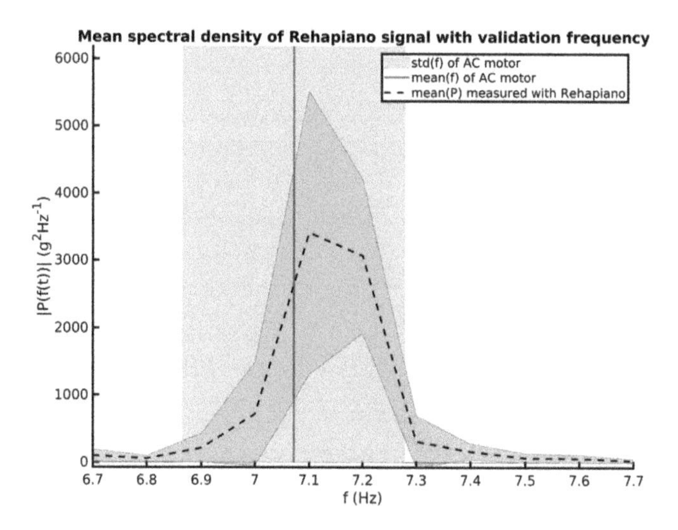

Figure 7. Mean frequency measured with optical encoder (red line), its standard deviation (green area), the Rehapiano mean power spectral density (black line), and its standard deviation (gray area) for high frequency experiments.

Subsequently, we calculated the mean and the variance of the measurements and from these values we computed the z-test. The z-test verifies our hypothesis that the measurement means of the Rehapiano and the optical encoder are the same at a significance level $\alpha = 0.05$. Table 2 shows the results of the hypothesis test for three frequencies. We cannot reject the null hypothesis for medium and high frequency at the 5% significance level. Therefore, we accept the null hypothesis for medium and high frequency, that $\mu_{oe} = \mu_r$. However, we reject the null hypothesis for low frequency $\mu_{oe} \neq \mu_r$.

Table 2. Mean frequencies measured by optical encoders and the Rehapiano and their variance, the result of the z-test, and the decision as to whether we can reject the null hypothesis.

Freq	$\mu_{oe}(Hz)$	σ_c	$\mu_r(Hz)$	s_r	z_{test}	H_0
Low	1.56	0.17	1.42	0.053	−2.51	1
Medium	2.96	0.09	2.94	0.021	−0.97	0
High	7.07	0.2	7.16	0.028	1.28	0

Because we did not succeed in verifying the Rehapiano for measuring the tremor frequencies lower than 3 Hz, in the following experiments, we considered only measurements of tremor frequencies between 3.5 and 7.5 Hz [9]. In this range, we considered the Rehapiano measurements to be valid for the purpose of detecting and evaluating PT and considered Hypothesis 1 to be proven.

We are confident that the Rehapiano produces valid measurements for lower frequencies and that the failure to verify it lies with the verification device we have used. More on this topic is in Section 7.

6.2. Distinction between Healthy Population and Patients with PT

Hypothesis:

- Measurements from the Rehapiano contain detectable tremor information. The performance metrics of a classifier that detects tremor should meet the following requirements.
  - Cross validation accuracy > 90%
  - Precision > 95%
  - Recall > 95%

We collected 490 measurements of 43 subjects from 49 sessions. Of those 43 subjects, 36 were healthy and seven were PD patients, with tremor rated on the FTMTRS (Table 1). All the healthy subjects completed the session once, producing 10 measurements each. Two PD patients repeated the exercise more than once (Table 1). Ninety-eight thumb measurements were uniformly excluded from the dataset, because of the PD patients' inability to maintain pressure on thumbs. We created the dataset with the following pipeline.

1. First, our algorithm filters the raw signal with an outlier filter that replaces values below the 1.25th percentile of the distribution using linear interpolation.
2. It then applies a band-pass filter that keeps frequencies between 3.5 Hz and 7.5 Hz.
3. Next, the algorithm calculates a one-sided amplitude spectrum of a 3 s signal, where the patient reached the desired force (Figure 8). After that, it resamples the result of the FT at 0.01 Hz between 3.5 and 7.5 Hz, creating a vector with 41 values describing the FT amplitude of the signal at specific frequencies.
4. Finally, we expect that FT amplitudes of the PD patients will be significantly different from the healthy population, and the data are labeled based on this assumption (PD patient: 1; Healthy Subject: 0).

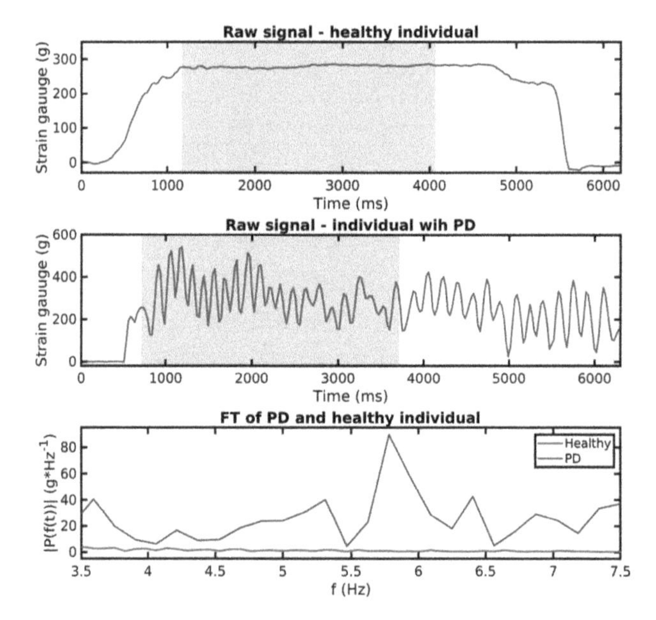

Figure 8. Comparison of healthy individual exercise and exercise for a patient with PD. Resampled values of FT transform (3rd) are used as input to our classifiers.

We trained four binary classifiers—support vector machines (SVM), naive Bayes (NB), decision tree (DT), and K-nearest neighbors (KNN). The target class of the classification was patients with PD. Classifiers had the following parameters: SVM: Gaussian kernel with $\gamma = 6.4$; NB: nonparametric Gaussian NB; DT: maximal number of nodes was set to 100, and the optimal feature for current node was selected by Gini's diversity index; KNN: $k = 13$, and the distance metric was cosine similarity. Input features of KNN and SVM were standardized. The mean of the feature was subtracted from every value in the column, and this value was divided by its standard deviation. We validated the classifiers with fivefold cross-validation; the cross-validation accuracy is shown in Table 3.

Table 3. Performance measures of different classifiers used to distinguish the healthy population from PD patients.

Class.	ValAccuracy	Sensitivity	Specificity	Precision	F_1
SVM	0.9311	0.798	0.9965	0.9965	0.8863
NB	0.9464	0.875	0.9722	0.9557	0.9136
DT	**0.9638**	**0.975**	**0.9861**	**0.9872**	**0.9811**
KNN	0.9285	0.9326	0.9756	0.9872	0.9537

Table 4 includes five metrics. Validation accuracy represents the average accuracy of classification from every K-fold classifier. Sensitivity shows the ability of the classifier to detect PD patients that truly have PT, whereas specificity is the ability to identify healthy people that do not have PT. Precision is the probability of making correct decision, when our classifier categorizes the measurement as a patient with PT. Subsequently, the F_1-score is the weighted average precision and sensitivity, and the formula is $F_1 = 2 * sensitivity * precision / (sensitivity + precision)$. All classifiers reach a validation accuracy higher than 90%. Although only the DT reaches sensitivity values higher than 95%, all four classifiers have a specificity higher than 95%. We highlight the DT that has the best F1 score.

Table 4. Comparison of different classifiers approaches on PT (LSTM: long short-term memory; GTB: gradient tree boosting; BCT: bagged classification tree; RF: random forest).

Name	Device	Scale	Method	Type	Accuracy	Sensitivity	Specificity
Our	Rehapiano	FTMTRS	DT	Binary	0.9638	0.975	0.9861
[41]	Gyroscope + Acc	UPDRS	LSTM/GTB	Multi	0.84/0.96*	-	-
[42]	Leap Motion	UPDRS	BCT	Binary	0.99	0.99	0.99
[43]	Smartphone	UPDRS	RF	Binary	-	0.90	0.82
[44]	Accelerometers	Binary	Welch (2)	Binary	0.95	0.98	0.69

* Correlation to UPDRS.

6.3. *Quantitative Assessment of the Tremor*

Hypothesis:

- Measurements from the Rehapiano provide quantitative information about tremor. Subsequent measurements of the same subject output the same tremor frequency.
  - Standard deviation of the measurements is less than 0.15 Hz.

We collected 48 measurements from one subject with PD. We repeated the first two steps from the data processing pipeline as described above. Subsequently, we applied Fourier transform on the signal and calculated the dominant frequency and peak power proportion from the power spectral density.

Table 5 contains the results of repeated measurements of a PD subject. Measurements are valid if the value of peak power proportion (Equation (2)) of the measurement is higher than the peak power proportion threshold. The peak power proportion threshold is a value above which we do not include measurement into the calculation of tremor frequency. We experimented with five peak power proportion thresholds from 0.5 to 0.9. The table is split into measurements of the right and the left hand. For each hand at a specific threshold, we provide the following information; the relative number of valid measurements, the absolute number of valid measurements, and the dominant frequency and its standard deviation. Absolute and relative number (absolute /total) of valid measurements both describe the amount of successful trials. The standard deviation of the measured PD tremor frequency for the right hand increased with the power peak ratio threshold, and we could obtain only two valid measurements at the threshold equal to 0.7. Because this was too small a statistical sample, we have not considered these measurements. We obtained nine valid measurements for the left hand even at the increased threshold equal to 0.8. The results indicate that the Rehapiano can be used to perform repeatable measurements. The mean of the calculated PD tremor frequency was 6.93 Hz with a standard deviation below 0.1 Hz.

Table 5. Results of repeated measurements with the Rehapiano of a PD subject.

	Left Hand Measurements			**Right Hand Measurements**		
V_f threshold	**Valid rel**	**Valid abs**	**$f \pm std(Hz)$**	**Valid rel**	**Valid abs**	**$f \pm std(Hz)$**
0.5	0.8571	24	6.7954 ± 0.4488	0.3928	11	6.27 ± 0.3193
0.6	0.6071	17	6.8688 ± 0.1471	0.2857	8	6.255 ± 0.3792
0.7	0.4285	12	$\mathbf{6.8925 \pm 0.126}$	0.0714	2	5.915 ± 0.7566
0.8	0.3214	9	$\mathbf{6.9344 \pm 0.0948}$	0	0	-
0.9	0	0	-	0	0	-

7. Discussion

We developed the Rehapiano device for the quick evaluation of action tremor. Tremor is measured during a steady state, when fingers are exerting the desired force. From a clinical viewpoint, action tremor, specifically kinetic (see Figure 1), appears during the targeted motion. Such movement is represented many times during everyday activities. Therefore, we decided to develop a device that measures this tremor, and based on the results of a particular activity, a physician can observe the development of PT.

Concerning Hypothesis 1, we rejected the null hypothesis based on the low frequency hypothesis test. Based on our observations, we still have confidence that the Rehapiano is capable of reliably measuring tremors of lower frequencies and that the problem of verifying it lies in the DC motor and the rubber band used for its verification. The difference between the measurements from the encoder and the Rehapiano may be due to the low torque of the DC motor at low voltage. The rubber band slows down the motor too much when being stretched and then, when being contracted, causes a peak in its angular velocity. This may distort the measurement. The authors will build a new verification device and repeat the procedure in the near future. For medium- and high-frequency tremors, we have confirmed Hypothesis 1.

Concerning Hypothesis 2, two factors influence the quality of the measurements and the resulting accuracy of the classifiers. The first factor is the subject's stress. According to the work in [16,45], the amplitude and frequency of tremor under psychological stress conditions increase significantly compared to a calm state. The second factor is the medication the subject receives. Based on the subjective opinions expressed by the subjects, they did not feel stressed during the measurement. All PD-affected subjects (except for the one measured in relation to Hypothesis 3) were taking medication to relieve PD symptoms. All the subjects completed the measurement protocol successfully. Two subjects had to repeat the measurement routine for one of their fingers due to wrong placement of the fingertip on the strain gauge. Based on the experimental results, we consider Hypothesis 2 to be confirmed.

As for Hypothesis 3, we subsequently measured both hands of a PD patient seven times. PT most often presents unilaterally and later progresses to include both sides of the body [46]. Both upper limbs of our subject were affected. Although the measurements of the left hand were valid, we had to ignore the measurements of the right hand. The patient was unable to keep the fingertips of the right hand in steady contact with the sensor, tapping the strain gauge more or less. From our perspective, this involuntary motion invalidated most of the right hand measurements. We discovered that, on multiple occasions, the signal contained several dominant frequencies. Therefore, we are currently working on several design and technological improvements to our Rehapiano device, primarily to improve the contact between the finger and strain gauge using neodymium magnets and gloves. Based on the experimental results of the PD subject's left hand measurements, we consider Hypothesis 3 to be confirmed.

8. Conclusions

This paper presents the Rehapiano system for measuring force applied by the fingertips, used here to detect and quantify hand tremors. We focused on measuring the action hand tremor of healthy subjects and PD patients in this study. We can confirm that the Rehapiano is capable of measuring tremors with frequencies higher than 3 Hz, that the Rehapiano measurements contain detectable and useful tremor information, and that the Rehapiano produces repeatable results. We have implemented a measurement protocol aided by a virtual nurse. Our device is noninvasive and non-wearable. It differs from the existing solutions due to its low measurement time, its comfort level, and the rapid processing and evaluation of the results. The average time of the measurement procedure was three minutes, and the subjects were not fatigued. We will be replacing the Hx711 converters to increase the sampling frequency, thus obtaining finer and higher quality data, which will also enable measurement of orthostatic tremor (between 14 Hz and 16 Hz).

We plan to expand the sample of the test subjects and to create measurement protocols for patients of specific groups. Future work also includes using the Rehapiano as a hand fine motor rehabilitation device. We are developing a rehabilitation computer game with the Rehapiano as the game controller. The game should motivate patients with motor disorders to exercise and improve their fine motor skills. Based on the field work and cooperation with the medical team, we learned that no PD patient barring one had problems in using the Rehapiano. The strain gauges' pods are equipped with neodymium magnets, although we have not yet used this feature. We plan to use fitting gloves with magnetic inserts to help keep the fingertips in place during the interaction. We have also learned that quantifying the tremor of PD patients at regular intervals could aid in determining the proper medication for them and in adjusting the dosage.

After carrying out all the planned improvements to the Rehapiano, we plan to use it to detect other types of tremor in clinical settings. Further, we want to compare Rehapiano measurements with reference measurements from IMUs and EMG. These experiments should provide conclusive evidence about the clinical feasibility of the Rehapiano.

Author Contributions: Conceptualization, N.F., M.J., M.B., and F.C.; Rehapiano hardware, N.F.; methodology, M.B., M.J., and F.C.; N.F. and M.J. proposed and described the algorithm and the methodology; software, M.J. and N.F.; protocol, N.F. and M.J.; investigation, M.B. and N.F.; N.F. designed and carried out the experiments with the patients; data processing, M.J.; writing—original draft preparation, N.F., M.J., M.B., and F.C.; editing, M.B. and F.C. All authors have read and agreed to the published version of the manuscript.

Funding: This publication is the result of the APVV grant Multimodal Human–Robot Interaction Using Cloud Resources (APVV-15-0731) (50%), H2020-MSCA-RISE-2018,824047—LIFEBOTS Exchange (25%), and VEGA, 2017-2020, 1/0663/17 Intelligent Cyber-Physical Systems in Heterogeneous Environments with the Support of IoE and Cloud Services (25%).

Acknowledgments: The authors gratefully acknowledge all the volunteers who participated in the trial. The study was performed in cooperation with the University Hospital of L. Pasteur in Košice, Svet Zdravia Hospital in Trebišov, and Retirement Home Trebišov, Slovakia. We thank Theodoz Molcanyi, Miriam Dziakova, Peter Mucha, Lukas Zbojovsky, Frantisek Mihalcik, and Nikola Harmadiova for their help and advice.

Conflicts of Interest: The authors declare no conflicts of interest.

References

1. Dorsey, E.R.; Elbaz, A.; Nichols, E.; Abd-Allah, F.; Abdelalim, A.; Adsuar, J.C.; Ansha, M.G.; Brayne, C.; Choi, J.Y.J.; Collado-Mateo, D. Global, regional, and national burden of Parkinson's disease, 1990–2016: A systematic analysis for the Global Burden of Disease Study 2016. *Lancet Neurol.* **2018**, *17*, 939–953. [CrossRef]
2. Moisan, F.; Kab, S.; Mohamed, F.; Canonico, M.; Le Guern, M.; Quintin, C.; Carcaillon, L.; Nicolau, J.; Duport, N.; Singh-Manoux, A.; et al. Parkinson disease male-to-female ratios increase with age: French nationwide study and meta-analysis. *J. Neurol. Neurosurg. Psychiatry* **2016**, *87*, 952–957. [CrossRef] [PubMed]
3. Politis, M.; Wu, K.; Molloy, S.; Bain, P.G.; Chaudhuri, K.R.; Piccini, P. Parkinson's disease symptoms: The patient's perspective. *Mov. Dis.* **2010**, *25*, 1646–1651. [CrossRef] [PubMed]

4. Chaudhuri, K.R.; Healy, D.G.; Schapira, A.H. Non-motor symptoms of Parkinson's disease: Diagnosis and management. *Lancet Neurol.* **2006**, *5*, 235–245. [CrossRef]
5. Abdo, W.F.; Van De Warrenburg, B.P.; Burn, D.J.; Quinn, N.P.; Bloem, B.R. The clinical approach to movement disorders. *Nat. Rev. Neurol.* **2010**, *6*, 29. [CrossRef]
6. Bendat, J.S.; Piersol, A.G. *Random Data: Analysis and Measurement Procedures*; John Wiley & Sons: Hoboken, NJ, USA, 2011; Volume 729.
7. Haubenberger, D.; Abbruzzese, G.; Bain, P.G.; Bajaj, N.; Benito-León, J.; Bhatia, K.P.; Deuschl, G.; Forjaz, M.J.; Hallett, M.; Louis, E.D. Transducer-based evaluation of tremor. *Mov. Dis.* **2016**, *31*, 1327–1336. [CrossRef]
8. Rovini, E.; Maremmani, C.; Cavallo, F. How wearable sensors can support Parkinson's disease diagnosis and treatment: A systematic review. *Front. Neurosci.* **2017**, *11*, 555. [CrossRef]
9. Dai, H.; Zhang, P.; Lueth, T. Quantitative assessment of parkinsonian tremor based on an inertial measurement unit. *Sensors* **2015**, *15*, 25055–25071. [CrossRef]
10. Samuel, M.; Torun, N.; Tuite, P.J.; Sharpe, J.A.; Lang, A.E. Progressive ataxia and palatal tremor (PAPT) Clinical and MRI assessment with review of palatal tremors. *Brain* **2004**, *127*, 1252–1268. [CrossRef]
11. Ahlskog, J.E. Slowing Parkinson's disease progression: Recent dopamine agonist trials. *Neurology* **2003**, *60*, 381–389. [CrossRef]
12. Sanger, T.D.; Chen, D.; Fehlings, D.L.; Hallett, M.; Lang, A.E.; Mink, J.W.; Singer, H.S.; Alter, K.; Ben-Pazi, H.; Butler, E.E. Definition and classification of hyperkinetic movements in childhood. *Mov. Dis.* **2010**, *25*, 1538–1549. [CrossRef]
13. Brittain, J.S.; Probert-Smith, P.; Aziz, T.Z.; Brown, P. Tremor suppression by rhythmic transcranial current stimulation. *Curr. Biol.* **2013**, *23*, 436–440. [CrossRef]
14. Choi, S.M. Movement disorders following cerebrovascular lesions in cerebellar circuits. *J. Mov. Dis.* **2016**, *9*, 80. [CrossRef] [PubMed]
15. Bötzel, K.; Tronnier, V.; Gasser, T. The differential diagnosis and treatment of tremor. *Deutsch. Ärztebl. Int.* **2014**, *111*, 225. [CrossRef] [PubMed]
16. Deuschl, G.; Bain, P.; Brin, M.; Committee, A.H.S. Consensus statement of the movement disorder society on tremor. *Mov. Dis.* **1998**, *13*, 2–23. [CrossRef] [PubMed]
17. Elble, R.J. Tremor. In *Neuro-Geriatrics*; Springer: Cham, Switzerland, 2017; pp. 311–326.
18. Zhang, J.; Xing, Y.; Ma, X.; Feng, L. Differential diagnosis of Parkinson disease, essential tremor, and enhanced physiological tremor with the tremor analysis of EMG. *Parkinson's Dis.* **2017**, *2017*, 1597907. [CrossRef]
19. Chen, W.; Hopfner, F.; Becktepe, J.S.; Deuschl, G. Rest tremor revisited: Parkinson's disease and other disorders. *Transl. Neurodegener.* **2017**, *6*, 16. [CrossRef]
20. Koller, W.C.; Vetere-Overfield, B.; Barter, R. Tremors in early Parkinson's disease. *Clin. Neuropharmacol.* **1989**, *12*, 293–297. [CrossRef]
21. Lance, J.W.; Schwab, R.S.; Peterson, E.A. Action tremor and the cogwheel phenomenon in Parkinson's disease. *Brain* **1963**, *86*, 95–110. [CrossRef]
22. Deuschl, G.; Krack, P.; Lauk, M.; Timmer, J. Clinical neurophysiology of tremor. *J. Clin. Neurophysiol.* **1996**, *13*, 110–121. [CrossRef]
23. Lee, H.J.; Lee, W.W.; Kim, S.K.; Park, H.; Jeon, H.S.; Kim, H.B.; Jeon, B.S.; Park, K.S. Tremor frequency characteristics in Parkinson's disease under resting-state and stress-state conditions. *J. Neurol. Sci.* **2016**, *362*, 272–277. [CrossRef] [PubMed]
24. Perumal, S.V.; Sankar, R. Gait and tremor assessment for patients with Parkinson's disease using wearable sensors. *ICT Express* **2016**, *2*, 168–174. [CrossRef]
25. Bhidayasiri, R. Differential diagnosis of common tremor syndromes. *Postgrad. Med. J.* **2005**, *81*, 756–762. [CrossRef] [PubMed]
26. Zach, H.; Dirkx, M.; Bloem, B.R.; Helmich, R.C. The clinical evaluation of Parkinson's tremor. *J. Parkinson's Dis.* **2015**, *5*, 471–474. [CrossRef]
27. Benito-León, J.; Domingo-Santos, Á. Orthostatic tremor: An update on a rare entity. *Tremor Other Hyperkinet. Mov.* **2016**, *6*, 411.
28. Sharott, A.; Marsden, J.; Brown, P. Primary orthostatic tremor is an exaggeration of a physiological response to instability. *Mov. Dis. Off. J. Mov. Dis. Soc.* **2003**, *18*, 195–199. [CrossRef]

29. Thenganatt, M.A.; Jankovic, J. Psychogenic tremor: A video guide to its distinguishing features. *Tremor Other Hyperkinet. Mov.* **2014**, *4*, 253.
30. Elble, R.J.; McNames, J. Using portable transducers to measure tremor severity. *Tremor Other Hyperkinet. Mov.* **2016**, *6*, 375.
31. Hurtado, J.M.; Gray, C.M.; Tamas, L.B.; Sigvardt, K.A. Dynamics of tremor-related oscillations in the human globus pallidus: a single case study. *Proc. Natl. Acad. Sci. USA* **1999**, *96*, 1674–1679. [CrossRef]
32. Elble, R.J.; Pullman, S.L.; Matsumoto, J.Y.; Raethjen, J.; Deuschl, G.; Tintner, R. Tremor amplitude is logarithmically related to 4-and 5-point tremor rating scales. *Brain* **2006**, *129*, 2660–2666. [CrossRef]
33. Niazmand, K.; Tonn, K.; Kalaras, A.; Fietzek, U.M.; Mehrkens, J.H.; Lueth, T.C. Quantitative evaluation of Parkinson's disease using sensor based smart glove. In Proceedings of the 2011 24th International Symposium on Computer-Based Medical Systems (CBMS), Bristol, UK, 27–30 June 2011; pp. 1–8.
34. Khan, F.M.; Barnathan, M.; Montgomery, M.; Myers, S.; Côté, L.; Loftus, S. A wearable accelerometer system for unobtrusive monitoring of parkinson's diease motor symptoms. In Proceedings of the 2014 IEEE International Conference on Bioinformatics and Bioengineering, Boca Raton, FL, USA, 10–12 November 2014; pp. 120–125.
35. Bhavana, C.; Gopal, J.; Raghavendra, P.; Vanitha, K.; Talasila, V. Techniques of measurement for Parkinson's tremor highlighting advantages of embedded IMU over EMG. In Proceedings of the 2016 International Conference on Recent Trends in Information Technology (ICRTIT), Chennai, India, 8–9 April 2016; pp. 1–5.
36. AVIA Semiconductors. 24-Bit Analog-to-Digital Converter (ADC) for Weigh Scales. 2006. Available online: https://www.mouser.com/datasheet/2/813/hx711_english-1022875.pdf (accessed on 23 January 2020).
37. Heida, T.; Wentink, E.C.; Marani, E. Power spectral density analysis of physiological, rest and action tremor in Parkinson's disease patients treated with deep brain stimulation. *J. NeuroEng. Rehabil.* **2013**, *10*, 70. [CrossRef]
38. Kumar, H.; Jog, M. A patient with tremor, part 1: Making the diagnosis. *CMAJ* **2011**, *183*, 1507–1510. [CrossRef]
39. Stacy, M.A.; Elble, R.J.; Ondo, W.G.; Wu, S.C.; Hulihan, J.; Group, T.S. Assessment of interrater and intrarater reliability of the Fahn–Tolosa–Marin Tremor Rating Scale in essential tremor. *Mov. Dis.* **2007**, *22*, 833–838. [CrossRef] [PubMed]
40. Western, D.G.; Neild, S.A.; Jones, R.; Davies-Smith, A. Personalised profiling to identify clinically relevant changes in tremor due to multiple sclerosis. *BMC Med. Inform. Decis. Mak.* **2019**, *19*, 1–18. [CrossRef] [PubMed]
41. Hssayeni, M.D.; Jimenez-Shahed, J.; Burack, M.A.; Ghoraani, B. Wearable Sensors for Estimation of Parkinsonian Tremor Severity during Free Body Movements. *Sensors* **2019**, *19*, 4215. [CrossRef] [PubMed]
42. Vivar-Estudillo, G.; Ibarra-Manzano, M.A.; Almanza-Ojeda, D.L. Tremor Signal Analysis for Parkinson's Disease Detection Using Leap Motion Device. In *Mexican International Conference on Artificial Intelligence*; Springer: Cham, Switzerland, 2018; pp. 342–353.
43. Kostikis, N.; Hristu-Varsakelis, D.; Arnaoutoglou, M.; Kotsavasiloglou, C. A smartphone-based tool for assessing parkinsonian hand tremor. *IEEE J. Biomed. Health Inform.* **2015**, *19*, 1835–1842. [CrossRef]
44. Manzanera, O.M.; Elting, J.W.; van der Hoeven, J.H.; Maurits, N.M. Tremor detection using parametric and non-parametric spectral estimation methods: A comparison with clinical assessment. *PLoS ONE* **2016**, *11*, e0156822. [CrossRef]
45. Raethjen, J.; Austermann, K.; Witt, K.; Zeuner, K.E.; Papengut, F.; Deuschl, G. Provocation of Parkinsonian tremor. *Mov. Dis. Off. J. Mov. Dis. Soc.* **2008**, *23*, 1019–1023. [CrossRef]
46. Jankovic, J. Distinguishing essential tremor from Parkinson's disease. *Pract. Neurol.* **2012**, 36–38.

 sensors

Article

Sleep in the Natural Environment: A Pilot Study

Fayzan F. Chaudhry [1,2,†], Matteo Danieletto [1,2,3,†], Eddye Golden [1,2,3], Jerome Scelza [2,3], Greg Botwin [2,3], Mark Shervey [2,3], Jessica K. De Freitas [1,2,3], Ishan Paranjpe [1], Girish N. Nadkarni [1,4,5], Riccardo Miotto [1,2,3], Patricia Glowe [1,2,3], Greg Stock [3], Bethany Percha [2,3], Noah Zimmerman [2,3], Joel T. Dudley [2,3,*] and Benjamin S. Glicksberg [1,2,3,*]

1 Hasso Plattner Institute for Digital Health at Mount Sinai, Icahn School of Medicine at Mount Sinai, New York, NY 10032, USA; fayzan.chaudhry@mssm.edu (F.F.C.); matteo.danieletto@mssm.edu (M.D.); eddye.golden@mssm.edu (E.G.); jessica.defreitas@icahn.mssm.edu (J.K.D.F.); ishan.paranjpe@icahn.mssm.edu (I.P.); girish.nadkarni@mountsinai.org (G.N.N.); riccardo.miotto@mssm.edu (R.M.); Patricia.Glowe@mssm.edu (P.G.)
2 Department of Genetics and Genomic Sciences, Icahn School of Medicine at Mount Sinai, New York, NY 10032, USA; scelzajr@gmail.com (J.S.); gbotwin@gmail.com (G.B.); markshervey@gmail.com (M.S.); bethany.percha@mssm.edu (B.P.); Noah.zimmerman@mssm.edu (N.Z.)
3 Institute for Next Generation Healthcare, Icahn School of Medicine at Mount Sinai, New York, NY 10032, USA; info@gregorystock.net
4 The Charles Bronfman Institute for Personalized Medicine, Icahn School of Medicine at Mount Sinai, New York, NY 10032, USA
5 Department of Medicine, Icahn School of Medicine at Mount Sinai, New York 10032, USA
* Correspondence: joel.dudley@mssm.edu (J.T.D.); benjamin.glicksberg@mssm.edu (B.S.G.)
† Authors contributed equally.

Received: 2 February 2020; Accepted: 29 February 2020; Published: 3 March 2020

Abstract: Sleep quality has been directly linked to cognitive function, quality of life, and a variety of serious diseases across many clinical domains. Standard methods for assessing sleep involve overnight studies in hospital settings, which are uncomfortable, expensive, not representative of real sleep, and difficult to conduct on a large scale. Recently, numerous commercial digital devices have been developed that record physiological data, such as movement, heart rate, and respiratory rate, which can act as a proxy for sleep quality in lieu of standard electroencephalogram recording equipment. The sleep-related output metrics from these devices include sleep staging and total sleep duration and are derived via proprietary algorithms that utilize a variety of these physiological recordings. Each device company makes different claims of accuracy and measures different features of sleep quality, and it is still unknown how well these devices correlate with one another and perform in a research setting. In this pilot study of 21 participants, we investigated whether sleep metric outputs from self-reported sleep metrics (SRSMs) and four sensors, specifically Fitbit Surge (a smart watch), Withings Aura (a sensor pad that is placed under a mattress), Hexoskin (a smart shirt), and Oura Ring (a smart ring), were related to known cognitive and psychological metrics, including the n-back test and Pittsburgh Sleep Quality Index (PSQI). We analyzed correlation between multiple device-related sleep metrics. Furthermore, we investigated relationships between these sleep metrics and cognitive scores across different timepoints and SRSM through univariate linear regressions. We found that correlations for sleep metrics between the devices across the sleep cycle were almost uniformly low, but still significant ($p < 0.05$). For cognitive scores, we found the Withings latency was statistically significant for afternoon and evening timepoints at $p = 0.016$ and $p = 0.013$. We did not find any significant associations between SRSMs and PSQI or cognitive scores. Additionally, Oura Ring's total sleep duration and efficiency in relation to the PSQI measure was statistically significant at $p = 0.004$ and $p = 0.033$, respectively. These findings can hopefully be used to guide future sensor-based sleep research.

Keywords: wearables; biosensors; sleep; Fitbit; Oura; Hexoskin; Withings; cognition

1. Introduction

Between 50 and 70 million Americans currently suffer from poor sleep [1]. A 2014 study from the Centers for Disease Control and Prevention found that over one third of Americans (34.8%) regularly sleep less than the recommended 7 hours per night [2]. Although the body has remarkable compensatory mechanisms for acute sleep deprivation, chronic poor sleep quality and suboptimal sleep duration are linked to many adverse health outcomes, including increased risk of diabetes [3], metabolic abnormalities [4], cardiovascular disease [5], hypertension [6], obesity [7], and anxiety and depression [8]. Chronic sleep deprivation also poses economical burdens to society, contributing to premature mortality, loss of working time, and suboptimal education outcomes that cost the US $280.6-411 billion annually [9]. However, the underlying mechanisms mediating the adverse effects of poor sleep remain unknown. Diverse factors and complex interactions govern the relationship between health and sleep, and there is likely substantial inter-individual variability. Pronounced gender [10], race [11], and ethnicity differences in sleep-related behaviors are well-established [2].

It is clear that broad, population-level studies of sleep are necessary to understand how lifestyle and environmental factors contribute to poor sleep and to link sleep abnormalities to their attendant negative health effects [12]. It is particularly important to capture individuals' sleep patterns in natural sleep settings (i.e., at home). However, traditional approaches to studying sleep do not permit these types of studies. Polysomnography (PSG), where brain waves, oxygen levels, and eye and leg movements are recorded, is the current "gold standard" approach to studying sleep. A PSG study typically requires the participant to sleep in a hospital or clinic setting with uncomfortable sensors placed on the scalp, face, and legs. These studies, which remove the participant from his/her natural sleep environment, are not well suited to longitudinal assessments of sleep. They also create issues such as the first night effect, which limit the translatability of laboratory sleep studies to real-life environments [13]. The recent development of clinical grade, at-home PSG tools has enabled quantification of the laboratory environment's effect on sleep [14]. Such studies have generally confirmed that participants sleep better at home than they do in a lab, although these findings are not universal [15].

Even with the availability of the at-home PSG, however, it is unlikely that the use of expensive, cumbersome, single-purpose equipment will promote the kinds of large-scale population studies that can quantify the diverse factors affecting sleep and its relationship to health outcomes. More user-friendly, lightweight, and unobtrusive sleep sensors are needed; ideally these would be embedded in devices that study participants already own. Recently, several companies have developed sub-clinical grade "wearable" technologies for the consumer market that passively collect high frequency data on physiological, environmental, activity, and sleep variables [16]. The Food and Drug Administration classifies these as general wellness products and they are not approved for clinical sleep studies. Due to their passivity, low risk, and growing ubiquity amongst consumers, it is clear that these devices present an intriguing new avenue for large-scale sleep data collection [17]. Combined with mobile application (app) software to monitor cognitive outcomes such as reaction time, executive function, and working memory, these devices could feasibly be used for large-scale, fully remote sleep studies.

This study aimed to determine the feasibility of monitoring sleep in a participant's natural environment with surveys completed electronically. Specifically, we performed a week-long pilot comparative study of four commercially available wearable technologies that have sleep monitoring capabilities. For the entire week, 21 participants were instrumented with all four devices, specifically Fitbit Surge (a smart watch), Withings Aura (a sensor pad that is placed under a mattress), Hexoskin (a smart shirt), and Oura Ring (a smart ring). To assess the feasibility of a fully remote study relating sleep features to cognition, we also assessed participants' daily cognitive function via a series of assessments on a custom-built mobile app. None of the four devices we compared in this study had been previously compared head-to-head for sleep and cognition research. Our results highlight some of the key difficulties involved in designing and executing large-scale sleep studies with consumer-grade wearable devices.

The rest of the paper is organized as follows. In Section 2, we describe the literature of related work including state-of-the-art research. In Section 3, we detail the materials and methods employed in this work, including the participant recruitment process, all metrics collected (e.g., device output), and the statistical tests performed. We detail the results from all assessments in Section 4. We discuss the implications of our work as well as limitations in Section 5 and finally conclude the paper in Section 6.

2. State-of-the-Art

This study built off of previous work that utilized comparisons of various devices and polysomnography [18,19]. For instance, de Zambotti et al. [19] directly compared the Oura ring with PSG. Correlation matrices from their study show poor agreement across different sleep stages, showing that tracking sleep stages was a problem for the Oura. However, this study concluded that the Oura's tracking of total sleep duration (TSD), sleep onset latency, and wake after sleep onset were not statistically different than that of PSG for these metrics. The Oura was found to track TSD in relative accordance with PSG in this regard. This suggests that many devices have trouble tracking TSD or participants had trouble wearing devices correctly outside of a monitored sleep lab.

The biggest question for these devices is, how well do they actually reflect sleep? The current consensus is mixed. For instance, de Zambotti et al. [20] found good overall agreement between PSG and Jawbone UP device, but there were over- and underestimations for certain sleep parameters such as sleep onset latency. Another study compared PSG to the Oura ring and found no differences in sleep onset latency, total sleep time, and wake after sleep onset, but the authors did find differences in sleep stage characterization between the two recording methods [19]. Meltzer et al. [21] concluded that the Fitbit Ultra did not produce clinically comparable results to PSG for certain sleep metrics. Montgomery-Downs et al. [22] found that Fitbit and actigraph monitoring consistently misidentified sleep and wake states compared to PSG, and they highlighted the challenge of using such devices for sleep research in different age groups. While such wearables offer huge promise for sleep research, there are a wide variety of additional challenges regarding their utility, including accuracy of sleep automation functions, detection range, and tracking reliability, among others [23]. Furthermore, comprehensive research including randomized control trials as well as interdisciplinary input from physicians and computer, behavioral, and data scientists will be required before these wearables can be ready for full clinical integration [24].

As there are many existing commercial devices, it is not only important to determine how accurate they are in capturing certain physiological parameters, but also the extent to which they are calibrated compared to one another. In this way, findings from studies that use different devices but measure similar outcomes can be compared in context. Murakami et al. [25] evaluated 12 devices for their ability to capture total energy expenditure against the gold standard and found that while most devices had strong correlation (greater than 0.8) compared to the gold standard, they did vary in their accuracy, with some significantly under- or overestimating energy expenditure. The authors suggested that most wearable devices do not produce a valid quantification of energy expenditure. Xie et al. [26] compared six devices and two smartphone apps regarding their ability to measure major health indicators (e.g., heart rate or number of steps) under various activity states (e.g., resting, running, and sleeping). They found that the devices had high measurement accuracy for all health indicators except energy consumption, but there was variation between devices, with certain ones performing better than others for specific indicators in different activity states. In terms of sleep, they found the overall accuracy for devices to be high in comparison to output from the Apple Watch 2, which was used as the gold standard. Lee et al. [27] performed a highly relevant study in which they examined the comparability of five devices total and a research-grade accelerometer to self-reported sleep regarding their ability to capture key sleep parameters such as total sleep time and time spent in bed, for one to three nights of sleep.

3. Materials and Methods

3.1. Research Setting

Participants were enrolled individually at the Harris Center for Precision Wellness (HC) and Institute for Next Generation Healthcare research offices within the Icahn School of Medicine at Mount Sinai. Monetary compensation in the form of a $100 gift card was provided to study participants upon device return. During the enrollment visit, participants met with an authorized study team member in a private office to complete the consent process, onboarding, and baseline procedures. The remainder of the study activities took place remotely with limited participant-team interaction. The study team maintained remote contact with each research participant throughout his/her participation via phone or email to answer any questions and provide technical support. The study was approved by the Mount Sinai Program for the Protection of Human Subjects (IRB #15-01012).

3.2. Recruitment Methods

To ensure a diverse population, the participants were recruited using a variety of methods, including flyers, institutional e-mails, social media, institution-affiliated websites, websites that help match studies with participants, and referrals.

3.3. Inclusion and Exclusion Criteria

Participants were eligible for the study if they were over 18 years old, had access to an iPhone, had basic knowledge of installing and using mobile applications and wearable devices, and were willing and able to provide written informed consent and participate in study procedures. Participants were ineligible for the study if they were colorblind, part of a vulnerable population, or unwilling to consent and participate in study activities.

3.4. Onboarding Questionnaires

During the initial study visit, participants were prompted to complete four questionnaires (see Supplemental S1–S4). All questionnaires were completed electronically via SurveyMonkey and the results were subsequently stored in the study team's encrypted and secured electronic database.

The Demographics Questionnaire (Supplemental S1) ascertained basic demographic information.

The 36-Item Short Form Health Survey (SF-36; Supplemental S2). The SF-36 evaluated eight domains: physical functioning, role limitations due to physical health, role limitations due to emotional problems, energy/fatigue, emotional well-being, social functioning, pain, and general health. The SF-36 takes roughly 5–10 min to complete.

The Morningness-Eveningness Questionnaire (MEQ; Supplemental S3) is a 19-question, multiple-choice instrument designed to detect when a person's circadian rhythm allows for peak alertness. The MEQ takes roughly 5–10 min to complete.

The Pittsburgh Sleep Quality Instrument (PSQI; Supplemental S4) is a nine-item, self-rated questionnaire that assesses sleep over the prior month. The PSQI has been shown to be sensitive and specific in distinguishing between good and poor sleepers. The PSQI utilizes higher numbers to indicate poorer sleep. The PSQI takes roughly 5–10 min to complete.

3.5. Technology Setup and Testing

After the initial screening visit, participants were asked to set up their devices and begin the week-long study at their leisure (Figure 1). The study team chose technologies based on performance and usability data obtained from HS#: 15-00292, "Pilot Evaluation Study on Emerging Wearable

Technologies." Each participant was assigned four sleep monitoring devices: a Fitbit Surge smart watch (Fitbit; first edition), a Hexoskin smart shirt (Hexoskin; male and female shirts and Classic device), a Withings Aura sleep pad/system (Withings; model number WAS01), and an Oura smart ring (Oura; first edition). Note that the form factors for the four devices were different; this was important to ensure that they could all be used at once and would not interfere with each other.

Setup for each device involved downloading the corresponding manufacturer's mobile application on the participant's iPhone and downloading the study team's custom HC App. Participants agreed to each manufacturer's software terms and conditions in the same manner as if they were to purchase and install the technologies themselves. In doing so, and as noted in the participant-signed consent document, participants acknowledged that the manufacturers would have access to identifiable information such as their names, email addresses, and locations. The HC App functioned as a portal to allow participants to authorize the sharing of data between the manufacturers' applications and the study team's database. During the initial setup period, the study team worked with participants to troubleshoot any issues and ensure proper data transmission to the database.

3.6. *Sleep Monitoring and Device-Specific Parameters*

Over a 7-day consecutive monitoring period of the participant's choosing, participants used the four different sleep monitoring technologies and completed daily assessments (Figure 1). The monitors measured physiological parameters (e.g., heart rate, heart rate variability, respiratory rate, temperature, and movement), activity parameters (e.g., number of steps per day), and sleep-related parameters, specifically time in each sleep stage, time in bed to fall asleep (latency), TSD, number of wakeups per night (wakeups), and standardized score of sleep quality (efficiency). The Withings and Oura both stage sleep as: (1) awake, (2) light, (3) deep, and (4) rapid eye movement (REM; Figure 1). The Hexoskin stages sleep as (1) awake, (2) non-REM (NREM), and (3) REM. The Fitbit stages sleep as (1) very awake, (2) awake, and (3) asleep.

3.7. *Daily Questionnaires and n-Back Tests*

Using the HC App, participants completed questionnaires and cognitive assessments on each day of the 7-day study. These included the n-back test and self-reported sleep metrics (SRSMs).

3.7.1. n-Back Tests

The n-back test [28] assesses working memory as well as higher cognitive functions/fluid intelligence. Participants were prompted to take the n-back test three times per day (morning, afternoon, and evening). In each test, participants were presented with a sequence of 20 trials, each of which consisted of a picture of one of eight stimuli: eye, bug, tree, car, bell, star, bed, anchor. The participant was asked whether the image was the same as the image n times back from the current image, where n = 1 or 2. The stimuli were chosen so that in the course of 20 trials, 10 would be congruent (the stimulus would match the n-back stimulus) and 10 would be incongruent. The participant had 500 milliseconds to enter a response. If no response was entered, the trial was counted as incorrect and a new trial was presented. The n-back tests took roughly 3 min each, for a total of under 10 min/day.

3.7.2. SRSMs

The participant was asked for an estimate of TSD, latency (i.e., time to fall asleep), and start to end sleep duration (i.e., TSD plus latency, referred to as Start-End). Participants self-reported these metrics electronically through the HC App at wakeup (1–2 min completion time).

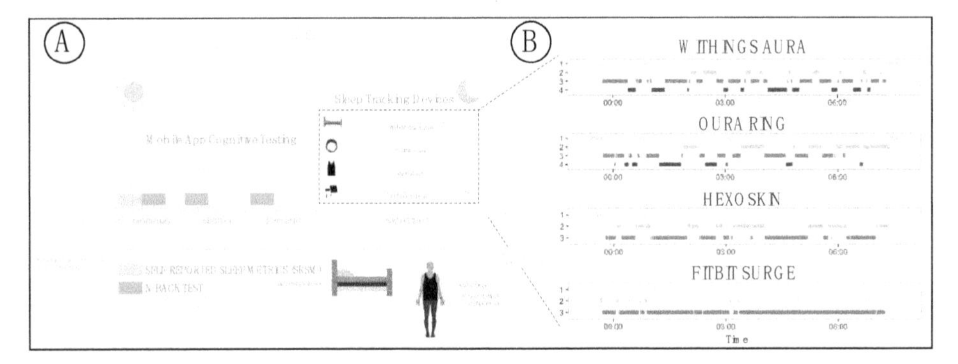

Figure 1. Study structure and data collection for our pilot sleep study. (**A**) Illustration of sleep study monitoring procedure and data collection strategies. (**B**) Example data showing a comparison of sleep staging of a single night for one study participant for all four devices.

3.8. n-Back Test Scoring

For each trial (i.e., each morning, afternoon, evening per study day), the participants' response time and the correctness/incorrectness of responses were recorded. We calculated four different scores for the n-back tests: median reaction time and percent correct, stratified by congruent vs. incongruent items. We treated all reaction times the same and did not segment or weigh based on items that the participants got correct vs. incorrect. Each participant was then given a cognitive score based on a self-created scoring function (Equation (1)) of the reaction time, degree of difficulty of question, and correctness. The metric accounts for variation across multiple elements of the n-back results leading to a greater representation of performance. The formula for the metric is

$$\frac{\sum\left(1-\frac{Reaction\ Time}{Max\ Reaction\ Time}\right)*Answer\ Correct*\frac{Steps\ Back}{2}}{n} \quad (1)$$

3.9. Inter-Device Comparisons for Sleep Staging and Metrics

We compared each pair of devices for overall correlation in sleep staging across all nights on a per-epoch basis. While the other three devices were used in this analysis, Fitbit was not included because it does not segment sleep by stages, rather measuring asleep vs. not asleep. Oura and Withings track four stages of sleep while the Hexoskin tracks three (see Section 3.6). Accordingly, the NREM sleep stages for Withings and Oura were combined into a single category (NREM) for this correlation analysis. After this transformation, these three devices had three stages of sleep used for this correlation analysis: (1) awake, (2) NREM, and (3) REM. We utilized Kendall's rank correlation for this analysis as sleep staging was ordinal. We performed Pearson correlation to compare the between-device correlation for specific device-produced sleep metrics, specifically TSD (all four devices) and REM (Oura, Hexoskin, and Withings), both in terms of total seconds. We also assessed the correlation of SRSMs, specifically TSD, to device-produced TSD (all four devices) across all nights per participant. We used Pearson's correlation for this analysis as density plots of these data did not reveal any outliers (Figure 2).

3.10. Statistical Models Linking Device Data to PSQI and n-Back Scores

We built a series of univariate linear models that regressed each individual sleep feature on either PSQI score or n-back score. The PSQI tracks quality of sleep, with higher values indicating poorer sleep. We performed a series of univariate linear regressions on the one-time reported PSQI against all available device and SRSMs (TSD and latency), taking the mean of each metric across all nights of sleep for each participant as a general representation of sleep quality. These device metrics include:

latency, TSD (in hours), wakeups (in number of events), efficiency, and REM (in hours). For these analyses, one participant was not included due to lack of data. Additionally, we used univariate linear regressions to compare n-back score against device and SRSM data. For each analysis, we regressed the n-back score of each timepoint (i.e., morning, afternoon, evening) against the mean of each device metric or SRSM feature by participant. In all of the regression models for the n-back scores, we analyzed only participants with two or more days of reported scores for each timepoint. This left us with 16, 19, and 18 participants out of the original 21 for morning, afternoon, and evening n-back tests, respectively.

3.11. Analysis of Missing Data

We analyzed the degree of missingness of each device-reported or self-reported field as measures of device reliability/quality or participant compliance, respectively. As the study progressed, some sleep features were also updated due to new advances in hardware and software on the device side, which resulted in missing data columns that were not included in the missing data plot.

4. Results

4.1. Summary of Study Population

Table 1 describes our study population, which consisted of 21 participants (11 female; 10 male). The median age of the cohort was 29 years (range: 23–41). The median PSQI score was 4 (range: 1–12). Sixteen of our participants were classified as normal sleepers, three were poor sleepers, and two were very poor sleepers. Median MEQ score was 52 (range: 35–73). We provide score summaries for all eight SF-36 subcategories at the bottom of Table 1. Additionally, the racial breakdown of the cohort was as follows: 17 Caucasian (white) and 4 Asian participants.

Table 1. Summary of the study population. The participant's gender (M/F/O), baseline assessment of sleep quality according to the Pittsburgh Sleep Quality Index (PSQI) (with higher values indicative of poorer sleep), age, SF-36 score (a measure of general health along eight axes), and MEQ time (optimal time of day) are included.

ID	Gender	Age	PSQI	MEQ	SF-36 Scores							
					Physical Functioning	Role Limitations (Physical)	Role Limitations (Emotional)	Energy	Emotional Well-Being	Social Functioning	Pain	General Health
1	F	23	1	50	100	100	100.0	50	68	87.5	100.0	55
2	F	26	4	47	90	100	66.7	45	72	100.0	100.0	60
3	F	27	5	52	100	100	100.0	45	56	87.5	90.0	50
4	F	27	2	36	100	100	100.0	65	80	75.0	100.0	55
5	F	27	4	58	100	100	100.0	50	76	87.5	90.0	55
6	F	28	3	52	100	100	100.0	55	76	75.0	100.0	60
7	F	28	3	40	90	100	33.3	50	72	87.5	67.5	55
8	F	29	12	35	100	100	0	15	36	50.0	67.5	55
9	F	31	4	49	95	100	100.0	60	84	100.0	100.0	55
10	F	39	4	49	60	50	100.0	45	44	87.5	77.5	55
11	F	41	5	53	100	100	100.0	95	96	100.0	100.0	60
12	M	25	10	55	100	100	66.7	85	76	100.0	100.0	60
13	M	29	5	52	100	100	100.0	50	88	100.0	100.0	50
14	M	29	4	41	100	100	100.0	50	76	100.0	100.0	60
15	M	31	3	56	95	100	100.0	65	80	75.0	90.0	50
16	M	34	12	73	100	100	66.7	50	52	62.5	100.0	55
17	M	35	6	52	100	100	100.0	75	80	100.0	100.0	50
18	M	37	3	61	90	100	66.7	50	80	87.5	90.0	55
19	M	39	8	72	100	100	100.0	80	88	100.0	100.0	55
20	M	41	6	55	95	100	100.0	50	84	100.0	80.0	55
21	M	41	9	52	95	100	66.7	35	52	87.5	70.0	60
MIN		23	1	35	60	50	0	15	36	50	67.5	50
MEDIAN		29	4	52	100	100	100	50	76	87.5	100	55
MAX		41	12	73	100	100	100	95	96	100	100	60

4.2. Inter-Device Comparisons for Sleep Stages and Metrics

Table 2 shows the summary statistics for all device-produced metrics and SRSMs. TSD was reported by all devices and by the participants themselves (i.e., as part of SRSM). Figure 2A shows a correlation matrix of TSD. The correlations were generally medium to weak ($\varrho < 0.7$ for all pairwise comparisons), although surprisingly the correlations of the SRSM with device estimates were on par with correlations among the devices themselves. Figure 2B shows a REM sleep (in sec) cycle correlations across the Oura, Hexoskin and Withings (Fitbit did not report an estimate of REM sleep). The correlation between Oura and Withings was highest at $\varrho = 0.44$, while Oura and Hexoskin had the lowest correlation ($\varrho = 0.22$). Figure 2C shows Kendall's rank correlation across overall sleep stages for Withings, Hexoskin, and Oura (see Section 3.9). All of these assessments were statistically significant at the $p < 0.05$ threshold. We report the p values from these analyses in Supplemental S5.

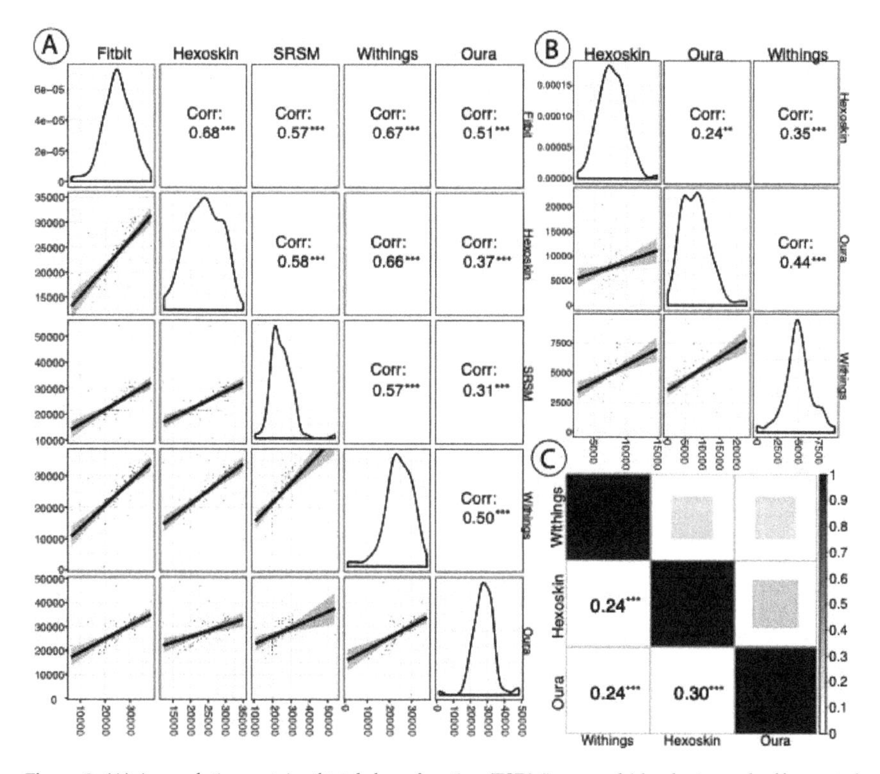

Figure 2. (**A**) A correlation matrix of total sleep duration (TSD) (in seconds) by device and self-reported estimation (i.e., self-reported sleep metrics (SRSMs)) with p value significance indication (* $p < 0.1$; ** $p < 0.05$; *** $p < 0.01$). Each point represents data from each night for each participant. The plots in the diagonals of A and B reflect the distribution of sleep metric of interest (TSD and REM, respectively). (**B**) A REM sleep (in sec) correlation across the Oura, Hexoskin, and Withings devices with p value significance indication (same as above). The Fitbit was excluded, as it does not track REM vs. NREM sleep. for each individual device. The plots in the bottom left of A and B show the trend line with 95% confidence intervals between devices. (**C**) A correlation matrix of overall sleep stages (awake, NREM, and REM) between Oura, Hexoskin, and Withings devices (Fitbit does not differentiate between NREM and REM) with p value significance indication (same as above).

Table 2. Summary metrics of device data and SRSMs. All units are in hours except wakeups which is in occurrences and efficiency (no units). Sleep efficiency is a metric to track percentage of time in bed while asleep. TSD is total sleep duration which is similar to start-end duration and similar features were utilized that included latency and other measures.

Device	Metric	n	Mean	St. Dev	Min	Pctl (25)	Pctl (75)	Max
Fitbit	Efficiency	129	94.70	15.70	31.00	94.00	97.00	193.00
	TSD All	129	7.47	1.47	3.78	6.50	8.43	11.40
	TSD	129	7.58	1.58	1.78	5.98	7.93	10.75
	Start-End	129	7.58	1.73	3.78	6.50	8.48	15.87
	Wakeups	129	1.60	1.20	0.00	1.00	2.00	8.00
Hexoskin	Efficiency	114	92.40	4.40	70.30	91.10	95.30	97.80
	TSD	114	6.72	1.31	3.45	5.78	7.81	9.69
	Start-End	135	7.57	1.42	3.93	6.57	8.58	11.43
	REM	123	2.15	0.57	0.69	1.77	2.53	4.12
	Latency	114	0.29	0.26	0.07	0.12	0.38	1.56
Oura	Efficiency	127	89.70	14.40	24.00	84.00	93.00	164.00
	TSD	128	7.69	1.72	0.42	6.73	8.75	13.48
	Start-End	130	10.67	11.63	4.62	6.97	9.55	117.60
	REM	127	2.17	1.11	0.00	1.29	2.81	6.38
	Deep	127	1.12	0.58	0.00	0.73	1.44	2.58
	Wakeups	127	2.40	1.90	0.00	1.00	4.00	7.00
	Latency	127	0.26	0.25	0.01	0.11	0.30	1.58
Withings	Efficiency	141	84.10	20.50	20.50	74.80	90.10	179.80
	TSD All	141	8.99	2.89	0.53	7.45	10.12	27.03
	TSD	141	6.97	1.75	0.33	5.95	8.15	10.97
	Start-End	141	9.30	4.45	0.42	7.08	9.73	34.55
	REM	141	1.40	0.46	0.00	1.15	1.67	2.63
	Deep	141	1.74	0.58	0.00	1.42	2.15	3.67
	Light	141	3.83	0.98	0.33	3.22	4.45	6.03
	Wakeups	141	2.40	2.60	0.00	0.00	3.00	13.00
	Latency	141	0.32	0.36	0.00	0.08	0.42	2.37
	Wakeup Duration	141	1.38	2.14	0.03	0.53	1.50	17.48
SRSMs	Start-End	122	7.34	1.45	4.50	6.35	8.24	12.33
	TSD	122	6.91	1.56	3.00	6.00	7.78	15.00
	Latency	122	0.24	0.23	0.02	0.08	0.33	2.00

4.3. PSQI, Cognitive Scores, and SRSMs vs. Device Data

Table 3 shows the results of a series of univariate linear models, each of which included either PSQI or cognitive score (morning, afternoon, and evening timepoints) as the dependent variable and the mean of the device metric per participant as the independent variable. The only statistically significant associations for PSQI at scale (i.e., significance threshold at alpha = 0.05) were for Oura's measurement of TSD and sleep efficiency ($p< 0.05$ for both). In both cases, an increase in TSD or sleep efficiency was associated with a significant decrease in PSQI score; since PSQI increases with poor sleep quality, these associations are in the expected direction (more sleep or more efficient sleep leads to better or lower PSQI). Withings latency was statistically significant for afternoon cognitive scores and evening cognitive scores at $p = 0.016$ and $p = 0.013$, respectively. We did not find any significant associations between SRSMs and cognitive scores or overall PSQI.

4.4. Cognitive Scores vs. Participant Summary Data

Table 4 shows the results of univariate linear models that regressed cognitive score on participant summary features. With regards to the morning cognitive score, there was a significant association with the SF-36 sub-category of physical functioning ($p = 0.014$); however, further analysis revealed that this was due to the presence of an outlier with very low physical functioning as well as a low cognitive score, and exclusion of this individual removed the significant association. The SF-36 sub-category of emotional well-being was trending towards significance ($p = 0.078$) with cognitive score that appears robust to the removal of individual data points. None of the other summary features were significantly associated with the morning cognitive score. Several other features were statistically significant ($p < 0.05$) across two or more cognitive score timepoints, and also of note is the consensus of significance of features across afternoon and evening cognitive score timepoints. (Table 4).

Table 3. Results of multiple univariate linear models for PSQI (left) and cognitive scores across all timepoints (right). For the PSQI-related models, the independent variables were the means of device data for each participant, and the dependent variable was PSQI. The higher the value is on the PSQI, the worse the sleep quality; thus, positive correlations suggest relation to poorer sleep quality. For the cognitive score-related models, the independent variables were the means of device data for each participant, and the dependent variable were the cognitive scores. We show the p values of each univariate regression for cognitive score by timepoint. Please see Supplemental S6–S8 for more statistics related to these regressions. All units are in hours with the exception of wakeups (number of occurrences) and efficiency (a standardized metric).

		PSQI				Cognitive scores (p Value)		
Device	**Feature**	**Coefficient**	**Std. Error**	**p-Value**	**R2**	**Morning**	**Afternoon**	**Evening**
Fitbit	TSD	−0.273	0.544	0.622	0.014	0.825	0.511	0.610
	Wakeups	1.570	1.005	0.136	0.119	0.329	0.672	0.857
Withings	TSD	−0.125	0.498	0.804	0.004	0.110	0.497	0.409
	Latency	−2.080	2.83	0.472	0.0291	0.869	0.016 **	0.013 **
	Efficiency	0.010	0.060	0.869	0.002	0.315	0.148	0.194
	Wakeups	0.352	0.427	0.421	0.036	0.888	0.361	0.378
	REM	0.260	1.962	0.896	0.001	0.342	0.617	0.557
Oura	TSD	−1.004	0.305	0.004 ***	0.376	0.265	0.197	0.221
	Latency	−7.311	4.445	0.117	0.131	0.366	0.499	0.563
	Efficiency	−0.092	0.040	0.033 **	0.228	0.285	0.332	0.301
	Wakeups	0.168	0.491	0.736	0.006	0.226	0.184	0.289
	REM	−0.526	0.715	0.471	0.029	0.656	0.732	0.713
Hexoskin	TSD	0.187	0.702	0.793	0.004	0.206	0.289	0.235
	Latency	1.249	4.444	0.782	0.004	0.995	0.481	0.718
	Efficiency	−0.226	0.272	0.417	0.037	0.530	0.527	0.798
	REM	0.397	1.833	0.831	0.003	0.128	0.186	0.180
SRSM	TSD	−0.725	0.558	0.210	0.086	0.725	0.361	0.273
	Latency	1.846	4.033	0.653	0.012	0.935	0.210	0.261
Observations			20			16	19	18

Note: * $p < 0.1$; ** $p < 0.05$; *** $p < 0.01$.

Table 4. In this collection of univariate linear models, the participants' summary data are the independent variables, and cognitive score is the dependent variable. We present the p values of each univariate regression for cognitive score by timepoint. Please see Supplemental S9–S11 for more statistics related to these regressions. These metrics all represent standardized scores.

		Cognitive Scores (p Value)	
Feature	**Morning**	**Afternoon**	**Evening**
PSQI	0.531	0.083 *	0.057 **
MEQ	0.529	0.057	0.120
Emotional Role Limitations	0.665	0.005 ***	0.003 ***
Energy	0.700	0.018 **	0.010 ***
General Health	0.769	0.823	0.961
Physical	0.014 **	0.745	0.597
Social	0.170	0.004 ***	0.002 ***
Well-being	0.078 *	0.005 ***	0.001 ***
Observations	16	19	18

Note: * $p < 0.1$; ** $p < 0.05$; *** $p < 0.01$.

4.5. Correlation between MEQ Preference and Cognitive Test Response Rates

We illustrate the rate of missingness for sleep-related metrics in Figure 3; in general, a large proportion of relevant data were missing due to noncompliance by users or device malfunctioning. We stratified response rate for morning, afternoon and evening test results, which are grouped by participants' MEQ segmentation into morning, intermediate, and night in Figure 4. We see that morning-preferred participants had the lowest response rate across all times. Furthermore, we see that afternoon response times were the highest for all MEQ groupings.

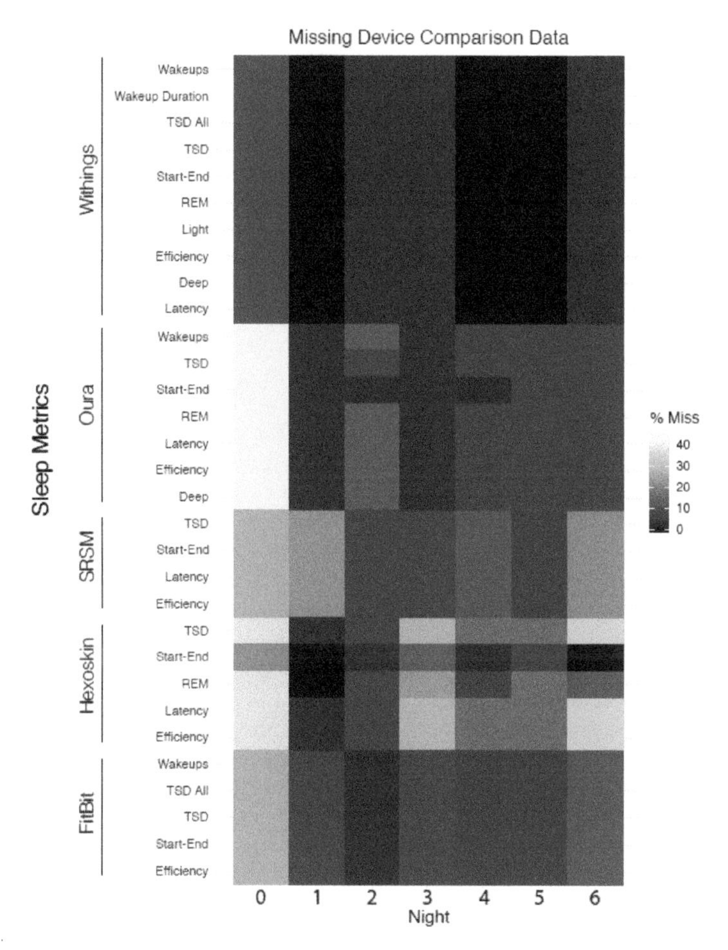

Figure 3. Plot of missing sleep-related data including SRSMs. Due to various device preferences, missing data are asymmetric across devices.

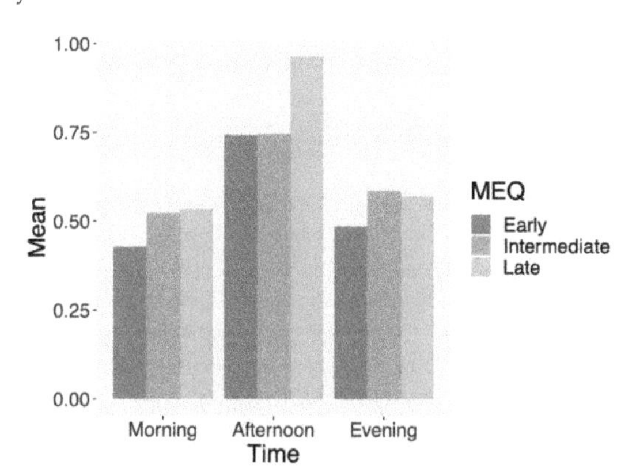

Figure 4. Average missing data for n-back tests by timepoint (morning, afternoon, and evening) and MEQ groupings (early, intermediate, or late).

5. Discussion

The results of our study reflect some general findings that are likely to impact most research involving wearable devices and mobile apps. First, because of low enrollment, our ability to detect effects was low; an effect would need to be highly pronounced to be detectable in a study population of this size. The effort involved in publicizing the study, enrolling participants, and ensuring they were able to complete the study (no device or app malfunctions, no devices running out of batteries, etc.) was substantial. Simple study designs with perhaps one or two devices that participants already own and are familiar with offer the greatest chance of success on a large scale. Second, there was substantial variability among the devices we tested, making the choice of device for any sleep study a material factor that can impact results. Even if it is impossible to assess which device is "preferred" for a given study design, this variability impacts cross-interpretability of results across different studies and will thwart attempts at meta-analyses. In this work, for TSD, we found that Oura, which has been shown to correlate strongly with PSG in prior work [19], had moderate correlations to Fitbit (0.51), Hexoskin (0.37), and Withings (0.50). Additionally, across the three devices that tracked REM, the maximum correlation was only 0.44 between Oura and the Withings. Finally, missingness and the presence of outliers were important considerations for all statistical analyses with this dataset. Although this was a pilot study, all of these issues are likely to translate to larger wearable device studies as well.

5.1. Study Limitations

There were several limitations to this study. First, we did not include other cognitive assessment tests such as the psychomotor vigilance test. Furthermore, while the n-back test is often used as an assessment of working memory in sleep-related research, the particular composite metric we derived to gauge performance has not been previously validated in this regard. A color-word association task based off the Stroop test was given, but we were not able to analyze results due to poor response rate. Additionally, as a result of using commercial sensors, we were unable to fully blind participants to the output of their sleep devices. While the participants were instructed not to check the nightly device sleep metric outputs when recording their estimated SRSMs, this could have led to biased responses if they did so. The biggest limitation of the current study was the lack of a gold standard for sleep metrics, namely PSG. It should be noted that sleep studies are extremely difficult to conduct with large numbers of participants due to the prohibitive cost of PSG. In the future, however, this field can face huge growth if some amalgamation of cheap, at home devices could reliably track various data and cross confirm results amongst themselves. This would be extremely beneficial in creating a mapping function of individual device metrics to PSG metrics, which in turn could allow these more simplistic sensors to accurately recreate conditions of PSG's at a low cost and in the comfort of participants' homes. This mapping function could increase recruitment of participants while decreasing cost for sleep studies.

5.2. Considerations Related to Cognitive Metrics and Self-Reported Sleep Quality Indices

PSQI has been shown to be a poor screening measure of PSG [29]. This may explain why self-reported one-time PSQI sleep quality variation was not well explained by much of the device data. However, the Oura ring's measurements of efficiency and sleep duration did explain variation in the one-time PSQI with statistical significance. These Oura tracking metrics may merit further investigation. Also, it is important to note that poor tracking metrics and a low number of participants could also be the reason more device data was not able to explain variation in PSQI. In terms of the SRSMs, specifically TSD, we found significant ($p < 0.05$), albeit low (range: 0.31–0.58) correlations between all devices.

Evidence of using the n-back test as a fluid intelligence metric is contentiously accepted, with some critics citing low correlation between n-back and other fluid intelligence tests [30]. The cognition metric, taken from the N-back results, and results from participant summary data had statistically significant associations. This provides a direction for further studies to investigate with larger samples. Ultimately, higher statistical power is needed to help understand these relationships. A recent study showed that poor sleep or deprivation may cause local deficits, specifically for tasks of an emotional nature [31]. This may suggest implementing a metric for wellbeing in addition to fluid intelligence tasks. Of particular note was Withings latency, which was statistically significant for the afternoon and evening cognitive scores ($p < 0.05$). Due to the low sample size, the importance of this is uncertain, but hopefully with subsequent studies could build on this work by further comparing latency with cognitive scores.

Insight from response rate based on MEQ segmentation into three categories (early, intermediate, and late preference) could help future study designs. Across all MEQ groups, n-back test response rates were highest in the afternoon. This suggests that crucial surveys should be administered around this time if possible. Another finding of note is that late-preferred participants had the highest n-back test response rate on average in the morning and afternoon timepoints. This finding suggests that participants who are not late-preferred may need extra motivating factors to increase their response rates.

6. Conclusions

We reported correlations among important sleep metrics for four different sleep tracker devices and correlated the results with self-reported questionnaires and cognitive metrics, specifically the n-back. Difficulty in participant enrollment and engagement led to new ideas about recruitment design and participant engagement design. Exploiting existing technology such as ReasearchKit or HealthKit from Apple can have a twofold benefit for recruiting people remotely (with an e-consent feature built into ResearchKit) and sharing electronic health records (EHR). By further combining this with additional data stores present in the HealthApp, participant eligibility screening can be improved [32]. In consideration of the missing data in the questionnaires and active tasks prescribed, we promote the use of as many passive collection procedures as possible. One such option is a smart mirror [33], which can be more passive than using a smartphone for data (e.g., imaging) collection. Finally, the weak correlation among devices opens new challenges for accurate interpretation and data portability for the end user. How will device-specific findings from various studies be taken in context to one another? The results from the current study can hopefully highlight the need for better standardization for sleep-related metrics across devices in order to make any robust and accurate conclusions.

Supplementary Materials: The following are available online at http://www.mdpi.com/1424-8220/20/5/1378/s1, S1: Demographic questionnaire, S2: SF-36 questionnaire, S3: MEQ questionnaire, S4: PSQI questionnaire, S5: Results table of all correlation analyses, S6: Results table of univariate analyses comparing morning cognitive scores to device data, S7: Results table of univariate analyses comparing afternoon cognitive scores to device data, S8: Results table of univariate analyses comparing evening cognitive scores to device data, S9: Results table of univariate analyses comparing morning cognitive scores to participant summary data, S10: Results table of univariate analyses comparing afternoon cognitive scores to participant summary data, S11: Results table of univariate analyses comparing evening cognitive scores to participant summary data.

Author Contributions: Conceptualization, G.S., N.Z., B.P., and J.T.D.; methodology, M.S., J.S., B.P., and N.Z.; software, J.S., D.S., M.S., and G.S.; formal analysis, F.F.C, M.D., and B.S.G.; investigation, G.B. and E.G.; validation, R.M., M.D., J.K.D.F., I.P., and G.N.N.; data curation, F.F.C., M.D., J.S., M.S., and N.Z.; writing—original draft preparation, F.F.C., M.D., E.G., B.P., N.Z., and B.S.G; writing—review and editing, all authors.; visualization, F.F.C., M.D., J.S., N.Z., and B.S.G.; supervision, J.T.D. and B.S.G.; project administration, G.B., E.G., and P.G.; funding acquisition, G.S. and J.T.D. All authors have read and agreed to the published version of the manuscript.

Funding: This research: and The Harris Center for Precision Wellness, was funded by generous gifts from Joshua and Marjorie Harris of the Harris Family Charitable Foundation and Julian Salisbury.

Acknowledgments: We acknowledge David E. Stark for assistance in designing the cognitive battery and Christopher Cowan for design and development of the HC mobile app.

Conflicts of Interest: The authors declare no conflict of interest. The funders had no role in the design of the study; in the collection, analyses, or interpretation of data; in the writing of the manuscript, or in the decision to publish the results.

References

1. National Center on Sleep Disorders Research. *National Institutes of Health Sleep Disorders Research Plan*; National Institutes of Health: Bethesda, MD, USA, 2011.
2. Liu, Y.; Wheaton, A.G.; Chapman, D.P.; Cunningham, T.J.; Lu, H.; Croft, J.B. Prevalence of Healthy Sleep Duration among Adults—United States, 2014. *MMWR Morb. Mortal. Wkly. Rep.* **2016**, *65*, 137–141. [CrossRef] [PubMed]
3. Shan, Z.; Ma, H.; Xie, M.; Yan, P.; Guo, Y.; Bao, W.; Rong, Y.; Jackson, C.L.; Hu, F.B.; Liu, L. Sleep duration and risk of type 2 diabetes: A meta-analysis of prospective studies. *Diabetes Care* **2015**, *38*, 529–537. [CrossRef]
4. Huang, T.; Redline, S. Cross-sectional and Prospective Associations of Actigraphy-Assessed Sleep Regularity With Metabolic Abnormalities: The Multi-Ethnic Study of Atherosclerosis. *Diabetes Care* **2019**, *42*, 1422–1429. [CrossRef] [PubMed]
5. Sanchez-de-la-Torre, M.; Campos-Rodriguez, F.; Barbe, F. Obstructive sleep apnoea and cardiovascular disease. *Lancet Respir. Med.* **2013**, *1*, 61–72. [CrossRef]
6. Liu, R.Q.; Qian, Z.; Trevathan, E.; Chang, J.J.; Zelicoff, A.; Hao, Y.T.; Lin, S.; Dong, G.H. Poor sleep quality associated with high risk of hypertension and elevated blood pressure in China: Results from a large population-based study. *Hypertens Res.* **2016**, *39*, 54–59. [CrossRef] [PubMed]
7. Chastin, S.F.; Palarea-Albaladejo, J.; Dontje, M.L.; Skelton, D.A. Combined Effects of Time Spent in Physical Activity, Sedentary Behaviors and Sleep on Obesity and Cardio-Metabolic Health Markers: A Novel Compositional Data Analysis Approach. *PLoS ONE* **2015**, *10*, e0139984. [CrossRef]
8. Alvaro, P.K.; Roberts, R.M.; Harris, J.K. A Systematic Review Assessing Bidirectionality between Sleep Disturbances, Anxiety, and Depression. *Sleep* **2013**, *36*, 1059–1068. [CrossRef]
9. Hafner, M.; Stepanek, M.; Taylor, J.; Troxel, W.M.; van Stolk, C. Why Sleep Matters-The Economic Costs of Insufficient Sleep: A Cross-Country Comparative Analysis. *Rand Health Q.* **2017**, *6*, 11.
10. O'Connor, C.; Thornley, K.S.; Hanly, P.J. Gender Differences in the Polysomnographic Features of Obstructive Sleep Apnea. *Am. J. Respir. Crit. Care Med.* **2000**, *161*, 1465–1472. [CrossRef]
11. Redline, S.; Tishler, P.V.; Hans, M.G.; Tosteson, T.D.; Strohl, K.P.; Spry, K. Racial differences in sleep-disordered breathing in African-Americans and Caucasians. *Am. J. Respir. Crit. Care Med.* **1997**, *155*, 186–192. [CrossRef]
12. Kiley, J.P.; Twery, M.J.; Gibbons, G.H. The National Center on Sleep Disorders Research—Progress and promise. *Sleep* **2019**, *42*. [CrossRef] [PubMed]
13. Newell, J.; Mairesse, O.; Verbanck, P.; Neu, D. Is a one-night stay in the lab really enough to conclude? First-night effect and night-to-night variability in polysomnographic recordings among different clinical population samples. *Psychiatry Res.* **2012**, *200*, 795–801. [CrossRef] [PubMed]
14. Means, M.K.; Edinger, J.D.; Glenn, D.M.; Fins, A.I. Accuracy of sleep perceptions among insomnia sufferers and normal sleepers. *Sleep Med.* **2003**, *4*, 285–296. [CrossRef]
15. Fry, J.M.; DiPhillipo, M.A.; Curran, K.; Goldberg, R.; Baran, A.S. Full Polysomnography in the Home. *Sleep* **1998**, *21*, 635–642. [CrossRef] [PubMed]
16. Peake, J.M.; Kerr, G.; Sullivan, J.P. A Critical Review of Consumer Wearables, Mobile Applications, and Equipment for Providing Biofeedback, Monitoring Stress, and Sleep in Physically Active Populations. *Front. Physiol.* **2018**, *9*. [CrossRef] [PubMed]
17. Bianchi, M.T. Sleep devices: Wearables and nearables, informational and interventional, consumer and clinical. *Metabolism* **2018**, *84*, 99–108. [CrossRef]
18. de Zambotti, M.; Goldstone, A.; Claudatos, S.; Colrain, I.M.; Baker, F.C. A validation study of Fitbit Charge 2™ compared with polysomnography in adults. *Chronobiol. Int.* **2018**, *35*, 465–476. [CrossRef]
19. de Zambotti, M.; Rosas, L.; Colrain, I.M.; Baker, F.C. The Sleep of the Ring: Comparison of the ŌURA Sleep Tracker Against Polysomnography. *Behav. Sleep Med.* **2019**, *17*, 124–136. [CrossRef]
20. de Zambotti, M.; Claudatos, S.; Inkelis, S.; Colrain, I.M.; Baker, F.C. Evaluation of a consumer fitness-tracking device to assess sleep in adults. *Chronobiol. Int.* **2015**, *32*, 1024–1028. [CrossRef]
21. Meltzer, L.J.; Hiruma, L.S.; Avis, K.; Montgomery-Downs, H.; Valentin, J. Comparison of a Commercial Accelerometer with Polysomnography and Actigraphy in Children and Adolescents. *Sleep* **2015**, *38*, 1323–1330. [CrossRef]
22. Montgomery-Downs, H.E.; Insana, S.P.; Bond, J.A. Movement toward a novel activity monitoring device. *Sleep Breath.* **2012**, *16*, 913–917. [CrossRef] [PubMed]

23. Liu, W.; Ploderer, B.; Hoang, T. In Bed with Technology: Challenges and Opportunities for Sleep Tracking. In Proceedings of the Annual Meeting of the Australian Special Interest Group for Computer Human Interaction, Parkville, Australia, 7–10 December 2015; pp. 142–151.
24. Piwek, L.; Ellis, D.A.; Andrews, S.; Joinson, A. The Rise of Consumer Health Wearables: Promises and Barriers. *PLoS Med.* **2016**, *13*, e1001953. [CrossRef] [PubMed]
25. Murakami, H.; Kawakami, R.; Nakae, S.; Nakata, Y.; Ishikawa-Takata, K.; Tanaka, S.; Miyachi, M. Accuracy of Wearable Devices for Estimating Total Energy Expenditure: Comparison with Metabolic Chamber and Doubly Labeled Water Method. *JAMA Intern. Med.* **2016**, *176*, 702–703. [CrossRef] [PubMed]
26. Xie, J.; Wen, D.; Liang, L.; Jia, Y.; Gao, L.; Lei, J. Evaluating the Validity of Current Mainstream Wearable Devices in Fitness Tracking Under Various Physical Activities: Comparative Study. *JMIR Mhealth Uhealth* **2018**, *6*, e94. [CrossRef] [PubMed]
27. Lee, J.-M.; Byun, W.; Keill, A.; Dinkel, D.; Seo, Y. Comparison of Wearable Trackers' Ability to Estimate Sleep. *Int. J. Environ. Res. Public Health* **2018**, *15*, 1265. [CrossRef] [PubMed]
28. Rosvold, H.E.; Mirsky, A.F.; Sarason, I.; Bransome Jr, E.D.; Beck, L.H. A continuous performance test of brain damage. *J. Consult. Psychol.* **1956**, *20*, 343. [CrossRef]
29. Buysse, D.J.; Hall, M.L.; Strollo, P.J.; Kamarck, T.W.; Owens, J.; Lee, L.; Reis, S.E.; Matthews, K.A. Relationships between the Pittsburgh Sleep Quality Index (PSQI), Epworth Sleepiness Scale (ESS), and clinical/polysomnographic measures in a community sample. *J. Clin. Sleep Med.* **2008**, *4*, 563–571. [CrossRef]
30. Kane, M.J.; Conway, A.R.A.; Miura, T.K.; Colflesh, G.J.H. Working memory, attention control, and the n-back task: A question of construct validity. *J. Exp. Psychol. Learn. Mem. Cognit.* **2007**, *33*, 615–622. [CrossRef]
31. Killgore, W.D.S. Effects of sleep deprivation on cognition. *Prog. Brain Res.* **2010**, *185*, 105–129.
32. Perez, M.V.; Mahaffey, K.W.; Hedlin, H.; Rumsfeld, J.S.; Garcia, A.; Ferris, T.; Balasubramanian, V.; Russo, A.M.; Rajmane, A.; Cheung, L. Large-scale assessment of a smartwatch to identify atrial fibrillation. *N. Engl. J. Med.* **2019**, *381*, 1909–1917. [CrossRef]
33. Miotto, R.; Danieletto, M.; Scelza, J.R.; Kidd, B.A.; Dudley, J.T. Reflecting health: Smart mirrors for personalized medicine. *NPJ Digit. Med.* **2018**, *1*, 1–7. [CrossRef] [PubMed]

Article

Hyperspectral Imaging for the Detection of Glioblastoma Tumor Cells in H&E Slides Using Convolutional Neural Networks

Samuel Ortega [1,2,*,†], Martin Halicek [1,3,†], Himar Fabelo [2], Rafael Camacho [4], María de la Luz Plaza [4], Fred Godtliebsen [5], Gustavo M. Callicó [2] and Baowei Fei [1,6,7,*]

1 Quantitative Bioimaging Laboratory, Department of Bioengineering, The University of Texas at Dallas, Richardson, TX 75080, USA; mth180000@utdallas.edu
2 Institute for Applied Microelectronics (IUMA), University of Las Palmas de Gran Canaria (ULPGC), 35017 Las Palmas de Gran Canaria, Spain; hfabelo@iuma.ulpgc.es (H.F.); gustavo@iuma.ulpgc.es (G.M.C.)
3 Department of Biomedical Engineering, Emory University and Georgia Institute of Technology, 1841 Clifton Road NE, Atlanta, GA 30329, USA
4 Department of Pathological Anatomy, University Hospital Doctor Negrin of Gran Canaria, Barranco de la Ballena s/n, 35010 Las Palmas de Gran Canaria, Spain; rcamgal@gobiernodecanarias.org (R.C.); mplaperb@gobiernodecanarias.org (M.d.l.L.P.)
5 Department of Mathematics and Statistics, UiT The Artic, University of Norway, Hansine Hansens veg 18, 9019 Tromsø, Norway; fred.godtliebsen@uit.no
6 Advanced Imaging Research Center, University of Texas Southwestern Medical Center, 5323 Harry Hine Blvd, Dallas, TX 75390, USA
7 Department of Radiology, University of Texas Southwestern Medical Center, 5323 Harry Hine Blvd, Dallas, TX 75390, USA
* Correspondence: sortega@iuma.ulpgc.es (S.O.); bfei@utdallas.edu (B.F.); Tel.: +34-928-451-220 (S.O.)
† These authors contributed equally to this work.

Received: 6 March 2020; Accepted: 28 March 2020; Published: 30 March 2020

Abstract: Hyperspectral imaging (HSI) technology has demonstrated potential to provide useful information about the chemical composition of tissue and its morphological features in a single image modality. Deep learning (DL) techniques have demonstrated the ability of automatic feature extraction from data for a successful classification. In this study, we exploit HSI and DL for the automatic differentiation of glioblastoma (GB) and non-tumor tissue on hematoxylin and eosin (H&E) stained histological slides of human brain tissue. GB detection is a challenging application, showing high heterogeneity in the cellular morphology across different patients. We employed an HSI microscope, with a spectral range from 400 to 1000 nm, to collect 517 HS cubes from 13 GB patients using 20× magnification. Using a convolutional neural network (CNN), we were able to automatically detect GB within the pathological slides, achieving average sensitivity and specificity values of 88% and 77%, respectively, representing an improvement of 7% and 8% respectively, as compared to the results obtained using RGB (red, green, and blue) images. This study demonstrates that the combination of hyperspectral microscopic imaging and deep learning is a promising tool for future computational pathologies.

Keywords: hyperspectral imaging; medical optics and biotechnology; optical pathology; convolutional neural networks; tissue diagnostics; tissue characterization; glioblastoma

1. Introduction

Traditional diagnosis of histological samples is based on manual examination of the morphological features of specimens by skilled pathologists. In recent years, the use of computer-aided technologies for aiding in this task is an emerging trend, with the main goal to reduce the intra and inter-observer variability [1]. Such technologies are intended to improve the diagnosis of pathological slides to be more reproducible, increase objectivity, and save time in the routine examination of samples [2,3]. Although most of the research carried out in computational pathology has been in the context of RGB (red, green, and blue) image analysis [4–7], hyperspectral (HS) and multispectral imaging are shown as promising technologies to aid in the histopathological analysis of samples.

Hyperspectral imaging (HSI) is a technology capable of capturing both the spatial and spectral features of the materials that are imaged. Recently, this technology has proven to provide advantages in the diagnosis of different types of diseases [8–10]. In the field of histopathological analysis, this technology has been used for different applications, such as the visualization of multiple biological markers within a single tissue specimen with inmunohistochemistry [11–13], the digital staining of samples [14,15], or diagnosis.

The analysis of HS images is usually performed in combination with machine learning approaches [16]. Traditionally, feature-based methods are used, such as supervised classifiers. Awan et al. performed automatic classification of colorectal tumor samples identifying four types of tissues: normal, tumor, hyperplastic polyp, and tubular adenoma with low-grade dysplasia. Using different types of feature extraction and band selection methods followed by support vector machines (SVM) classification, the authors found that the use of a higher number of spectral bands improved the classification accuracy [17]. Wang et al. analyzed hematoxylin and eosin (H&E) skin samples to facilitate the diagnosis of melanomas. The authors proposed a customized spatial-spectral classification method, which provided an accurate identification of melanoma and melanocytes with high specificity and sensitivity [18]. Ishikawa et al. presented a method for pancreatic tumor cell identification using HSI. They first proposed a method to remove the influence of the staining in the HS data, and then they applied SVM classification [19].

Although these authors have proven the feasibility of the feature learning methods for the diagnosis of histopathological samples using HS information, the performance of these approaches may be improved by using deep learning (DL) schemes. DL approaches automatically learn from the data in which features are optimal for classification, potentially outperforming handcrafted features [20]. In the case of HS images, both the spatial and spectral features are exploited simultaneously. Recently, only a few researchers employed DL for the classification of HS images for histopathological applications. Malon et al. proposed the use of a convolutional neural network (CNN) for the detection of mitotic cells within breast cancer specimens [21]. Haj-Hassan et al. also used CNNs for the classification of colorectal cancer tissue, showing performance improvements compared to traditional feature learning approaches [22].

In this paper, we propose the use of CNNs for the classification of hematoxylin and eosin (H&E) stained brain tumor samples. Specifically, the main goal of this work was to differentiate between high-grade gliomas (i.e., glioblastoma (GB)) and non-tumor tissue. In a previous study, we presented a feature learning approach for this type of disease [23]. Although such research was shown as a useful proof-of-concept on the possibilities of HS for histopathological analysis of GB, it presented some limitations, such as poor spatial and spectral resolution, and the lack of a rigorous experimental design. In this work, the image quality and spectral range have been significantly improved, resulting in a more appropriate experimental design for realistic clinical applications.

2. Materials and Methods

2.1. Acquisition System

The instrumentation employed in this study consists of an HS camera coupled to a conventional light microscope (Figure 1). The microscope is an Olympus BX-53 (Olympus, Tokyo, Japan). The HS camera is a Hyperspec® VNIR A-Series from HeadWall Photonics (Fitchburg, MA, USA), which is based on an imaging spectrometer coupled to a CCD (Charge-Coupled Device) sensor, the Adimec-1000m

(Adimec, Eindhoven, Netherlands). This HS system works in the visual and near-infrared (VNIR) spectral range from 400 to 1000 nm with a spectral resolution of 2.8 nm, sampling 826 spectral channels and 1004 spatial pixels. The push-broom camera performs spatial scanning to acquire an HS cube with a mechanical stage (SCAN, Märzhäuser, Wetzlar, Germany) attached to the microscope, which provides accurate movement of the specimens. The objective lenses are from the LMPLFLN family (Olympus, Tokyo, Japan), which are optimized for infra-red (IR) observations. The light source is a 12 V, 100 W halogen lamp.

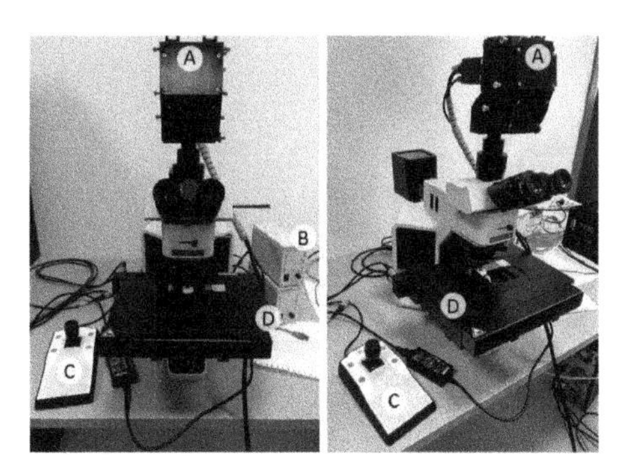

Figure 1. Microscopic hyperspectral (HS) acquisition system. (**A**) HS camera. (**B**) Halogen light source. (**C**) Positioning joystick. (**D**) XY linear stage.

To ensure high quality acquisitions, the methodology proposed in a previous research work to maximize the quality of HS images acquired with a push-broom microscope [24] was followed. This methodology includes the optimal speed determination of the scanning, a dynamic range configuration, an appropriate alignment, and the correct focusing procedure. We developed custom software for synchronizing the scanning movement and the camera acquisition. Although we are not focused on collecting a whole-slide HS image of the specimens, the software was developed to allow the acquisition of consecutive HS cubes in a row to save time in the acquisition of the images, thus reducing the human intervention in the process. Due to the challenges imposed by the high dimensionality of the HS images, we decided to collect images with a spatial size of 800 lines, producing HS cubes of 800 × 1004 × 826, i.e., number of lines × number of rows × number of bands.

2.2. GB Histological Samples

The specimens investigated in this research work consist of human biopsies extracted during brain tumor resection procedures. The pathological slides in this study were processed and analyzed by the Pathological Anatomy Department of the University Hospital Doctor Negrín at Las Palmas of Gran Canaria (Spain). The study protocol and consent procedures were approved by the Comité Ético de Investigación Clínica-Comité de Ética en la Investigación (CEIC/CEI) of the same hospital. After the resection, the samples were dehydrated and embedded in paraffin blocks. The blocks were then mounted in microtomes and sliced in 4 µm thick slices. Finally, the slices were rehydrated and stained with H&E. After routine examination of the samples, every sample was diagnosed by pathologists as GB, according to the World Health Organization (WHO) classification of tumors of the nervous system [25]. After the pathologist confirmed the GB diagnosis, macroscopic annotations of the GB locations were made on the physical glass slides using a marker-pen. Non-tumor areas are defined as areas in the pathological slide where there is no discrete presence of tumor cells. Within the areas annotated by a pathologist, we selected regions of interest (ROI) that were subsequently digitized using HS instrumentation. Within each ROI, different numbers of HS images were acquired for analysis. Figure 2 shows an example of the annotations within the pathological slide, the selection of different ROIs (shown at 5×), and the HS images (imaged at 20×) that are used in this study for classification. In this case, red color annotations indicate areas diagnosed as GB, while non-tumor areas were annotated in blue marker. In this feasibility study, a total of 13 patients were analyzed.

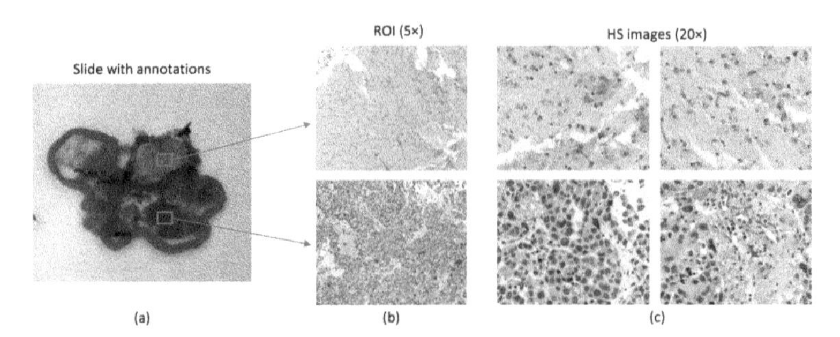

Figure 2. Pathological samples used in this study. (**a**) Macroscopic annotations performed in pathological slides after diagnosis. Blue squares denote regions of interest (ROIs) within annotations; (**b**) ROIs from (**a**) shown at 5×; (**c**) Examples of HS images used in this study for classification (imaged at 20×).

2.3. Hyperspectral Dataset

Using the aforementioned instrumentation, some of the areas highlighted by pathologists from each slide were imaged. The positioning joystick of the microscope was used to select the initial position of the first HS image within a ROI to be captured. Then, we configured in the software the number of images to be captured consecutively. This number of images should keep relatively low to avoid the focus worsening of the images throughout the specimen. In this case, a maximum of 10 HS images were extracted consecutively from a ROI. We used a 20× magnification for image acquisition, producing a HS image size of 375 × 299 μm. This magnification was chosen because it allowed the visualization of the cell morphology; hence, the classifier was able to exploit both the spatial and the spectral features of data. In Figure 3, we show some examples of HS images used in this study, together with the spectral signatures of representative tissue components, i.e., cells and background for both tumor and non-tumor regions.

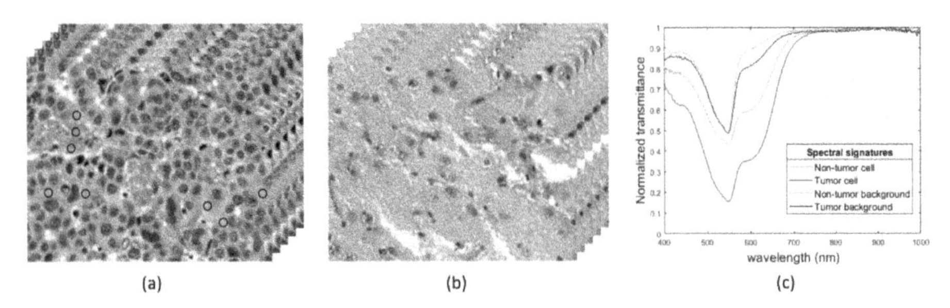

Figure 3. HS histopathological dataset. (**a**,**b**) HS cubes from tumor and non-tumor samples, respectively. (**c**) Spectral signatures of different parts of the tissue: tumor cells (red), non-tumor cells (blue), tumor background tissue (black), and non-tumor background tissue (green).

In this research, we used a CNN to perform the classification of the samples. Due to the nature of the data for this study, the ground truth assignment into tumor or non-tumor is shared across each selected ROI; thus, each HS image is assigned within a certain class. For this reason, it was decided to perform the classification in a patch-based approach because a fully-convolutional design was not feasible. There are two motivations on the selection of the patch size. Firstly, the patch should be large enough to contain more than one cell, but if the patch is too large, then the CNN could learn that the tumor is located only in dense cell patches. Secondly, the smaller the patches, the higher the quantity of patches will be extracted from a single HS image, so the number of samples to train the CNN will be increased. Finally, we choose a patch size of 87 × 87 pixels. In principle, from a spectral cube of size 800 × 1004, 99 patches can be extracted. However, there are some situations where most parts of the patches consisted only of a blank space of light. For this reason, we decided to reject patches that were composed by more than 50% of light, i.e., half of the patch is empty.

The method to reject the patches which presented high amount of light is as follows. Firstly, the RGB image is extracted from the HS cube and is transformed to the hue-saturation-value color representation. Then, the hue value of each image is extracted and binarized using a threshold empirically configured to separate the pixels belonging to the specimen and the pixels containing background light. The generation of the patches can be observed in Figure 4, where the last row of the patches (in Figure 4c) represents patches that have been rejected in the database due to high content of background light pixels.

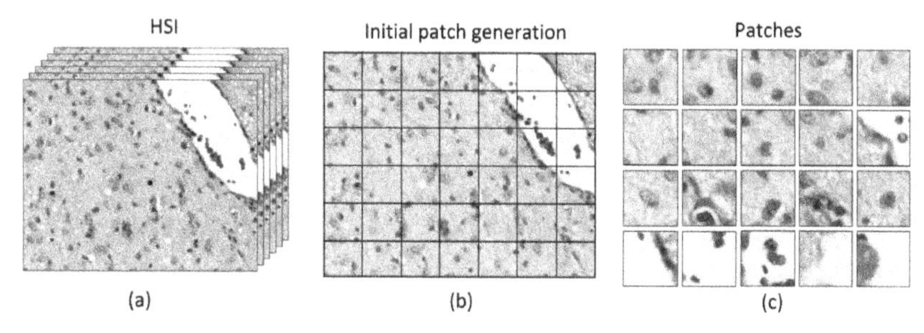

Figure 4. Generation of patches. (**a**) Original HS image; (**b**) grid of patches within the HS image; (**c**) patches of size 87 × 87 used in the classification. The last row contained patches that were rejected for the dataset for having more than 50% of empty pixels. HSI: hyperspectral imaging.

The database used in this work consists of 527 HS images, where 337 are non-tumor brain samples and 190 were diagnosed as GB. It should be highlighted that only the biopsies from 8 patients presented both non-tumor and tumor samples; the other 5 patients only presented tumor samples. The summary of the employed dataset is detailed in Table 1. After extracting the patches that were valid to be processed, we had a total of 32,878 patches from non-tumor tissue and 16,687 from tumor tissue.

Table 1. HS histopathological dataset summary.

Patient ID	Images		Patches	
	Non-Tumor	Tumor	Non-Tumor	Tumor
P1	48	12	4595	1090
P2	36	12	3563	1188
P3	31	12	3058	1178
P4	40	12	3779	1158
P5	66	12	5675	1165
P6	48	12	4586	1188
P7	44	12	4289	1184
P8	24	36	3333	2260
P9	0	22	0	1695
P10	0	12	0	1094
P11	0	12	0	1169
P12	0	12	0	1137
P13	0	12	0	1181
Total	337	190	32,878	16,687

2.4. Processing Framework

The processing framework applied to each HS cube is composed by the following steps. First, a standard flat field correction is applied to the images. To this end, the images are transformed from radiance to normalized transmittance by using a reference image that is captured from a blank area of the pathological slide [23]. Then, due to the high correlation of spectral information between adjacent spectral bands, a reduced-band HS image is generated by averaging the neighbors' spectral bands, reducing the number of spectral bands from 826 bands to 275 and slightly reducing the white Gaussian noise. This band reduction is also beneficial for alleviating computational cost in the subsequent image processing. Finally, each image is divided into patches, which will train the CNN. In this section, we will detail the architecture of the proposed neural network, the metrics that are used for performance evaluation, and the proposed data partition scheme.

2.4.1. Convolutional Neural Network

We employed a custom 2D-CNN for the automatic detection of non-tumor and tumor patches. As mentioned previously, these types of networks are able to exploit together the spatial and spectral features of the sample. The performance of DL approaches for the classification of HS data has been proven both for medical and for non-medical applications [26]. We used the TensorFlow implementation of the Keras Deep Learning API [27,28] for the development of this network. This selection was made because it allows effective development of CNN architectures, training paradigms, and efficient deployment between the Python programming language and GPU deployment of training/testing. The architecture of this CNN is mainly composed by 2D convolutional layers. We detail the description of the network in Table 2, where the input size of each layer is shown in each row, and the output size is the input size of the subsequent layer. All convolutions and the dense layer were performed with ReLU (rectified linear unit) activation functions with a 10% dropout. The optimizer used was stochastic gradient descend with a learning rate of 10^{-3}.

Table 2. Schematic of the proposed convolutional neural network (CNN).

Layer	Kernel Size	Input Size
Conv2D	3×3	$87 \times 87 \times 275$
Conv2D	3×3	$85 \times 85 \times 256$
Conv2D	3×3	$83 \times 83 \times 256$
Conv2D	3×3	$81 \times 81 \times 512$
Conv2D	3×3	$79 \times 79 \times 512$
Conv2D	3×3	$77 \times 77 \times 1024$
Conv2D	3×3	$75 \times 75 \times 1024$
Conv2D	3×3	$73 \times 73 \times 1024$
Global Avg. Pool	25×25	$73 \times 73 \times 1024$
Dense	256 neurons	1×1024
Dense	Logits	1×256
Softmax	Classifier	1×2

2.4.2. Evaluation Metrics

The metrics for measuring the classification performance of the proposed CNN were overall accuracy, sensitivity, and specificity. Overall accuracy measures the overall performance of the classification; sensitivity measures the proportion of true positives that are classified correctly; and specificity measures the ability of the classifier for identifying false negatives. The equations for these metrics according to the false positives (FP), false negatives (FN), true positives (TN), and true negatives (TN) are shown in Equations (1)–(3). Additionally, we used the area under the curve (AUC) of the receiver operating curve (ROC) of the classifier as an evaluation metric. The AUC has been proven to be more robust compared to overall accuracy. AUC is decision threshold independent, shows a decreasing standard error when the number of test samples increases, and is more sensitive to Analysis of Variance (ANOVA) test [29].

$$Sensitivity = \frac{TP}{TP + FN}, \quad (1)$$

$$Specificity = \frac{TN}{TN + FP}, \quad (2)$$

$$Accuracy = \frac{TN + TP}{TN + TP + FP + FN}. \quad (3)$$

2.4.3. Data Partition

In this work, we split data into training, validation and test sets. We were targeting a real clinical application, and, for this reason, data partition is intended to minimize bias, where the patients used for train, validation and test are independent. We were limited to 13 patients, where five of them only had samples belonging to tumor class. For this reason, we decided to perform the data partition in 4 different folds, where every patient should be part of the test set across all the folds. We proposed the use of three folds with 9 training patients, a single validation patient, and 3 test patients. The remaining fold is composed by 8 training patients, a single validation patient, and 4 test patients. Regarding the distribution of the classes in each fold, the patient selected for validation in each fold should have samples from both types of classes (non-tumor and tumor). The initial data partition scheme is shown in Table 3, where data from patients who only have tumor samples has been highlighted (‡).

Table 3. Data partition design (patients with only tumor samples are marked with ‡).

Fold ID	Training Patients	Validation Patients	Test Patients
F1	9 (5 + 4‡)	1	3 (2 + 1‡)
F2	9 (5 + 4‡)	1	3 (2 + 1‡)
F3	9 (5 + 4‡)	1	3 (2 + 1‡)
F4	8 (5 + 3‡)	1	4 (2 + 2‡)

We decided to make the patient assignment randomly within the different folds. However, the distribution of patients in fold F4 was different from the others and required some minor manual adjustments in data partitioning. Nonetheless, the rest of assignments were performed randomly. Fold F4 required assigning two tumor-only specimens for testing, so we decided to manually assign the tumor-only samples that have the least number of patches (i.e., P10 and P12). Furthermore, because fold F4 had fewer training patients compared to the other folds, we decided to assign the patient with the most patches (i.e., P5) to train this fold. The final data partition into the different folds is shown in Table 4.

Table 4. Final data partition (patients with only tumor samples are marked with ‡).

Fold ID	Training Patients	Validation Patients	Test Patients
F1	P2, P3, P4, P5, P8 P9‡, P10‡, P12‡, P13‡	P6	P1, P7, P11‡
F2	P1, P2, P5, P7, P8 P9‡, P10‡, P11‡, P12‡	P3	P4, P6, P13‡
F3	P1, P3, P4, P6, P8 P10‡, P11‡, P12‡, P13‡	P7	P2, P5, P9‡
F4	P2, P4, P5, P6, P7 P9‡, P11‡, P13‡	P1	P3, P8, P10‡, P12‡

3. Experimental Results

3.1. Validation Results

We trained different CNNs using the data from each fold, and, using the validation data, we selected the aforementioned CNN architecture (Table 2) as the best candidate for the classification of the samples. As can be observed in Table 1, the data between tumor and non-tumor classes are not balanced: the number of non-tumor samples is twice the number of tumor samples. For this reason, we performed data augmentation on the tumor data to balance the data during training, creating twice the number of tumor patches to train the CNN than cited in Table 1. Such data augmentation consisted in a single spatial rotation of tumor patches.

At the beginning of the validation phase, some of the folds presented problems when they were trained, showing poor performance metrics in the validation set. For this reason, we carefully examined the tumor HS images from each patient, and we detected that accidentally some necrosis areas were included in the dataset as tumor samples. These necrosis areas (found in P8) were excluded from the dataset. After excluding the necrosis areas, we got competitive results for all the folds in the validation set. These results are shown in Table 5. The models for each fold were selected because they all presented high AUC, higher than 0.92, and the results in terms of accuracy, sensitivity and specificity were balanced, indicating that the models identified correctly both non-tumor and tumor tissue.

Table 5. Classification results on the validation dataset, across all four folds (F). AUC: area under the curve.

Partition	HSI				RGB			
	AUC	Accuracy (%)	Sensitivity (%)	Specificity (%)	AUC	Accuracy (%)	Sensitivity (%)	Specificity (%)
F1	0.92	84	84	85	0.88	77	71	88
F2	0.97	93	91	94	0.95	87	83	93
F3	0.95	88	90	88	0.93	87	91	79
F4	0.95	89	87	91	0.92	92	93	89
Avg.	0.95	88	88	89	0.92	86	84	87
Std.	0.02	3.70	3.16	3.87	0.03	6.29	9.98	5.91

In order to provide a comparison of performance between HSI and RGB imagery, we performed the classification of synthetic RGB images using the same CNN. Such RGB images were extracted from the HS data, where each color channel was generated equalizing the spectral information to match the spectral response of the human eye [30]. After separately training the CNN with RGB patches, the models selected after the validation were found to be competitive. Nevertheless, the validation performance when using HSI data was more accurate in each fold and presented more balanced sensitivity and specificity values (Table 5).

3.2. Test Results

After the model selection in the validation phase, we applied them to independent patients for the test set. These results are shown in Table 6. Some results show good discrimination between non-tumor and tumor tissues, i.e., patients P1, P3, and P8. For these patients, the AUC, sensitivity and specificity are comparable to the values obtained during validation. The tumor detection in patients P9 to P13 was also highly accurate. However, there are some patients where the classification performance was poor. Although the sensitivity is high in patients P2 and P5, the specificity is low, which indicates there may be an issue classifying non-tumor patches. There are also some patients with poor accuracy, namely patients P4 and P7, which have results slightly better than random guessing. Finally, the results obtained for patient P6 are suspicious, being substantially worse than random guessing.

Table 6. Initial classification results on the test dataset.

Patient ID	HSI				RGB			
	AUC	Accuracy (%)	Sensitivity (%)	Specificity (%)	AUC	Accuracy (%)	Sensitivity (%)	Specificity (%)
P1	0.97	92	90	94	0.92	90	97	61
P2	0.75	77	99	69	0.98	85	99	80
P3	0.95	85	91	80	0.96	92	97	78
P4	0.62	57	57	58	0.69	77	98	7
P5	0.81	69	81	64	0.66	59	59	60
P6	0.35	37	38	36	0.21	67	81	7
P7	0.64	59	64	57	0.51	45	36	76
P8	0.98	96	96	96	0.99	97	97	97
P9	N.A.	99	99	N.A.	N.A.	89	89	N.A.
P10	N.A.	89	89	N.A.	N.A.	43	43	N.A.
P11	N.A.	92	92	N.A.	N.A.	98	98	N.A.
P12	N.A.	92	92	N.A.	N.A.	84	84	N.A.
P13	N.A.	99	99	N.A.	N.A.	88	88	N.A.
Avg.	0.76	80	84	69	0.74	78	82	58
Std.	0.22	19	19	20	0.28	19	22	34

The selection of the models in the validation phase was performed using independent patients for validation; for this reason, such inaccuracies on test data was unexpected. To determine the reasons for the misclassifications, we used the CNN models to generate heat maps for all the patients, and we carefully examined them. After this analysis, we found that some HS images presented problems; hence, the results were worsened for these reasons. As mentioned before, we performed a careful inspection of tumor HS images in the validation set. However, upon inspection after the test outcomes, we discovered there were also problems in non-tumor samples. There were four main sources of errors in the images: (1) some HS images were contaminated with the ink used by pathologists to delimitate the diagnosed regions ($n = 15$); (2) some images were unfocused ($n = 13$); (3) some samples presented artifacts from histopathological processing ($n = 2$); and (4) other images were composed mainly by red blood cells ($n = 2$). Examples of these images can be observed in Figure 5.

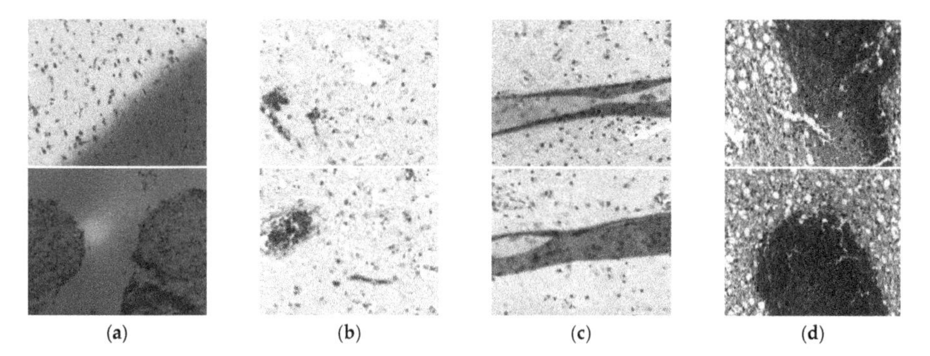

Figure 5. Example of image defects detected in the test dataset. (**a**) Ink contamination; (**b**) unfocused images; (**c**) artifacts in the specimens; (**d**) samples mainly composed of red blood cells.

Furthermore, due to the suspicious results obtained on patient P6, the specimen was examined again by a pathologist for reassessing the initial diagnosis. After this examination, the pathologist realized a problem with the selection of ROIs in the HS acquisition for the non-tumor areas. In Figure 6, we show the initial evaluation of the sample, where the tumor area was annotated by using a red marker contour, and the rest of the sample was considered as non-tumor. Figure 6b corresponds to the second evaluation of the sample. The original annotation of tumor was technically correct, but the yellow markers indicate the location of the highly invasive malignant tumor, i.e., GB. Although the other tumor areas correspond to tumor, their cells are atypical and cannot be considered a high-grade GB. In both Figure 6a,b, the ROIs selected for HS acquisitions are highlighted with squared boxes, where red and blue boxes indicate tumor and non-tumor ROIs, respectively. As can be observed in Figure 6b, the non-tumor areas selected for our experiments were located too close to areas where the infiltrating GB was identified; thus, they contain extensive lymphocytic infiltration and cannot be considered strictly non-tumor samples. Furthermore, it was found that the GB of this patient was not typical, presenting low cellular density in the tumor areas. Finally, the ROI selected from the tumor area was located where the diagnosis is tumor but cannot be considered a high-grade glioma, i.e., GB. These reasons explain the seemingly inaccurate results obtained in the classification. Nevertheless, such bad results helped us to find an abnormality in the sample.

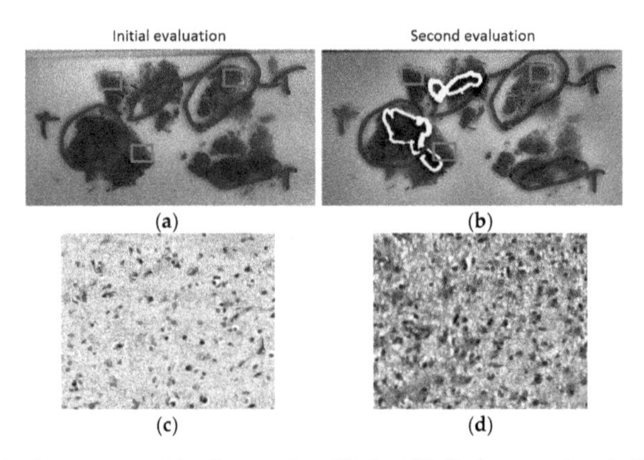

Figure 6. Evaluation assessment for the samples of Patient P6. Red pen markers indicate the initial evaluation of tumor regions. Regions without pen contour were considered as non-tumor. Red squares indicate the ROIs of tumor samples. Blue squares indicate the ROIs of non-tumor samples. (**a**) Initial evaluation of the sample; (**b**) second evaluation of the sample, where a yellow marker is used for the updated tumor areas; (**c**) example of HSI from tumor ROI; (**d**) example of HSI from non-tumor ROI.

In order to quantify the influence of the inclusion of incorrect HS images in the classification, we evaluated again the classifiers when the corrupted HS images were excluded from the dataset. These HS images were only removed from the test. The CNN was not trained again to avoid introducing bias in our experiments. These results are shown in Table 7. Patients where data exclusion was performed are indicated with an asterisk (*), and the results of patient P6 were removed due to the diagnosis reasons explained before. The results of the classification after data exclusion improved significantly for patients P2 and P7, while the results of other patients keep constant after the exclusion of some HS images. This data removal also boosts the overall metrics across the patients, due to the improvement in the classification in some patients and because of the removal of patient P6 due to justifiable clinical reasons.

Table 7. Final classification results on the test set after excluding incorrect HS images.

Patient	HSI				RGB			
	AUC	Accuracy (%)	Sensitivity (%)	Specificity (%)	AUC	Accuracy (%)	Sensitivity (%)	Specificity (%)
P1*	0.98	93	91	96	0.93	90	97	61
P2*	0.99	89	99	83	0.99	87	79	99
P3*	0.95	85	91	80	0.96	92	97	78
P4	0.62	57	57	58	0.69	77	98	7
P5*	0.81	69	81	64	0.66	58	57	60
P6†	-	-	-	-	-	-	-	-
P7*	0.74	66	71	63	0.68	58	50	77
P8	0.98	96	96	96	0.99	97	97	97
P9	N.A.	99	99	N.A.	N.A.	89	89	N.A.
P10	N.A.	89	89	N.A.	N.A.	43	43	N.A.
P11	N.A.	92	92	N.A.	N.A.	98	98	N.A.
P12	N.A.	92	92	N.A.	N.A.	84	84	N.A.
P13	N.A.	99	99	N.A.	N.A.	88	88	N.A.
Avg.	0.87	85	88	77	0.84	80	81	68
Std.	0.15	14	13	16	0.16	18	20	31

* Data exclusion; † Data removed.

Regarding the classification performance of HSI compared to RGB, the results suggest the superiority of HSI (see Tables 6 and 7). The average metrics on the whole datasets are worse for RGB images, especially in terms of specificity and sensitivity. We consider good performance in classification when all the metrics are high, with balanced specificity and sensitivity. For example, P4 presents a better AUC for RGB but really poor specificity (7%). For this reason, the HSI classification for such patient presents a better performance. Only for P2, P3, and P8, the performance of RGB is approximately equivalent to HSI. P11 is the only patient where RGB substantially outperforms HSI. For patients where the performance is the most promising (e.g., P1, P2, P3, and P8), RGB classification is also accurate. However, the sensitivity and specificity are not as balanced compared to HSI. Furthermore, the standard deviation in specificity and sensitivity are higher for RGB classification, which show a wider spread of the classification results compared to HSI. The decrease of performance of RGB images compared to HSI is more evident in patients with only tumor samples, where HSI classification was shown to be really accurate (e.g., P9, P10, P12, and P13). Finally, in patients where the classification of HSI was found poor (e.g., P4, P5, and P7), HSI performance is still shown to be more competitive than the RGB counterpart. On average, the accuracy of the classification is improved 5% when using HSI instead of RGB imaging, and particularly, the specificity and specificity are increased achieving 7% and 9% of improvement, respectively (Table 7).

3.3. Heat Map Results

Beyond the results obtained for the analysis of the patches, we also qualitatively evaluated the outcomes of the classification by generating classification heat maps from the HS images. In these maps, the probability of each pixel to be classified as tumor is represented, where red values indicate high probability and blue values indicate low probability. The inputs of our CNN are patches of 87 × 87 pixels, for this reason the resolution of the heat maps cannot contain pixel-level details. To provide them with resolution enough for a useful interpretation, we generated classification results for a 23-pixels sliding window length. We show two different types of heat maps in Figures 7 and 8.

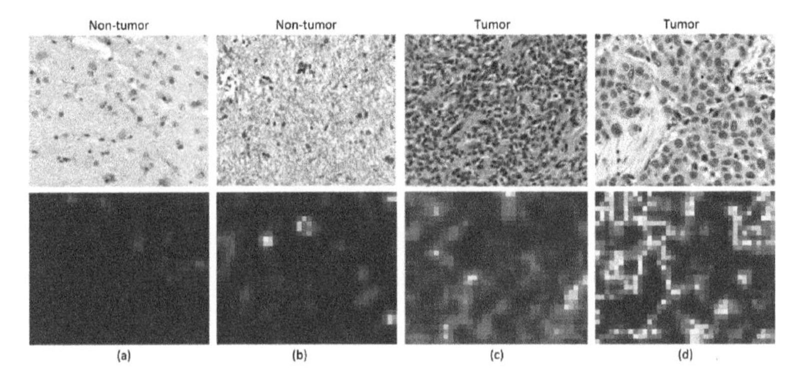

Figure 7. Heat maps from good performance patients. (**a**) Non-tumor tissue with no false positive; (**b**) non-tumor tissue with some false positives; (**c**) tumor tissue with no false negative; (**d**) tumor tissue with some false negatives.

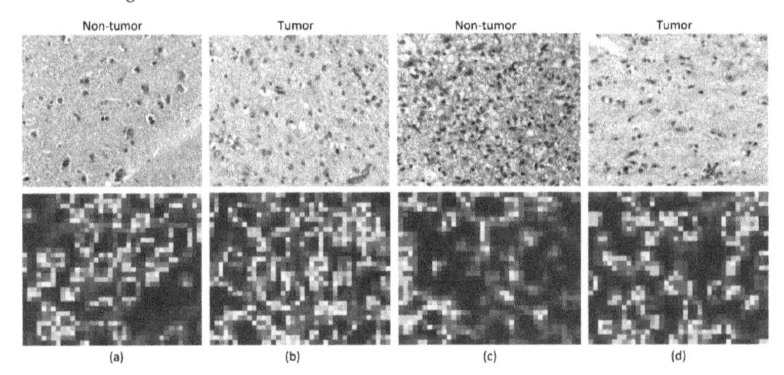

Figure 8. Heat maps from bad performing patients. (**a**,**b**) Non-tumor and tumor maps from Patient P4; (**c**,**d**) non-tumor and tumor maps from Patient P6.

On the one hand, in Figure 7, we illustrate the different types of results that are obtained in patients where the models were proven to classify accurately the samples. Figure 7a,c show examples of non-tumor and tumor images that were classified correctly, with no presence of neither false positives nor false negatives, respectively. In Figure 7b, we can see the presence of some false positives in a non-tumor tissue image, but such false positives are located in an area where there is a cluster of cells, indicating that it is a suspicious region. Finally, Figure 7d shows a tumor image where there are some regions classified as non-tumor tissue. Nevertheless, such false negatives are located in areas where there are no cells. The FP in Figure 7b suggests the CNN perceives areas with high cell density as tumor. The FN shown in Figure 7d has a clinical interpretation, but it is computed as a bad result in the quantitative evaluation of the classification. Furthermore, a more detailed ground truth scheme for classification may improve the classification performance, e.g., the inclusion of brain background tissue, blood vessels, or blood cells.

On the other hand, we show, in Figure 8, the heat maps from patients that present the worst performance in the quantitative evaluation of the results. Firstly, Figure 8a,b show the results for Patient P4. It can be observed that the heat maps for each kind of tissue are similar, presenting false positives for the non-tumor image and false negatives for the tumor image. For this patient, the heat maps and the quantitative results are coherent, showing that the CNN is not able to accurately classify the samples from this patient. Secondly, Figure 8c,d show the heat maps for Patient P6. As mentioned before, the non-tumor tissue of this patient was proven to be adjacent to the tumor area, and hence cannot be considered as non-tumor tissue. In this case, Figure 8c shows that the non-tumor area has been classified as tumor, which is in fact correct. Finally, it was also discussed that Patient P6 presented a non-typical GB with low cellularity. A heat map from a tumor image from this patient (Figure 8d) shows that tumor cells are highlighted as tumor, but the areas with low cellular presence are diagnosed as non-tumor.

4. Discussion and Conclusion

In this research work, we present a hyperspectral microscopic imaging system and a deep learning approach for the classification of hyperspectral images of H&E pathological slides of brain tissue samples with glioblastoma of human patients.

As described in the introduction, this research is the continuation of a previous research that presented some drawbacks [23]. Firstly, the total number of HS cubes used in our previous work was limited to 40, only 4 HS cubes per patient. Secondly, the instrumentation used in this previous work presented limitations in both the spectral and spatial information. Regarding the spectral information, the spectral range was restricted to 419–768 nm due to limitations of the microscope optical path. The spatial information was limited due to the use of a scanning platform unable to image the complete scene, so the analysis of the HS images was restricted to a low magnification (5×), which was not sufficient to image the morphological features of the sample. Additionally, the main goal of such previous work was to develop a preliminary proof-of-concept on the use of HSI for the differentiation of tumor and non-tumor samples, showing promising results.

In this work, an improved acquisition system capable of capturing high-quality images in a higher magnification (20×), and with a higher spectral range (400–1000 nm) has been used to capture a total amount of 517 HS cubes. The use of 20× magnification allows the classifier to exploit both the spectral and the spatial differences of the samples to make a decision.

Such dataset was then used to train a CNN and to perform the classification between non-tumor and tumor tissue. Due to a limited number of patients involved in this study and with the aim to provide a data partition scheme with minimum bias, we decided to split the dataset in four different folds where the training, validation, and testing data belonged to different patients. Each fold was trained with 9 patients, where only 5 of them presented both types of samples, i.e., tumor and non-tumor tissue.

After selecting models with high AUC and balanced accuracy, sensitivity and specificity in the validation phase, some results on the test set were not accurate at all. For this reason, we carefully inspected the heat maps generated by the classifiers for each patient in order to find a rationale about the inaccurate results. After this, we detected four types of problems in the images that could worsened the results, namely the presence of ink or artifacts in the images, unfocused images, or excess of red blood cells. We reported both results, before and after cleaning wrong HS images, for a fair experimental design. We consider that the test results after removing such defective HS images are not biased because the rationale of removing the images from the test set is justifiable and transparent. These corrupted images were part of the training set, but it is unknown if the training process of the CNN was affected.

We also found a patient, P6, where the results were really inaccurate. For this reason, the regions of interest that were analyzed by HS were re-examined by the pathologists. After examining the sample, an atypical subtype of GB was found, and examination revealed that the ROIs selected as normal samples were close to the tumor area, which cannot be considered as non-tumor. Although the classification results were not valid for this patient, by using the outcomes of the CNN, we were able to identify a problem with the prior examination and ROI selection within the sample. Additionally, although this patient was used both as part of the training and as a patient used for validation, the results are not significantly affected by this fact. These results highlight the robustness of the CNN for tumor classification. Firstly, although the validation results of fold 1 were good when evaluating patient P6, the model from this fold was also capable of accurately classifying patients P1 and P11. Secondly, although patient P6 was used as training data for fold 3 and fold 4, the outcomes of these models were not proven to be significantly affected by contaminated training data.

Although the results are not accurate in every patient, after excluding incorrectly labeled and contaminated HS images, nine patients showed accurate classification results (P1 to P3 and P8 to P13). Two patients provided acceptable results (P5 and P7), and only a single patient presented results that were slightly better than random guessing (P4). Nevertheless, these results can be considered promising for two main reasons. First, a limited number of samples were used for training, especially for the non-tumor class, which was limited to only five training patients for each CNN. Second, the high inter-patient variability shows significant differences between tumor samples among the different patients. As can be observed in the analysis of heat maps (Figures 7 and 8), there is a significant heterogeneity in cellular morphology in different patients' specimens, which makes GB detection an especially challenging application. To handle these challenges, the number of patients should be increased in future works, and to deal with the high inter-patient variability, HS data from more than a single patient should be used to validate the models.

Finally, we found that HSI data perform slightly better than RGB images for the classification. Such improvement is more evident when the classification is performed on challenging patients (e.g., P5 or P7) or in patients with only tumor samples. Furthermore, the classification results of HSI are shown to provide more balanced sensitivity and specificity, which is the goal for clinical applications, improving the average sensitivity and specificity by 7% and 9% with respect to the RGB imaging results, respectively. Nevertheless, more research should be performed to definitively demonstrate the superiority of HSI over conventional RGB imagery.

Author Contributions: Conceptualization, S.O., M.H., H.F., G.M.C. and B.F.; software, S.O. and M.H.; validation, S.O. and M.H.; investigation, S.O., M.H., and H.F.; resources, G.M.C. and B.F.; data curation, R.C. and M.d.l.L.P.; writing—original draft preparation, S.O., and M.H.; writing—review and editing, H.F., F.G., R.C., M.d.l.L.P., G.M.C., and B.F.; supervision, R.C., M.d.l.L.P., G.M.C. and B.F.; project administration, G.M.C., and B.F.; funding acquisition, G.M.C., and B.F. All authors have read and agreed to the published version of the manuscript.

Funding: This research is supported in part by the Cancer Prevention and Research Institute of Texas (CPRIT) grant RP190588. This work has been supported by the Spanish Government through PLATINO project (TEC2017-86722-C4-4-R), and the Canary Islands Government through the ACIISI (Canarian Agency for Research, Innovation and the Information Society), ITHaCA project "Hyperspectral identification of Brain tumors" under Grant Agreement ProID2017010164. This work was completed while Samuel Ortega was beneficiary of a pre-doctoral grant given by the "*Agencia Canaria de Investigacion, Innovacion y Sociedad de la Información (ACIISI)*" of the "*Conserjería de Economía, Industria, Comercio y Conocimiento*" of the "*Gobierno de Canarias*", which is part-financed by the European Social Fund (FSE) (*POC 2014-2020, Eje 3 Tema Prioritario 74 (85%)*).

Conflicts of Interest: The authors declare no conflict of interest. The founding sponsors had no role in the design of the study; in the collection, analyses, or interpretation of data; in the writing of the manuscript, and in the decision to publish the results.

Ethical Statements: Written informed consent was obtained from all of the participant subjects, and the study protocol and consent procedures were approved by the Comité Ético de Investigación Clínica-Comité de Ética en la Investigación (CEIC/CEI) of the University Hospital Doctor Negrin: IdenTificacion Hiperespectral de tumores CerebrAles (ITHaCA), Code: 2019-001-1.

References

1. Van Es, S.L. Digital pathology: Semper ad meliora. *Pathology* **2019**, *51*, 1–10. [CrossRef]
2. Flotte, T.J.; Bell, D.A. Anatomical pathology is at a crossroads. *Pathology* **2018**, *50*, 373–374. [CrossRef] [PubMed]
3. Madabhushi, A.; Lee, G. Image analysis and machine learning in digital pathology: Challenges and opportunities. *Med. Image Anal.* **2016**, *33*, 170–175. [CrossRef] [PubMed]
4. Song, Y.; Zhang, L.; Chen, S.; Ni, D.; Lei, B.; Wang, T. Accurate Segmentation of Cervical Cytoplasm and Nuclei Based on Multiscale Convolutional Network and Graph Partitioning. *IEEE Trans. Biomed. Eng.* **2015**, *62*, 2421–2433. [CrossRef] [PubMed]
5. Rezaeilouyeh, H.; Mollahosseini, A.; Mahoor, M.H. Microscopic medical image classification framework via deep learning and shearlet transform. *J. Med. Imaging* **2016**, *3*, 044501. [CrossRef] [PubMed]

6. Mishra, M.; Schmitt, S.; Wang, L.; Strasser, M.K.; Marr, C.; Navab, N.; Zischka, H.; Peng, T. Structure-based assessment of cancerous mitochondria using deep networks. In Proceedings of the 2016 IEEE 13th International Symposium on Biomedical Imaging (ISBI), Prague, Czech Republic, 13–16 April 2016; pp. 545–548.
7. Liu, Y.; Kohlberger, T.; Norouzi, M.; Dahl, G.E.; Smith, J.L.; Mohtashamian, A.; Olson, N.; Peng, L.H.; Hipp, J.D.; Stumpe, M.C. Artificial Intelligence–Based Breast Cancer Nodal Metastasis Detection: Insights Into the Black Box for Pathologists. *Arch. Pathol. Lab. Med.* **2019**, *143*, 859–868. [CrossRef] [PubMed]
8. Halicek, M.; Fabelo, H.; Ortega, S.; Callico, G.M.; Fei, B. In-Vivo and Ex-Vivo Tissue Analysis through Hyperspectral Imaging Techniques: Revealing the Invisible Features of Cancer. *Cancers (Basel)* **2019**, *11*, 756. [CrossRef]
9. Johansen, T.H.; Møllersen, K.; Ortega, S.; Fabelo, H.; Garcia, A.; Callico, G.M.; Godtliebsen, F. Recent advances in hyperspectral imaging for melanoma detection. *Wiley Interdiscip. Rev. Comput. Stat.* **2019**, *12*, e1465. [CrossRef]
10. Ortega, S.; Fabelo, H.; Iakovidis, D.; Koulaouzidis, A.; Callico, G.; Ortega, S.; Fabelo, H.; Iakovidis, D.K.; Koulaouzidis, A.; Callico, G.M. Use of Hyperspectral/Multispectral Imaging in Gastroenterology. Shedding Some–Different–Light into the Dark. *J. Clin. Med.* **2019**, *8*, 36. [CrossRef]
11. Campbell, M.J.; Baehner, F.; O'Meara, T.; Ojukwu, E.; Han, B.; Mukhtar, R.; Tandon, V.; Endicott, M.; Zhu, Z.; Wong, J.; et al. Characterizing the immune microenvironment in high-risk ductal carcinoma in situ of the breast. *Breast Cancer Res. Treat.* **2017**, *161*, 17–28. [CrossRef]
12. Jiang, C.; Huang, Y.-H.; Lu, J.-B.; Yang, Y.-Z.; Rao, H.-L.; Zhang, B.; He, W.-Z.; Xia, L.-P. Perivascular cell coverage of intratumoral vasculature is a predictor for bevacizumab efficacy in metastatic colorectal cancer. *Cancer Manag. Res.* **2018**, *10*, 3589–3597. [CrossRef] [PubMed]
13. Feng, Z.; Bethmann, D.; Kappler, M.; Ballesteros-Merino, C.; Eckert, A.; Bell, R.B.; Cheng, A.; Bui, T.; Leidner, R.; Urba, W.J.; et al. Multiparametric immune profiling in HPV– oral squamous cell cancer. *JCI Insight* **2017**, *2*, e93652. [CrossRef]
14. Bautista, P.A.; Abe, T.; Yamaguchi, M.; Yagi, Y.; Ohyama, N. Digital Staining of Unstained Pathological Tissue Samples through Spectral Transmittance Classification. *Opt. Rev.* **2005**, *12*, 7–14. [CrossRef]
15. Bayramoglu, N.; Kaakinen, M.; Eklund, L.; Heikkila, J. Towards Virtual H&E Staining of Hyperspectral Lung Histology Images Using Conditional Generative Adversarial Networks. In Proceedings of the 2017 IEEE International Conference on Computer Vision Workshops (ICCVW), Venice, Italy, 22–29 October 2017; pp. 64–71.
16. Ghamisi, P.; Plaza, J.; Chen, Y.; Li, J.; Plaza, A.J. Advanced Spectral Classifiers for Hyperspectral Images: A review. *IEEE Geosci. Remote Sens. Mag.* **2017**, *5*, 8–32. [CrossRef]
17. Awan, R.; Al-Maadeed, S.; Al-Saady, R. Using spectral imaging for the analysis of abnormalities for colorectal cancer: When is it helpful? *PLoS ONE* **2018**, *13*, e0197431. [CrossRef]
18. Wang, Q.; Li, Q.; Zhou, M.; Sun, L.; Qiu, S.; Wang, Y. Melanoma and Melanocyte Identification from Hyperspectral Pathology Images Using Object-Based Multiscale Analysis. *Appl. Spectrosc.* **2018**, *72*, 1538–1547. [CrossRef]
19. Ishikawa, M.; Okamoto, C.; Shinoda, K.; Komagata, H.; Iwamoto, C.; Ohuchida, K.; Hashizume, M.; Shimizu, A.; Kobayashi, N. Detection of pancreatic tumor cell nuclei via a hyperspectral analysis of pathological slides based on stain spectra. *Biomed. Opt. Express* **2019**, *10*, 4568. [CrossRef]
20. Litjens, G.; Kooi, T.; Bejnordi, B.E.; Setio, A.A.A.; Ciompi, F.; Ghafoorian, M.; van der Laak, J.A.W.M.; van Ginneken, B.; Sánchez, C.I. A survey on deep learning in medical image analysis. *Med. Image Anal.* **2017**, *42*, 60–88. [CrossRef]
21. Malon, C.; Cosatto, E. Classification of mitotic figures with convolutional neural networks and seeded blob features. *J. Pathol. Inform.* **2013**, *4*, 9. [CrossRef]
22. Haj-Hassan, H.; Chaddad, A.; Harkouss, Y.; Desrosiers, C.; Toews, M.; Tanougast, C. Classifications of multispectral colorectal cancer tissues using convolution neural network. *J. Pathol. Inform.* **2017**, *8*, 1.
23. Ortega, S.; Fabelo, H.; Camacho, R.; de la Luz Plaza, M.; Callicó, G.M.; Sarmiento, R. Detecting brain tumor in pathological slides using hyperspectral imaging. *Biomed. Opt. Express* **2018**, *9*, 818. [CrossRef] [PubMed]
24. Ortega, S.; Guerra, R.; Diaz, M.; Fabelo, H.; Lopez, S.; Callico, G.M.; Sarmiento, R. Hyperspectral Push-Broom Microscope Development and Characterization. *IEEE Access* **2019**, *7*, 122473–122491. [CrossRef]

25. Louis, D.N.; Perry, A.; Reifenberger, G.; von Deimling, A.; Figarella-Branger, D.; Cavenee, W.K.; Ohgaki, H.; Wiestler, O.D.; Kleihues, P.; Ellison, D.W. The 2016 World Health Organization Classification of Tumors of the Central Nervous System: A summary. *Acta Neuropathol.* **2016**, *131*, 803–820. [CrossRef] [PubMed]
26. Li, S.; Song, W.; Fang, L.; Chen, Y.; Ghamisi, P.; Benediktsson, J.A. Deep Learning for Hyperspectral Image Classification: An Overview. *IEEE Trans. Geosci. Remote Sens.* **2019**, *57*, 1–20. [CrossRef]
27. Abadi, M.; Agarwal, A.; Barham, P.; Brevdo, E.; Chen, Z.; Citro, C.; Corrado, G.S.; Davis, A.; Dean, J.; Devin, M.; et al. TensorFlow: Large-Scale Machine Learning on Heterogeneous Distributed Systems. *arXiv* **2016**, arXiv:1603.04467. Available online: https://arxiv.org/abs/1603.04467 (accessed on 29 March 2020).
28. Chollet, F. Keras: Deep learning for humans. GitHub Repos. 2015. Available online: https://github.com/keras-team/keras (accessed on 29 March 2020).
29. Bradley, A.P. The use of the area under the ROC curve in the evaluation of machine learning algorithms. *Pattern Recognit.* **1997**, *30*, 1145–1159. [CrossRef]
30. Halicek, M.; Dormer, J.; Little, J.; Chen, A.; Fei, B. Tumor detection of the thyroid and salivary glands using hyperspectral imaging and deep learning. *Biomed. Opt. Express* **2020**, *11*, 1383–1400. [CrossRef] [PubMed]

Article

Contactless Vital Signs Measurement System Using RGB-Thermal Image Sensors and Its Clinical Screening Test on Patients with Seasonal Influenza

Toshiaki Negishi [1], Shigeto Abe [2], Takemi Matsui [3], He Liu [4], Masaki Kurosawa [1], Tetsuo Kirimoto [1] and Guanghao Sun [1,*]

1. Graduate School of Informatics and Engineering, The University of Electro-Communications, 1-5-1 Chofugaoka, Chofu, Tokyo 182-8585, Japan; negishi@secure.ee.uec.ac.jp (T.N.); kurosawa@uec.ac.jp (M.K.); kirimoto@ee.uec.ac.jp (T.K.)
2. Takasaka Clinic, Fukushima 973-8407, Japan; rsh71841@nifty.com
3. Graduate School of System Design, Tokyo Metropolitan University, Tokyo 191-0065, Japan; tmatsui@tmu.ac.jp
4. School of Materials Science and Engineering, Harbin University of Science and Technology, Harbin 150000, China; he.liu@hrbust.edu.cn

* Correspondence: guanghao.sun@uec.ac.jp; Tel.: +81-42-443-5412

Received: 18 March 2020; Accepted: 10 April 2020; Published: 13 April 2020

Abstract: *Background:* In the last two decades, infrared thermography (IRT) has been applied in quarantine stations for the screening of patients with suspected infectious disease. However, the fever-based screening procedure employing IRT suffers from low sensitivity, because monitoring body temperature alone is insufficient for detecting infected patients. To overcome the drawbacks of fever-based screening, this study aims to develop and evaluate a multiple vital sign (i.e., body temperature, heart rate and respiration rate) measurement system using RGB-thermal image sensors. *Methods:* The RGB camera measures blood volume pulse (BVP) through variations in the light absorption from human facial areas. IRT is used to estimate the respiration rate by measuring the change in temperature near the nostrils or mouth accompanying respiration. To enable a stable and reliable system, the following image and signal processing methods were proposed and implemented: (1) an RGB-thermal image fusion approach to achieve highly reliable facial region-of-interest tracking, (2) a heart rate estimation method including a tapered window for reducing noise caused by the face tracker, reconstruction of a BVP signal with three RGB channels to optimize a linear function, thereby improving the signal-to-noise ratio and multiple signal classification (MUSIC) algorithm for estimating the pseudo-spectrum from limited time-domain BVP signals within 15 s and (3) a respiration rate estimation method implementing nasal or oral breathing signal selection based on signal quality index for stable measurement and MUSIC algorithm for rapid measurement. We tested the system on 22 healthy subjects and 28 patients with seasonal influenza, using the support vector machine (SVM) classification method. *Results*: The body temperature, heart rate and respiration rate measured in a non-contact manner were highly similarity to those measured via contact-type reference devices (i.e., thermometer, ECG and respiration belt), with Pearson correlation coefficients of 0.71, 0.87 and 0.87, respectively. Moreover, the optimized SVM model with three vital signs yielded sensitivity and specificity values of 85.7% and 90.1%, respectively. *Conclusion*: For contactless vital sign measurement, the system achieved a performance similar to that of the reference devices. The multiple vital sign-based screening achieved higher sensitivity than fever-based screening. Thus, this system represents a promising alternative for further quarantine procedures to prevent the spread of infectious diseases.

Keywords: contactless measurement; vital signs; RGB-thermal image processing; infection diseases

---

1. Introduction

Emerging infectious diseases are serious threats to global health. During the last two decades, there have been travel-related outbreaks of infectious diseases, such as severe acute respiratory syndrome and novel Coronavirus (2019-nCoV), around the world in 2003 and 2019 [1,2]. To contain the outbreak of emerging viral diseases, infrared thermography (IRT) has been applied for fever screening of passengers with suspected infection in many international quarantine stations [3–5]. IRT is an effective method for measuring elevated body temperature. However, monitoring body temperature alone is insufficient for accurate detection of infected patients, as IRT monitoring facial surface temperature can be affected by many factors such as antipyretic consumption [6]. The positive predictive values of fever-based screening using IRT vary from 3.5% to 65.4%, indicating the limited efficacy for detecting symptomatic passengers [7].

To overcome the drawbacks of fever-based screening, we previously proposed a screening method based on simultaneously measuring three vital signs—body temperature, heart rate (HR) and respiration rate (RR)—using multiple sensors, that is, medical radar, thermograph, photo-sensor and RGB cameras [8–10]. These three vital signs were included in the criteria of the systemic inflammatory response syndrome [11]. Symptoms of the most infectious diseases tend to include an elevated HR and RR; hence, a screening that combines these three vital signs will improve the precision of detecting patients with such symptoms. Therefore, we developed contact and contactless vital sign measurement systems to investigate the feasibility of our screening method (Figure 1). In brief, the contact-type system (Ver.1.0) comprises three sensors, that is, medical radar, photo-sensor and thermograph [8]. The medical radar detects tiny body surface movements caused by respiration, the thermograph measures the highest temperature of the face and the photo-sensor monitors pulse waves to calculate the HR. To enable a completely contactless system (Ver.2.0), we combined RGB and the thermal image to extract multiple vital signs from the facial image [10]. The RR can be measured by monitoring the temperature changes around the nasal and oral areas accompanying inspiration and expiration. The RGB camera measures the blood volume pulse (BVP) through variations in the light absorption from the human facial area. We tested the systems on patients with seasonal influenza and dengue fever and the results indicate a sensitivity ranging from 81.5–98% [12].

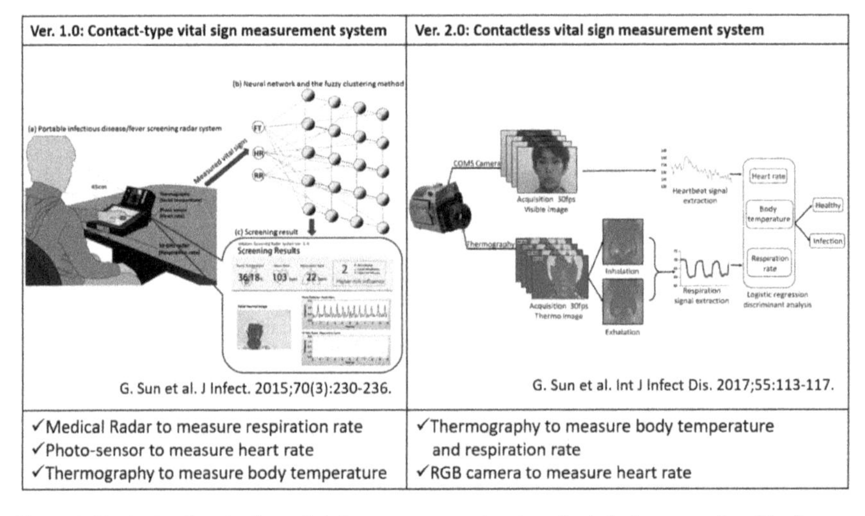

Figure 1. Contact and contactless vital sign measurement systems for infection screening. The figures were with copyright permission [8,10].

In this study, to promote the widespread use of our vital sign-based infection screening method, we enhanced the function of the Ver.2.0 contactless system to enable a stable, reliable and real-time system. We improved the stability of HR and RR measurement with the RGB-thermal image fusion approach for a highly reliable facial region-of-interest (ROI) tracking [13]. Moreover, we focused on improving the robustness of extracting BVP and respiration signal from the RGB camera and IRT. We proposed a signal processing method for reconstructing the BVP waveform using all RGB channels and selecting nasal or oral breathing based on signal quality index (SQI), for improving the signal-to-noise ratio. To enable a real-time system, we implemented a multiple signal classification (MUSIC) algorithm to estimate the pseudo-spectrum from limited time-domain BVP and respiration signals within 15 s [14]. Finally, we tested the system on 22 healthy subjects and 41 patients with influenza-like symptoms (28 diagnosed influenza patients and 13 undiagnosed patients).

The remainder of this paper is organized as follows. In the Section "Materials and Methods," we describe an overview of our system and proposed signal and image processing methods. The Section "Results" contains the results of comparison between our contactless system with contact-type reference devices and screening performance on detecting influenza patients using a support vector machine (SVM). In the Section "Discussion and Conclusion," we discuss our findings and draw conclusions.

2. Materials and Methods

2.1. Related Work on Vision based Clinical Screening

Vision-based clinical screening using RGB and thermal image sensors have recently attracted increasing attention in academia and industry. Ming-Zher Poh et al. developed a robust method for measuring HR and HRV from digital RGB video recording of skin color changes [15]. He Liu et al. proposed a novel method using dual cameras to estimate arterial oxygen saturation [16]. Philips Research has been launching an app called "*Vital Signs Camera*" in 2012. Moreover, the thermal camera-based approaches have been widely applied in clinical screening and research, such as fever screening and human pose estimation [5]. To enable such specific applications, image processing method for keypoint detection has been proposed using a stacked hourglass network and feature boosting networks [17–19].

2.2. Overview of Infectious Screening System using RGB-thermal Image Sensors

In our previous work, a dual image sensor-based infectious screening system was developed for predicting the possibility of infection [10]. It comprises an RGB camera and an IRT for measuring HR, RR and body temperature. We used DFK23U618 (The Imaging Source Co. Ltd., Germany) as the RGB camera and FLIR A315 (FLIR Systems, Inc., USA) as the IRT. The visible video was recorded at a speed of 15 frames per second (fps) with a pixel resolution of 640 × 480 and the thermal video was recorded at a speed of 15 fps with a pixel resolution of 320 × 240. An RGB camera senses fluctuations in hemoglobin absorption derived from the volumetric change in facial blood vessels and obtains heartbeat signals. An IRT detects temperature changes between inhalation and exhalation in the nasal or oral area. In addition, the facial skin temperature is measured by the IRT. Multiple vital signs distinguish between patients with influenza and healthy subjects. Figure 2 shows an overview of an infectious screening system.

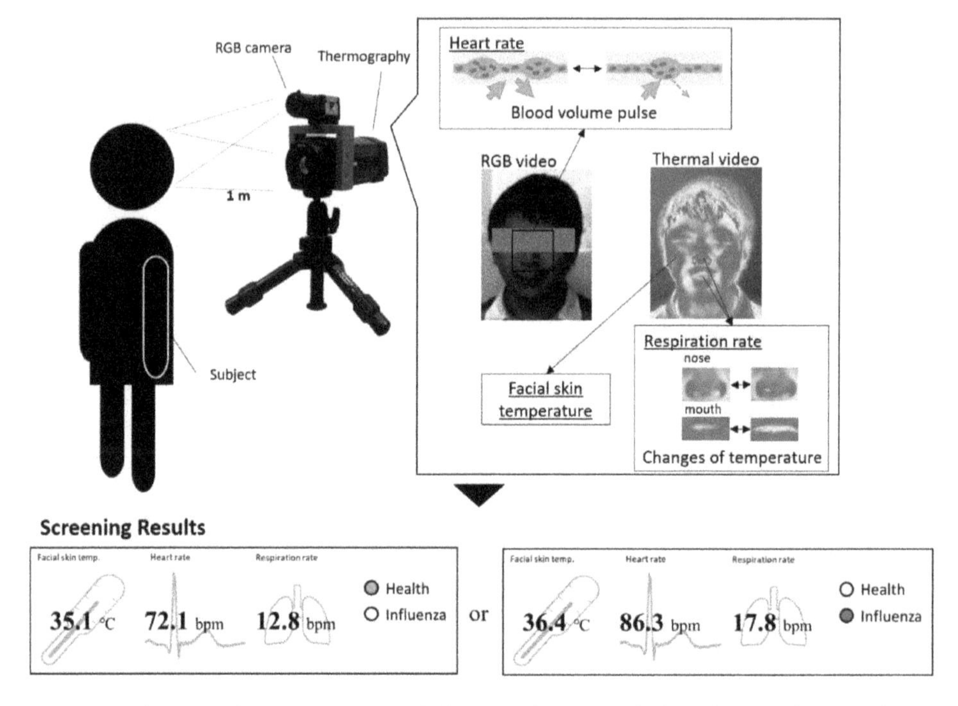

Figure 2. Overview of measurement principle that remotely senses multiple vital signs and an example of screening result.

2.3. Sensor Fusion Combining RGB sensor and IRT for ROI Detection

A stable measurement of the body temperature and RR using an IRT needs a detailed ROI detection of facial landmarks (i.e., face, nose and mouth) because temperature is estimated at the facial area and respiration occurs at the nose and mouth. An RGB camera can detect facial landmarks finely using previous methods [20]. Therefore, we introduced a sensor fusion method to obtain facial landmarks in a thermal video determined by an RGB video.

The facial landmarks in a thermal video are detected by homography of the RGB image coordinates of the nose and mouth, detected by "dlib" of an open-source library to thermal image coordinates. The homography between the images is represented by equation (1) and the homography matrix H is represented as

$$H = \begin{pmatrix} h_{11} & h_{12} & h_{13} \\ h_{21} & h_{22} & h_{23} \\ h_{31} & h_{32} & h_{33} \end{pmatrix}, \quad x_{thermo} = \frac{h_{11}x_{RGB}+h_{12}y_{RGB}+h_{13}}{h_{31}x_{RGB}+h_{32}y_{RGB}+h_{33}}, \quad y_{thermo} = \frac{h_{21}x_{RGB}+h_{22}y_{RGB}+h_{23}}{h_{31}x_{RGB}+h_{32}y_{RGB}+h_{33}}, \tag{1}$$

where x_{RGB}, y_{RGB}, x_{thermo} and y_{thermo} are image coordinates in the RGB and thermal images. Each h_{ij} $(i, j = 1, 2, 3)$ in Equation (1) is an element of the homography matrix H. Figure 3 shows a flowchart of image processing conducted to estimate the homography matrix H. Its standard is the face profile between the RGB and thermal images using pattern matching. First, from the RGB and thermal images shown in Figure 3a,b, the profile part is abstracted using the "grabcut" method [21] of OpenCV, to obtain the profile images shown in Figure 3c. The combination of coordinates between the images is found by obtaining the oriented fast and rotated BRIEF (ORB) characteristics of the two profile images and by performing a full search of the corresponding points from the characteristic points of each image obtained [22]. The homography matrix for the combination of image coordinates obtained is estimated using the random sample consensus method [23]. Finally, the facial landmarks in the thermal image (Figure 3e) are detected by applying the homography matrix to RGB's facial landmarks (Figure 3d).

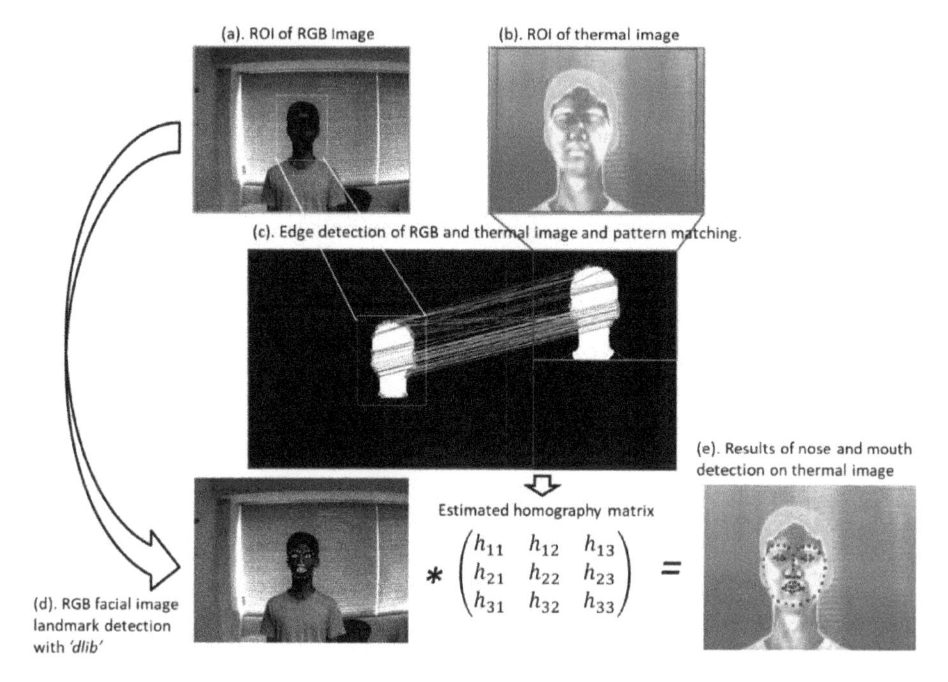

$$* \begin{pmatrix} h_{11} & h_{12} & h_{13} \\ h_{21} & h_{22} & h_{23} \\ h_{31} & h_{32} & h_{33} \end{pmatrix} =$$

Figure 3. Feature matching for region-of-interest (ROI) detection in thermal image. The figure reproduced with copyright permission from Reference [14].

2.4. RGB Sensor Processing for HR Estimation Using Tapered Window, Signal Reconstruction based on Softsig and MUSIC Algorithm

The fundamental method of HR estimation using an RGB camera has been described previously [15]. The RGB camera senses tiny color fluctuations in the facial skin with other noise. To remove the noise components, methods such as independent component analysis (ICA) and soft signature-based extraction (Softsig) [24] are used. In this study, we introduce the tapered window and signal reconstruction method into HR estimation for a stable measurement, which achieved an infection screening system. The observed RGB time-series data have components of heartbeat, motion artifact and noise from other light sources. The tapered window and signal reconstruction method is based on the Softsig demix heartbeat signal. Figure 4 shows an overview of HR estimation in this system.

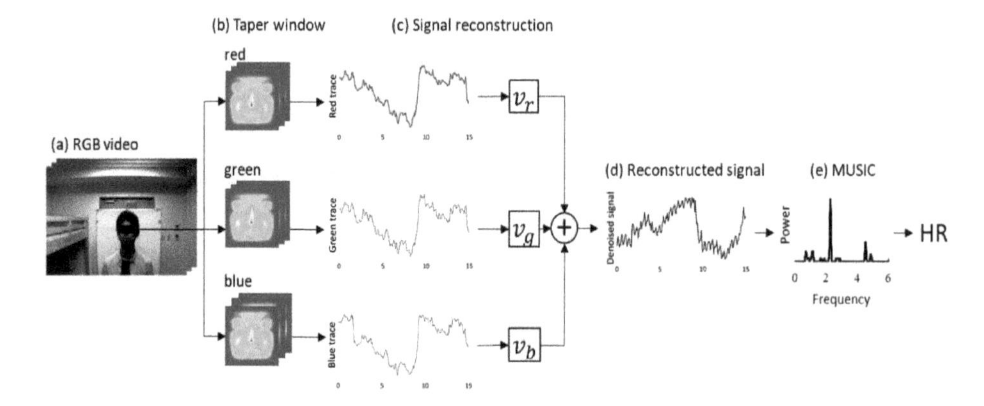

Figure 4. Block diagram of signal processing for HR estimation. (**a**) RGB video with ROI detected by OpenCV. (**b**) RGB ROI image applied to tapered window. (**c**) Raw RGB time-series data and reconstruction vector $V = \left(v_r, v_g, v_b\right)$ determined by kurtosis of spectra. (**d**) Reconstructed signal using V. (**e**) Power spectra obtained by MUSIC.

Tapered window, which is a general window function, was applied to the detected facial ROI (Figure 4b). In facial ROI, the edge area suffers from the lag affected by the face tracker. On the other hand, the ROI center can achieve a stable tracking of the facial skin. Therefore, we adopted tapered window to weighted ROI to reduce the noise raised by facial tracking. A 1d-tapered window is represented as

$$tapaer_{1d}(i) = \begin{cases} 0.5x(i)\left(1 - \cos\left(\frac{2\pi i}{2m}\right)\right) & (i = 0, 1, 2, \ldots, m-1) \\ 0.5x(i)\left(1 - \cos\left(\frac{2\pi(n-i-1)}{2m}\right)\right) & (i = n-m, \ldots, n) \\ x(i) & (otherwise), \end{cases} \tag{2}$$

where m indicates the tapered portion and has a value of $0.05 \cdot n$. To apply the tapered window to a 2d-image, the 2d-tapered window is expressed as

$$tapaer_{2d}(x, y) = taper_{1d}(x) \cdot taper_{1d}(y), \tag{3}$$

where x and y are the x-coordinates and y-coordinates of ROI, respectively.

The aim of signal reconstruction is to find a reconstruction vector $V = \left(v_r, v_g, v_b\right)$ for extracting the heartbeat signal by utilizing the difference among RGB absorption. Reconstructing a BVP signal using three RGB channels to optimize a linear function for improving the signal-to-noise ratio. According to a previous study, the reflection strength of the heartbeat is referred to as the relation in G>B>R order among the RGB channels. Using this relation, signal reconstruction can be expressed as

$$y(t) = v_r x_r(t) + v_g x_g(t) + v_b x_b(t), \tag{4}$$

where v_r, v_g, and v_b are the reconstruction vector. While this method is based on the Softsig method, we improved the determined method for vector V. To recover the pulse signal, we selected V to maximize the kurtosis of the spectra in the HR range of [0.75–4.0 Hz] (Figure 4c).

Finally, the MUSIC method was introduced to realize HR and RR measurements within a short time period. This method permits the realization of high-resolution HR and RR frequency estimation based on short-period measurement data Equation (5) expresses the spectrum estimation formula of the MUSIC method [14]:

$$S_{MUSIC}(f) = \frac{1}{\sum_{k=M+1}^{p} \left| e^{T}(f) W_k \right|^2} \times \frac{1}{\delta f}, \tag{5}$$

where $e(f_i)$ represents a complex sinusoidal wave vector and W_k represents the eigenvector of the correlation matrix. This system applies the MUSIC method separately to the HR and RR time-series data obtained from the video. In the case of heartbeat, the peak of 0.75–3.0 Hz (45–180 beats per minute (bpm)) of the obtained spectrum was assumed to be the HR.

2.5. IRT Sensor Processing for RR Estimation Using Nasal and Oral Breathing Decision based on SQI and MUSIC Algorithm and Body Temperature Estimation

The current approach of respiration measurement using an IRT is based on nasal temperature change. However, mouth breathing is reported in 17% of the total population [25]. For a stable RR measurement using an IRT, we must also measure oral temperature changes and select nasal or oral temperature changes dependent on strongly including respiration. To choose nasal or oral breathing, we quantified temperature traces via nasal and oral areas using SQI. Moreover, the MUSIC algorithm achieved rapid measurement for RR estimation. Figure 5 shows an overview of the respiration measurement that introduces nasal and oral breathing measurement method and MUSIC algorithm.

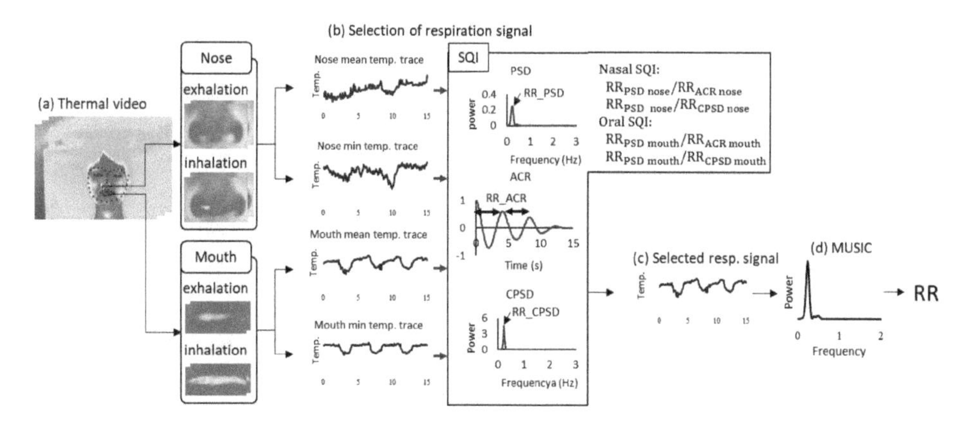

Figure 5. Block diagram of signal processing for respiration rate (RR) estimation. (**a**) Thermal video frame with facial landmark detected by the fusion sensor system described in Section 2. (**b**) Time-series data extracted from nasal and oral areas. (**c**) Respiration signal that chooses from four signals (b) based on SQI. (**d**) Power spectra obtained by MUSIC.

First, the nasal and oral areas were detected using the fusion sensor system described in Section 2. The possible respiration signals were extracted by the two areas. The mean temperature fluctuation $x_{mean}(t)$ in each ROI and the min temperature fluctuation $x_{min}(t)$ in each ROI are expressed as

$$x_{mean}(t) = \frac{1}{mn} \sum_{x=0}^{m-1} \sum_{y=0}^{n-1} I(x, y, t) x_{min}(t) = \min_{0<x<m-1,\, 0<y<n-1} I(x.y, t), \tag{6}$$

where *I(x,y,t)* is the pixel temperature at the image coordinate (*x*, *y*) in the ROI and time *t*, *m* is the width of the ROI and n is the height of the ROI. $x_{mean}(t)$ and $x_{min}(t)$ include the respiration signals.

Second, the respiration signal is selected from nasal and oral temperature traces using the four extracted signals: $x_{mean\ nose}(t)$, $x_{min\ nose}(t)$, $x_{mean\ mouth}(t)$ and $x_{minmouth}(t)$. Selection of the proposed respiration signal is conducted using the nasal SQI and oral SQI, based on the agreement of frequency estimated by power spectral density (PSD), autocorrelation (ACR) and cross-power spectral density (CPSD). The frequency of PSD using $x_{mean}(t)$ was estimated from the peak of power spectra from 0.1–0.75 Hz, to provide the range of RR measurement. The frequency of ACR using $x_{mean}(t)$ was estimated from the average peak interval. The frequency of CPSD using $x_{mean}(t)$ and $x_{min}(t)$ was estimated from the peak of cross-power spectra ranging from 0.1–0.75 Hz. If the temperature change in the nasal or oral area includes dominant respiration frequency, CPSD indicates the frequency by strengthening the respiration frequency between $x_{mean}(t)$ and $x_{min}(t)$ in the ROI. The following two rules are adopted sequentially:

1. Rule 1 (nasal SQI): If the ratio of $RR_{PSD\ nose}$ to $RR_{ACR\ nose}$ and that of $RR_{PSD\ nose}$ to $RR_{CSPD\ nose}$ obtained by the nasal area lie between 0.85 and 1.15, we select the nasal temperature change as the respiration signal. (This index shows that the nasal area includes the respiration signal because a ratio close to 1 indicates that the respiration frequency is dominant)
2. Rule 2 (oral SQI): If the ratio of $RR_{PSD\ mouth}$ to $RR_{ACR\ mouth}$ and that of $RR_{PSD\ mouth}$ to $RR_{CSPD\ mouth}$ obtained by the oral area lie between 0.85 and 1.15, we select the oral temperature change as the respiration signal. (This index shows that the oral area includes the respiration signal because a ratio close to 1 indicates that the respiration frequency is dominant)

If the two rules are not satisfied, we select nasal area as the respiration signal.

This system applies the MUSIC method separately to the HR and RR time-series data obtained from the video. In the case of respiration, the peak of 0.1– 0.75 Hz (6–45 bpm) of the spectrum obtained was assumed to be the RR. Temperature was also determined as the max facial temperature in the detected facial ROI using the sensor fusion technique.

2.6. SVM Discriminant Analysis to Predict Patients with Seasonal Influenza based on the Three Vital Signs Measured

Aiming at screening using features of HR, RR and body temperature of patients with infection, we proposed a classification model based on SVM. SVM is a method that predicts the separating hyperplane to maximize the margin between the two classes and achieves a high generalization capability. The SVM discriminant function is defined as

$$\begin{aligned} &\min_{w,\ w_0,\ \xi} \left(\frac{1}{2}\|w\|^2 + C\sum_{i=0}^{N} \xi_i \right) \\ &\text{subject to} \begin{cases} y_i f(x_i) \geq 1 - \xi_i \\ \xi_i \geq 0 \end{cases}, \end{aligned} \tag{7}$$

where w is a constant that indicates the SVM coefficients corresponding to HR, RR and temperature; y_i is a category of health or infection; C is the penalty parameter and ξ_i is the slack parameter; $f(x_i)$ is linear discriminant function formula $w \cdot x_i + w_0$. The calculation of SVM is performed using the MATLAB software.

2.7. Evaluation of the System in Laboratory and Clinical Settings

Laboratory and clinical testing of the system was conducted in 2019. Twenty-two healthy control subjects with no symptoms of fever (23.4 years of average age) participated in the laboratory test at the University of Electro-Communications. A total of 41 patients (45.0 years of average age) with symptoms such as influenza were included, who visited Takasaka Clinic, Fukushima, Japan. Their RR, HR and body temperature were measured using the contactless system; reference measurements were simultaneously obtained using a contact-type electrocardiogram (ECG) (LRR-03, GMS Co. Ltd., Tokyo, Japan) or pulse oximeter (SAT-2200 Oxypal mini, NIHONKOHDEN Co., Tokyo, Japan), clinical thermometer (TERUMO electric thermometer C230, TERUMO Co., Tokyo, Japan) and a respiration effort belt (DL-231, S&ME Inc.,Tokyo, Japan). It should be noted that, some patients may show increased heart rate due to white-coat hypertension. This study was approved by the Committee on

Human Research of the Faculty of System Design, Tokyo Metropolitan University and the University of Electro-Communications. All subjects gave their informed written consent.

2.8. Statistical Analysis

The Bland–Altman plot and scatter plot were utilized for statistical and graphical proof of the agreement between the proposed method and reference method [26]. The reference vital signs were measured by ECG or a pulse oximeter for HR, respiration effort belt for RR and electronic thermometer for axillary temperature. The results from the SVM classification model were used to calculate the sensitivity, specificity negative predictive value (NPV) and positive predictive value (PPV). A leave-one-out cross-validation was performed to avoid overfitting.

3. Results

3.1. HR Measurements Using RGB Sensor in a Laboratory and Clinical Setting

Figure 6 presents an example of signal recovery applied using the proposed method, by employing the tapered window and signal reconstruction based on Softsig. Raw traces of RGB color (Figure 6a) contained a dominant frequency of noise components, which can be observed by their spectra (Figure 6b), because the ground truth of HR measured by the pulse oximeter is 1.83 Hz. However, applying the proposed method, we can observe a clear peak of the HR frequency component in Figure 6e. This example shows the advantage of the proposed HR estimation.

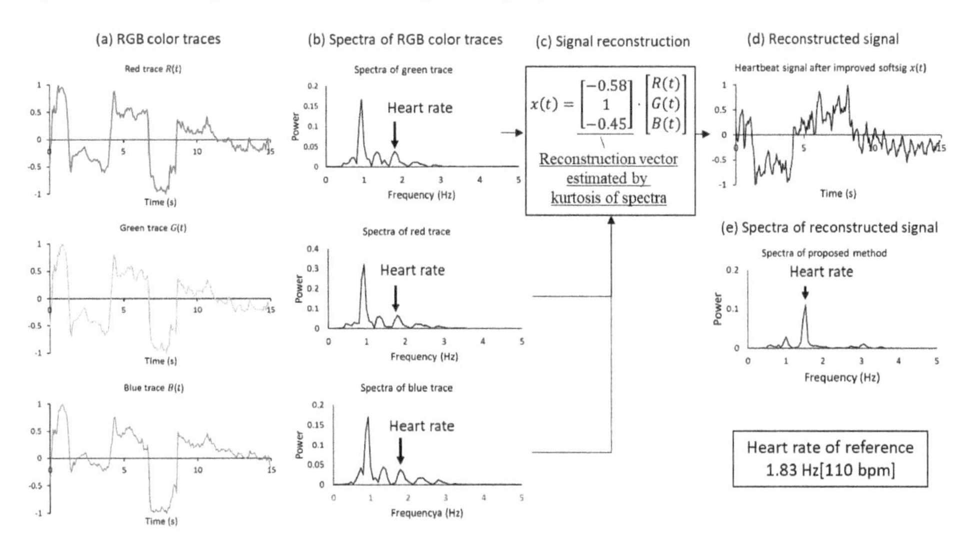

Figure 6. Recovery of heartbeat signal by applying tapered window and signal reconstruction. (**a**) RGB color traces obtained by RGB video. (**b**) Spectra estimated by Fast Fourier Transform (FFT). (**c**) Signal reconstruction determined through kurtosis of the spectra. (**d**), (**e**) Reconstructed signal and its spectra.

To evaluate the tapered window, signal reconstruction and MUSIC, we compared the proposed method to raw green trace, which uses only green channel and Fast Fourier Transform (FFT). The green trace method is a general method for estimating HR using an RGB camera. The ground truth of HR was measured by ECG and the pulse oximeter. We performed 15 s measurement four times against healthy control subjects and obtained 128 pairs of HRs from all subjects, which included 22 healthy control subjects and 41 patients with influenza-like symptoms. A comparison of HR estimation is shown in Figure 7. Figure 7a shows the Bland–Altman plot of green trace applying FFT. The 95% limits of agreement ranged from -23.5 to 33.4 bpm (standard deviation $\sigma = 14.5$) and the root mean square error (RMSE) was 15.3. Figure 7c shows the scatter plot of the green trace method; the Pearson correlation coefficient was 0.48. Figure 7b shows the Bland–Altman plot of the proposed method, which applies the tapered window, signal reconstruction and MUSIC. The 95% limits of agreement ranged from -10.4 to 12.6 bpm (standard deviation $\sigma = 5.85$) and RMSE was 5.93. Figure 7d shows the scatter plot of the proposed method; the Pearson correlation coefficient was 0.87. The results showed that the proposed method can reduce the 95% limits of agreement from [−23.5, 33.4] to [−10.4, 12.6] bpm. Especially, the result of patients with influenza-like illness (red circle) was improved because the experiment at a clinic is close to a real-world setting.

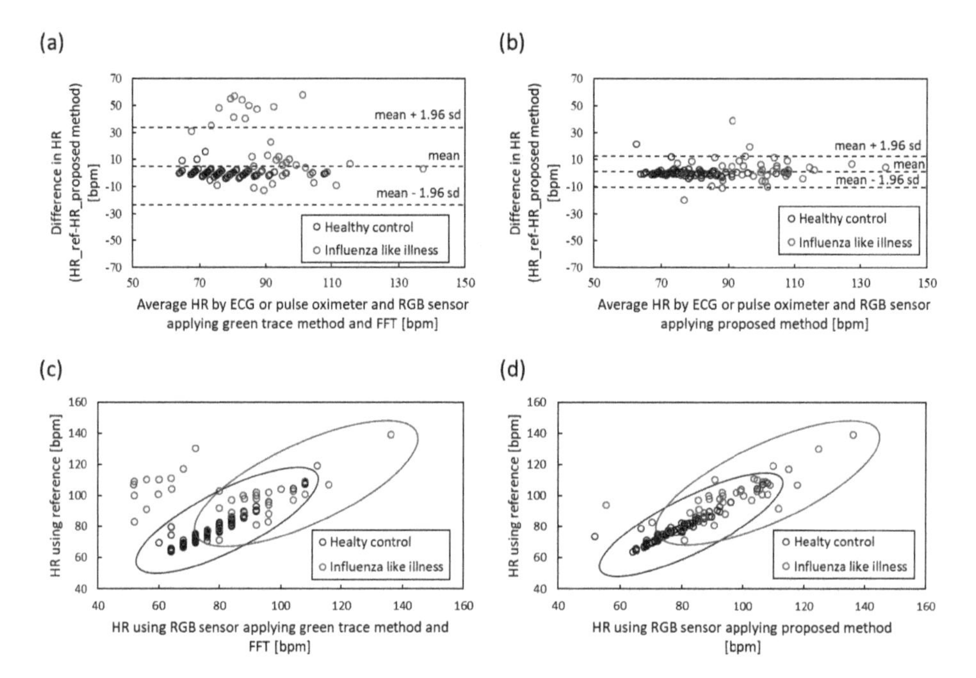

Figure 7. Bland–Altman plots and scatter plots of heart rate (HR) obtained by RGB sensor and electrocardiogram (ECG) or pulse oximeter. (**a**) Bland–Altman plot of raw green trace method applying FFT. (**b**) Bland–Altman plot of the proposed method applying tapered window, signal reconstruction and MUSIC. (**c**) Scatter plot of raw green trace. (**d**) Scatter plot of proposed method.

3.2. RR and Body Temperature Measurements Using IRT at a Laboratory and Clinical Settings

Figure 8 shows an example of the signal selection applied by the proposed method, which is detailed in Section 2. The mean and minimum temperature changes in each ROI are shown in Figure 8b,d. To determine the respiration signal from four signals, we calculated the SQI parameters, which included the PSD, ACR and CPSD of each signal (Figure 8c,e). Using the SQI parameters, we chose the respiration signal.

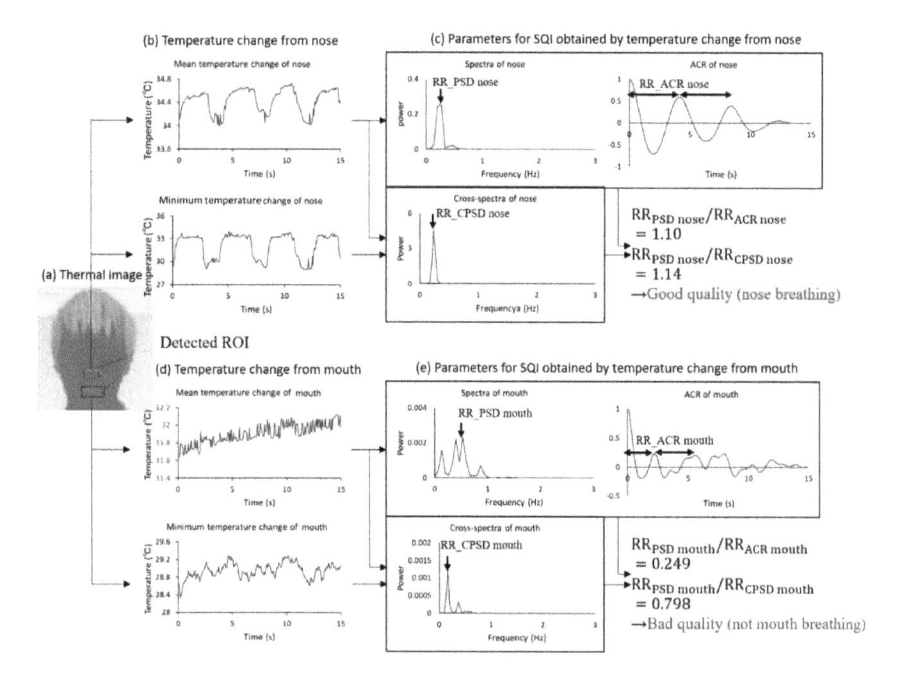

Figure 8. Determination of respiration signal applying nasal and oral breathing decision based on SQI. (**a**) Thermal facial image with ROI. (**b**) Mean and minimum temperature fluctuations in nasal area. (**c**) SQI parameter obtained by power spectral density (PSD), autocorrelation (ACR) and cross-power spectral density (CPSD) of nasal temperature changes. (**d**) Mean and minimum temperature fluctuations in oral area. (**e**) SQI parameter obtained by PSD, ACR and CPSD.

To evaluate the nasal or oral breathing decision based on SQI and MUSIC, we compared the proposed method with the raw temperature change in the nasal area applied to FFT, which is a general method for estimating RR using IRT. The ground truth of RR was measured using the respiratory effort belt. We performed 15 s measurement four times and obtained 88 pairs of RRs from 22 healthy control subjects, including 6 subjects with nose clip for instructing subjects to mouth breathing. A comparison of RR estimation is shown in Figure 9. Figure 9a shows the Bland–Altman plot of nasal temperature change. The 95% limits of agreement ranged from -7.60 to 7.99 bpm (standard deviation $\sigma = 3.98$) and the RMSE was 3.98. Figure 9c shows the scatter plot of nasal temperature change; the Pearson correlation coefficient was 0.53. Figure 9b shows the Bland–Altman plot of the proposed method. The 95% limits of agreement ranged from -2.97 to 3.67 bpm (standard deviation $\sigma = 1.68$) and the RMSE was 1.73. Figure 9d shows the scatter plot of the proposed method; the Pearson correlation coefficient was 0.87. The results showed that the proposed method can reduce the 95% limits of agreement from [−7.60, 7.99] bpm to [−2.97, 3.67] bpm.

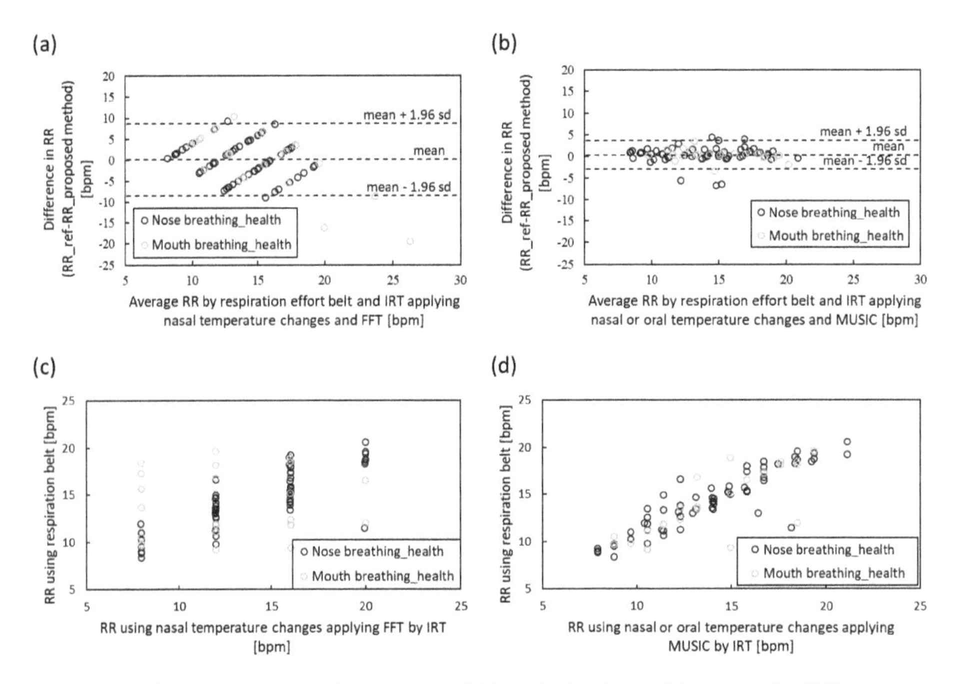

Figure 9. Bland–Altman plots and scatter plots of RR obtained by infrared thermography (IRT) sensor and respiratory effort belt. (**a**) Bland–Altman plot of nasal temperature change under the application of FFT. (**b**) Bland–Altman plot of the proposed method applying nasal or oral signal selection using SQI and MUSIC. (**c**) Scatter plot of nasal temperature change under FFT application. (**d**) Scatter plot of the proposed method.

Facial temperature, which is estimated by ROI detection using sensor fusion, was also evaluated. The ground truth of the temperature was measured using an electric thermometer. From all subjects, which included 22 healthy control subjects and 41 patients with influenza-like symptoms, a comparison of temperature estimation is shown in Figure 10. Figure 10a shows the Bland–Altman plot of temperature. The 95% limits of agreement ranged from -0.45 to 2.56 °C (standard deviation $\sigma = 0.77$) and the RMSE was 1.30. Figure 10b shows the scatter plot; the Pearson correlation coefficient was 0.71.

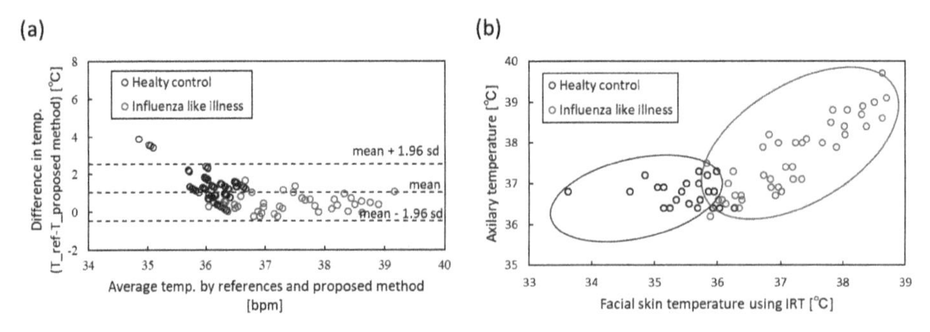

Figure 10. Bland–Altman plots and scatter plots of body temperature obtained by IRT sensor and electric thermometer. (**a**) Bland–Altman plot. (**b**) Scatter plot.

3.3. *Classification of Healthy Control Subjects and Influenza Patients*

SVM established a classification model using three vital signs, including HR, RR and temperature, estimated by RGB and IRT sensors. The vital signs were measured for 22 healthy control subjects and 28 influenza patients (45.5 years of average age) diagnosed as influenza using virus isolation from all 41 patients with influenza-like symptoms. Figure 11a illustrates the distribution of the vital signs (22 blue dots: healthy control subjects, 28 red dots: influenza patients) and the separating hyperplane obtained by SVM using all data. SVM classification using the three vital signs achieved more accurate screening than fever-based classification (Figure 11b). Figure 11c presents the result obtained through leave-one-out cross-validation. The sensitivity, specificity, NPV and PPV were 85.7%, 90.1%, 83.3% and 92.3%, respectively. The fever-based screening using an electric thermometer was adopted to compare SVM classification. The sensitivity and specificity were 60.7% and 86.4%, respectively.

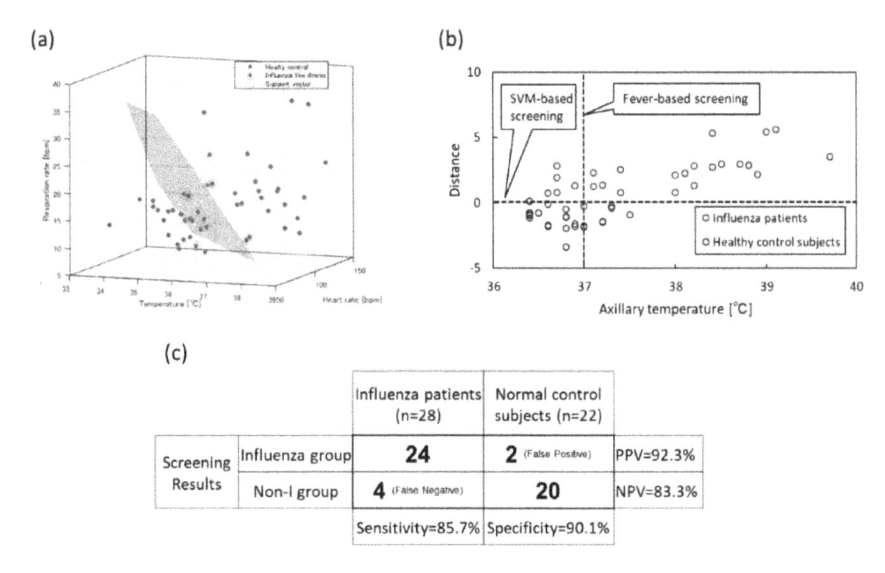

		Influenza patients (n=28)	Normal control subjects (n=22)	
Screening Results	Influenza group	24	2 (False Positive)	PPV=92.3%
	Non-I group	4 (False Negative)	20	NPV=83.3%
		Sensitivity=85.7%	Specificity=90.1%	

Figure 11. Classification model based on Support Vector Machine (SVM). (**a**) SVM classification. (**b**) Confusion matrix.

4. Discussion and Conclusions

The outbreak of 2019-nCoV was first reported in Wuhan, China, in December 2019 and was confirmed to have spread to more than 110 countries as of March 2020. When such a novel virus outbreaks, enhanced public health quarantine and isolation is essential. For this purpose, we developed a multiple vital sign measurement system for the mass screening of infected individuals in places of mass gathering. In this study, we focused on developing our system to measure three vital signs, to achieve automation, stability and swiftness for practical use in real-world settings. From a technical perspective, we proposed specific signal and image processing methods for highly reliable vital sign measurements and compared them with conventional methods (Tables 1 and 2). Tapered window, RGB signal reconstruction and MUSIC were applied for HR measurement. Automatic ROI tracking using sensor fusion and nasal or oral breathing selection using SQI and MUSIC were applied for HR measurement. The proposed method showed agreement with their reference devices (HR: [−10.4, 12.6] bpm, RR: [−2.97, 3.67] bpm, temperature: [−0.449, 2.56] °C). The reliability and stability of our system on vital sign measurement were significantly improved.

Table 1. Comparison of proposed RGB signal reconstruction method with conventional green trace method on HR measurement.

HR	RMSE	Bland–Altman	Pearson Correlation
RGB signal reconstruction and MUSIC	5.93	The 95% limits of agreement -10.4 to 12.6 bpm ($\sigma = 5.85$)	0.87
Green trace alone and FFT	15.30	The 95% limits of agreement -23.5 to 33.4 bpm ($\sigma = 14.5$)	0.48

Table 2. Comparison of proposed Nasal/oral SQI method with conventional nasal alone method on RR measurement.

RR	RMSE	Bland–Altman	Pearson Correlation
Nasal or oral SQI and MUSIC	1.73	The 95% limits of agreement -2.97 to 3.67 bpm ($\sigma = 1.68$)	0.89
Nasal and FFT	3.98	The 95% limits of agreement -7.60 to 7.99 bpm ($\sigma = 3.98$)	0.53

Moreover, we tested multiple vital sign-based screening in a laboratory and a clinic. The proposed method's sensitivity and specificity (85.7%, 90.1%) were found to be higher than those of fever-based screening (60.7%, 86.4%). The tendency of the three vital signs measured by healthy control subjects and influenza patients is shown in Figure 12. The medians of facial skin temperature of influenza patients and healthy control subjects were 37.3 and 35.5 °C, respectively. The medians of HR of influenza patients and healthy control subjects were 99.3 and 76.4 bpm. The medians of RR of influenza patients and healthy control subjects were 18.9 and 14.0 bpm. Each vital sign of patients with influenza was found to be elevated. This contributed to improvement in SVM classification based on the three vital signs.

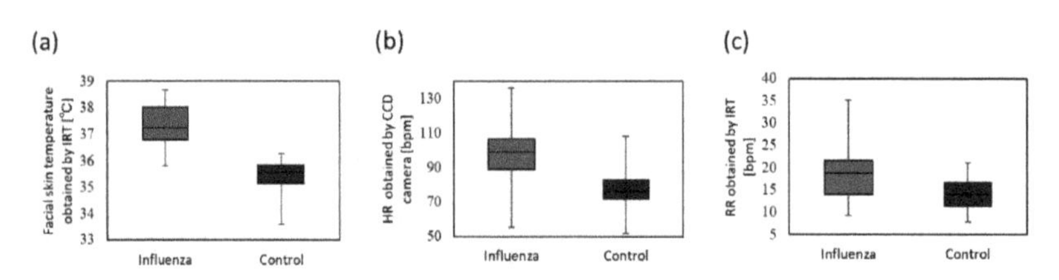

Figure 12. Box plot of vital signs between influenza patients and healthy control subjects. (**a**) Facial skin temperature. (**b**) HR. (**c**) RR.

However, the proposed method has some limitations. The ROI detection of sensor fusion may fail when the background has the color of skin or hair. In terms of the classification test based on SVM, the facial skin temperature may include the influence of the ambient environment. The measurement environment at a laboratory is different from that at a clinic, even at the same ambient temperature. This causes a difference in facial skin temperature regardless of the seasonal influenza. Therefore, we need to develop environment-invariant temperature estimation using an IRT.

In conclusion, we proposed automatic, stable and rapid HR, RR and body temperature measurements using an RGB-thermal sensor and its application for the screening of infectious diseases. This method introduces (1) the sensor fusion approach for the detection of detailed facial landmarks in a thermal image, (2) HR estimation, which introduces tapered window, signal reconstruction and MUSIC and (3) RR estimation, which implements nasal or oral breathing selection using SQI and MUSIC.

Moreover, we demonstrated a classification model based on SVM using healthy control subjects and patients with seasonal influenza. The results indicate that the proposed method is indispensable for the high performance of contactless multiple vital sign measurements for infection screening.

Author Contributions: Conceptualization, G.S., S.A. and T.M.; methodology, T.N., T.M., H.L., M.K., T.K. and G.S.; software, T.N., G.S.; validation, T.N., G.S., S.A. and T.M.; formal analysis, T.N., G.S.; investigation, G.S.; resources, G.S.; data curation, T.N., G.S., S.A.; writing—original draft preparation, T.N., G.S.; visualization, T.N., G.S.; supervision, G.S.; project administration, G.S.; funding acquisition, G.S. All authors have read and agreed to the published version of the manuscript.

Funding: This research was funded by in part by the JSPS KAKENHI Grant-in-Aid for Scientific Research (B) under Grant 19H02385, The Okawa Foundation for Information and Telecommunications and in part by the National Science Foundation Program of China under Grant 61801149.

Conflicts of Interest: The authors declare no conflict of interest.

References

1. Parashar, U.D.; Anderson, L.J. Severe acute respiratory syndrome: Review and lessons of the 2003 outbreak. *Int. J. Epidemiol.* **2004**, *33*, 628–634. [CrossRef] [PubMed]
2. Hui, D.S.; IAzhar, E.; Madani, T.A.; Ntoumi, F.; Kock, R.; Dar, O.; Ippolito, G.; Mchugh, T.D.; Memish, Z.A.; Drosten, C.; et al. The continuing 2019-nCoV epidemic threat of novel coronaviruses to global health—The latest 2019 novel coronavirus outbreak in Wuhan, China. *Int. J. Infect. Dis.* **2020**, *91*, 264–266. [CrossRef] [PubMed]
3. Ng, E.Y.; Kaw, G.J.; Chang, W.M. Analysis of IR thermal imager for mass blind fever screening. *Microvasc. Res.* **2004**, *68*, 104–109. [CrossRef] [PubMed]
4. Chiang, M.F.; Lin, P.W.; Lin, L.F.; Chiou, H.Y.; Chien, C.W.; Chu, S.F. Mass screening of suspected febrile patients with remote-sensing infrared thermography: Alarm temperature and optimal distance. *J. Formos. Med. Assoc.* **2008**, *107*, 937–944. [CrossRef]
5. Sun, G.; Matsui, T.; Kirimoto, T.; Yao, Y.; Abe, S. Applications of infrared thermography for noncontact and noninvasive mass screening of febrile international travelers at airport quarantine stations. In *Application of Infrared to Biomedical Sciences*; Ng, E.Y.K., Etehadtavakol, M., Eds.; Springer: Singapore, 2017; pp. 347–358.
6. Nishiura, H.; Kamiya, K. Fever screening during the influenza (H1N1-2009) pandemic at Narita International Airport, Japan. *BMC Infect Dis.* **2011**, *11*, 111. [CrossRef] [PubMed]
7. Bitar, D.; Goubar, A.; Desenclos, J.C. International travels and fever screening during epidemics: A literature review on the effectiveness and potential use of non-contact infrared thermometers. *Eurosurveillance* **2009**, *12*, 19115.
8. Sun, G.; Matsui, T.; Hakozaki, Y.; Abe, S. An infectious disease/fever screening radar system which stratifies higher-risk patients within ten seconds using a neural network and the fuzzy grouping method. *J. Infect.* **2015**, *70*, 230–236. [CrossRef] [PubMed]
9. Yao, Y.; Sun, G.; Matsui, T.; Hakozaki, Y.; van Waasen, S.; Schiek, M. Multiple vital-sign-based infection screening outperforms thermography independent of the classification algorithm. *IEEE Trans. Biomed. Eng.* **2016**, *63*, 1025–1033. [CrossRef] [PubMed]
10. Sun, G.; Nakayama, Y.; Dagdanpurev, S.; Abe, S.; Nishimura, H.; Kirimoto, T.; Matsui, T. Remote sensing of multiple vital signs using a CMOS camera-equipped infrared thermography system and its clinical application in rapidly screening patients with suspected infectious diseases. *Int. J. Infect. Dis.* **2017**, *55*, 113–117. [CrossRef] [PubMed]
11. Kaukonen, K.M.; Bailey, M.; Pilcher, D.; Cooper, D.J.; Bellomo, R. Systemic inflammatory response syndrome criteria in defining severe sepsis. *N. Engl. J. Med.* **2015**, *372*, 1629–1638. [CrossRef] [PubMed]
12. Sun, G.; Trung, N.V.; Matsui, T.; Ishibashi, K.; Kirimoto, T.; Furukawa, H.; Hoi, L.T.; Huyen, N.N.; Nguyen, Q.; Abe, S.; et al. Field evaluation of an infectious disease/fever screening radar system during the 2017 dengue fever outbreak in Hanoi, Vietnam: A preliminary report. *J. Infect.* **2017**, *75*, 593–595. [CrossRef] [PubMed]
13. Negishi, T.; Sun, G.; Liu, H.; Sato, S.; Matsui, T.; Kirimoto, T. Stable contactless sensing of vital signs using RGB-thermal image fusion system with facial tracking for infection screening. *Conf. Proc. IEEE Eng. Med. Biol. Soc.* **2018**, 4371–4374. [CrossRef]

14. Negishi, T.; Sun, G.; Sato, S.; Liu, H.; Matsui, T.; Abe, S.; Nishimura, H.; Kirimoto, T. Infection screening system using thermography and CCD camera with good stability and swiftness for non-contact vital-signs measurement by feature matching and MUSIC algorithm. *Conf. Proc. IEEE Eng. Med. Biol. Soc.* **2019**, 3183–3186. [CrossRef]
15. Poh, M.Z.; McDuff, D.J.; Picard, R.W. Advancements in noncontact, multiparameter physiological measurements using a webcam. *IEEE Trans. Biomed. Eng.* **2011**, *58*, 7–11. [CrossRef] [PubMed]
16. Liu, H.; Ivanov, K.; Wang, Y.; Wang, L. A novel method based on two cameras for accurate estimation of arterial oxygen saturation. *BioMed. Eng. Online* **2015**, *14*, 52. [CrossRef] [PubMed]
17. Newell, A.; Yang, K.; Deng, J. Stacked hourglass networks for human pose estimation. In Proceedings of the European Conference on Computer Vision, Amsterdam, The Netherlands, 8–16 October 2016.
18. Liu, J.; Ding, H.; Shahroudy, A.; Duan, L.; Jiang, X.; Wang, G.; Kot, A. Feature boosting network for 3D pose estimation. *IEEE Trans. Pattern Anal. Mach. Intell.* **2016**, *42*, 494–501. [CrossRef] [PubMed]
19. Nibali, A.; He, Z.; Morgan, S.; Prendergast, L. 3D human pose estimation with 2D marginal heatmaps. In Proceedings of the IEEE Winter Conference on Applications of Computer Vision (WACV), Waikoloa Village, HI, USA, 8–10 January 2019.
20. Kazemi, V.; Sulivan, J. One millisecond face alignment with an ensemble of regression trees. In Proceedings of the IEEE Conference on Computer Vision and Pattern Recognition, Columbus, OH, USA, 23–28 June 2014.
21. Rother, C.; Kolmogorov, V.; Blake, A. "GrabCut"—Interactive foreground extraction using iterated graph cuts. In *ACM Transactions on Graphics Siggraph*; Association for Computing Machinery: New York, NY, USA, 2004; pp. 309–314.
22. Rublee, E.; Rabaud, V.; Konolige, K.; Bradski, G. ORB: An efficient alternative to SIFT or SURF. In Proceedings of the 2011 International Conference on Computer Vision, Barcelona, Spain, 6–13 November 2011; pp. 2564–2571.
23. Raguram, R.; Chum, O.; Pollefeys, M.; Matas, J.; Frahm, J.M. USAC: A universal framework for random sample consensus. *IEEE Trans. Pattern Anal. Mach. Intel.* **2012**, *35*, 2022–2038. [CrossRef] [PubMed]
24. Wang, W.; den Brinker, A.C.; de Haan, G. Single element remote-PPG. *IEEE Trans. Biomed. Eng.* **2018**. [CrossRef] [PubMed]
25. Izuhara, Y.; Matsumoto, H.; Nagasaki, T.; Kanemitsu, Y.; Murase, K.; Ito, I.; Oguma, T.; Muro, S.; Asai, K.; Tabara, Y.; et al. Mouth breathing another risk factor for asthma: The Nagahama study. *Eur. J. Allergy Clin. Immunol.* **2016**, *71*, 1031–1036. [CrossRef] [PubMed]
26. Bland, J.M.; Altman, D.G. Statistical methods for assessing agreement between two methods of clinical measurement. *Lancet* **1986**, *1*, 307–310. [CrossRef]

 *sensors*

Article

Contactless Real-Time Heartbeat Detection via 24 GHz Continuous-Wave Doppler Radar Using Artificial Neural Networks

Nebojša Malešević [1,*], Vladimir Petrović [2,*], Minja Belić [3], Christian Antfolk [1], Veljko Mihajlović [3] and Milica Janković [2,*]

[1] Department of Biomedical Engineering, Faculty of Engineering, Lund University, Box 118, 221 00 Lund, Sweden; christian.antfolk@bme.lth.se

[2] School of Electrical Engineering, University of Belgrade, Bulevar kralja Aleksandra 73, 11120 Belgrade, Serbia

[3] Novelic, Veljka Dugoševića 54/A3, 11000 Belgrade, Serbia; minja.belic@novelic.com (M.B.); veljko.mihajlovic@novelic.com (V.M.)

* Correspondence: nebojsa.malesevic@bme.lth.se (N.M.); petrovicv@etf.bg.ac.rs (V.P.); piperski@etf.bg.ac.rs (M.J.)

Received: 6 March 2020; Accepted: 19 April 2020; Published: 21 April 2020

Abstract: The measurement of human vital signs is a highly important task in a variety of environments and applications. Most notably, the electrocardiogram (ECG) is a versatile signal that could indicate various physical and psychological conditions, from signs of life to complex mental states. The measurement of the ECG relies on electrodes attached to the skin to acquire the electrical activity of the heart, which imposes certain limitations. Recently, due to the advancement of wireless technology, it has become possible to pick up heart activity in a contactless manner. Among the possible ways to wirelessly obtain information related to heart activity, methods based on mm-wave radars proved to be the most accurate in detecting the small mechanical oscillations of the human chest resulting from heartbeats. In this paper, we presented a method based on a continuous-wave Doppler radar coupled with an artificial neural network (ANN) to detect heartbeats as individual events. To keep the method computationally simple, the ANN took the raw radar signal as input, while the output was minimally processed, ensuring low latency operation (<1 s). The performance of the proposed method was evaluated with respect to an ECG reference ("ground truth") in an experiment involving 21 healthy volunteers, who were sitting on a cushioned seat and were refrained from making excessive body movements. The results indicated that the presented approach is viable for the fast detection of individual heartbeats without heavy signal preprocessing.

Keywords: artificial neural network; Doppler radar; heart rate; real-time processing

---

1. Introduction

The contactless monitoring of heart rate has numerous advantages over conservative methods that use contact sensors, such as electrocardiogram (ECG) monitors, conventional photoplethysmography (PPG) sensors or piezoresistive sensors [1]. Contactless sensors offer improved mobility and obviate the need for attaching or cleaning electrodes, but also have the unique ability to be used on patients who suffer from skin irritations, painful skin damage like lacerations or burns, as well as patients who exhibit anxiety or allergic reactions to contact sensors. Furthermore, some contactless instruments, such as radar-based sensors [2], can be used for heart rate monitoring through clothes or other obstacles.

The real-time operation of heart rate monitors is required for the timely detection of potentially dangerous conditions in hospitals or in-home health care applications. The heart rate and its variability can be used for emotion, stress [3,4] or drowsiness detection [5] and real-time operation is often necessary for these applications.

In recent years, significant progress has been made in the development of radar-based heart rate monitors [6–32]. The potential for the production of compact low-power sensors, which are completely non-obstructive and harmless to human health, placed radar technology as one of the most promising options for contactless vital signs monitoring. Radar sensors are used for the detection of sub-millimeter movements of chest wall skin surface that occur due to heartbeats, whereas various signal processing methods are employed for heart rate extraction from discretized radar signals. Radar technology has shown not only great potential for heart rate estimation but also the potential for extracting ventricular ejection timing using nonlinear filtering methods [33].

The most frequent radar architectures used in heart rate estimation sensors are continuous-wave (CW) Doppler radars [6–24], frequency-modulated continuous-wave (FMCW) radars [3,25], and impulse radio ultra-wideband (IR UWB) radars [26–30]. CW Doppler and FMCW radars mostly outperform IR UWB radars in terms of power consumption and sensitivity [2]. The tracking of fine chest wall motion can be obtained by measuring the phase shift of the reflected signals of continuous-wave radars. The higher the frequency of the transmitted radar signal, the higher sensitivity can be obtained. While FMCW radars can detect both the absolute and relative displacement of the chest wall surface, CW Doppler radars are only capable of tracking relative displacement. This means that FMCW radars could be applied in multiple person heart rate estimation [25]. However, CW Doppler radars have a simpler hardware architecture and lower power consumption, and in single person applications, the relative displacement information obtained by the CW Doppler radar can be enough for a good heart rate estimation.

Many research groups have extensively investigated the monitoring of heart rate using CW Doppler radars. Most of the previous research was based on the experimental data monitored in studies with healthy participants lying or sitting in a controlled environment. The early published methods were based on (1) the simple filtering of heartbeat-related signals and applying a threshold to the filtered signals for extraction of heartbeat locations [6–8,10,12], or (2) heart rate frequency estimation using spectral analysis [9,13,15–17]. These approaches were hardly capable of fast and real-time performance and high-accuracy estimation at the same time. Simple band pass filtering would provide small latency, but the filtered output signals need further processing in order to automatically extract heartbeats. The robustness of these methods is hence very limited. The research in [12] showed that the error of the heart rate extraction can be drastically increased just when the subject changes their position from supine to still sitting. On the other hand, the frequency domain approaches would need a long window (5–30 s) of data for achieving sufficient frequency resolution for the detection of the heart rate harmonic. Additionally, they usually focus on high accuracy harmonic extraction and do not necessarily offer methods for distinguishing the heart rate harmonic from breathing and intermodulation harmonics. When the heart rate harmonic extraction is applied, the achieved accuracy is moderate (mean relative error of heat rate estimation around 10% [15]). Additionally, the testing set was limited to data recorded on a small number of human subjects (1 or 2, except in [12] where 10 subjects participated in the study).

However, more recent studies used data from more subjects (up to 10 participants in sitting position) and presented promising results in terms of detection accuracy. In [18], the ensemble empirical mode decomposition (EEMD) was used for the extraction of heartbeat information and in [22] the autocorrelation and frequency-time phase regression (FTPR) provided an algorithm robust to noisy conditions, but both of them used relatively long data windows for the heart rate assessment (10–15 s), which produced a large delay.

High-accuracy approaches capable of real-time operation with CW Doppler radar architectures, which achieve a relatively small delay, have recently been presented [19–21,23,24]. Authors in [19] and [20] used a dynamic variation of the time window for processing via the fast Fourier transform (FFT) [19] or the Wavelet transform (WT) [20]. In [21], the polyphase-basis discrete cosine transform has been used for heart rate estimation. All these methods improved heart rate detection accuracy in the frequency domain when shorter data windows were used (2–5 s). Specific heartbeat signal has been obtained using the short-time Fourier transform (STFT) analysis in [23], which was further filtered through an adaptive band pass filter for improved quality. The control of the adaptive band pass filter was done using the information extracted from the time domain analysis of the heartbeat signal on windows of 2–3 s. In [24] it has been shown that the analysis of the frequency domain only did not give satisfactory results. Therefore, the heart rate information was extracted using frequency domain analysis (window length: 3.5 s) for the coarse estimation and time domain processing using a band pass filter bank for the refinement of the results. This approach resulted in small algorithm delay (~2.5 s) and high accuracy.

Recently, new approaches based on supervised or unsupervised machine-learning algorithms [31,32] were introduced in the CW Doppler radar systems, and the first results have shown promising advantages in terms of heartbeat detection delay and source separation capabilities (robustness of heartbeat detection to respiration motion or random body motion) compared to traditional approaches. Convolutional neural networks (CNN) were applied in [29] to estimate heart rate from UWB radar signals. However, due to the lack of training data, this approach was person-specific since the CNN needed to be trained for each subject separately. To the best of our knowledge, there is no published paper that used artificial neural networks for heartbeat detection using the CW Doppler radar technology. This paper focused on the development of a system for instantaneous heart rate estimation (delay of less than 1 s from heartbeat occurrence) using a shallow artificial neural network (ANN) as a main signal processing element. Additionally, the goal was to develop a detection algorithm that was person-independent. The contribution of this work is in the development of the system for detecting individual heartbeats considering the following requirements: (1) a low-complexity time domain-based algorithm (without relying on periodic occurrences of heartbeat-related chest displacements as in the case of traditional spectral approaches), (2) suitable for real-time human presence detection, (3) calibration-free (no need for I/Q imbalance, offset compensation or the usage of any demodulation techniques) and 4) testing on a separate group of subjects from those whose heart rate signals were used in the ANN model selection and training process.

2. Materials and Methods

2.1. Basics of CW Doppler Radar Operation

A typical architecture of quadrature continuous-wave Doppler radar is shown in Figure 1a. The radar transmitted sinusoidal electromagnetic waves generated in the local oscillator (LO) and amplified in the power amplifier (PA). The transmitted signal reflects from the target, where the reflected signal is modulated in phase. The transmitted signal can be expressed as

$$T(t) = A_T \cos(2\pi f t + \theta(t)), \tag{1}$$

where A_T and f are the amplitude and the frequency of the transmitted signal, respectively, and $\theta(t)$ is the phase noise of the local oscillator. The received signal can be expressed as

$$R(t) = A_R \cos\left(2\pi f t - \frac{4\pi d_0}{\lambda} - \frac{4\pi x(t)}{\lambda} + \theta\left(t - \frac{2d_0}{c}\right)\right), \tag{2}$$

where A_R, f and λ are the amplitude, the frequency and the wavelength of the carrier signal, respectively, c is the speed of light, d_0 is the nominal radar–target distance and $x(t)$ is the target's relative displacement [7]. The received signal is demodulated using a quadrature demodulator as shown in Figure 1a. The resulting signals are two baseband signals: the in-phase signal (I), which is in phase with the carrier and the quadrature signal (Q), which is phase-shifted from the carrier by 90°. These signals are expressed as [10]

$$I(t) = A_I \cos\left(\theta_0 + \frac{4\pi x(t)}{\lambda} + \Delta\theta(t)\right) + DC_I, \tag{3}$$

$$Q(t) = A_Q \sin\left(\theta_0 + \frac{4\pi x(t)}{\lambda} + \Delta\theta(t) + \Delta\varphi\right) + DC_Q, \tag{4}$$

where A_I and A_Q are amplitudes ($A_I \neq A_Q$ due to the I/Q amplitude imbalance), DC_I and DC_Q are DC offsets, θ_0 is the constant phase shift due to the constant nominal distance d_0 from (2), $\Delta\varphi$ is the phase shift due to the I/Q phase imbalance and $\Delta\theta(t)$ is the total residual phase noise which can be neglected in vital signs detection applications since the distance between the target and the radar system is small [7]. Baseband signals are further digitized using analog-to-digital converters (ADCs) and processed in a digital signal processing unit.

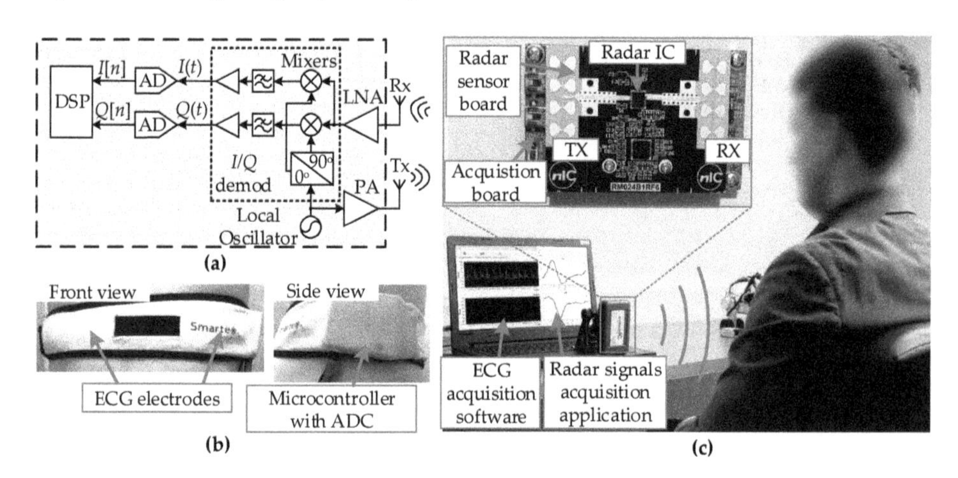

Figure 1. (**a**) Architecture of the quadrature Doppler radar used for the measurements; (**b**) Photographs of the Smartex Wearable Wellness System (WWS) used for the electrocardiogram (ECG) measurement; (**c**) Experiment setup.

2.2. *Instrumentation*

In this study, a CW Doppler radar with a carrier frequency of 24 GHz was used for heart rate estimation. The recording of the reference ("ground truth") signal for the ANN training and the validation of the estimation accuracy was done using a wearable cardiorespiratory monitoring system (Smartex Wearable Wellness System (WWS), Pisa, Italy) including a single lead ECG and a piezo band as shown in Figure 1b. The ECG sensor was chosen since its accuracy was higher than other heart rate monitors such as pulse oximeters or photoplethysmographs. The system contained a microcontroller for data acquisition and Bluetooth connection for wireless data transmission to the personal computer (PC). It was a CE (Conformité Européenne) certified system. Additionally, the electrodes were connected to the skin using the wet textile fabric, which eliminated any potential irritation that could come from the sticky adhesive electrodes, particularly if applied on non-glabrous skin [34]. The sample rate for the ECG recording was 250 Hz.

The radar system used in the experiment was a Novelic Radar Module, NRM24 [24]. Figure 1c shows the radar module placed in the experimental setup. It was a DC-coupled Doppler radar sensor. The module was compact (8 × 5 × 1 cm) and portable and consisted of two stacked printed circuit boards (PCBs). The radar sensor PCB included the main part of the analog frontend: antennas, an integrated radar transceiver and a phase-locked loop integrated circuit. Antenna beamwidths (BW) were $BW_{\theta,\,\text{3-dB}} = 25°$, $BW_{\theta,\,\text{6-dB}} = 33°$, $BW_{\theta,\,\text{10-dB}} = 43°$ (elevation) and $BW_{\varphi,\,\text{3-dB}} = 44°$, $BW_{\varphi,\,\text{6-dB}} = 65°$, $BW_{\varphi,\,\text{10-dB}} = 90°$ (azimuth), whereas the antenna gain was 12 dBi. The maximum power at the transmit antenna input was 10 dBm. The second PCB was the acquisition board, which included baseband amplifiers and filters, a power supply circuitry, an ARM Cortex-M4 based microcontroller (MCU) that integrated a multichannel 12-bit ADC and serial-to-USB converters for data transfer. The baseband filter had a cut-off frequency of 100 Hz, which was considered high enough for the vital sign detection application. The sampling rate was set to f_S = 1 kHz, while the data logging on the PC was performed using a custom-made application that communicated via serial connection with the MCU.

The alignment of the datapoints from the Smartex WWS system and the radar system was done by matching timestamps in the logged data.

2.3. Database Recordings

The radar recordings, as well as the ECG data used as reference, were obtained from 21 healthy human volunteers who took part in the experiments (14 males and 7 females, aged 26.1 ± 5.1, with a height of 179.5 ± 11.6 cm and a weight of 74.2 ± 16.4 kg). Subjects were free of any diagnosed acute/chronic cardiac or respiratory problems, based on their self-report. Participants were acquainted with the protocol in advance and gave informed consent. The study was approved by the ethical committee of the University of Belgrade—School of Electrical Engineering, Serbia. The subjects had the wearable ECG strapped around their thorax and were instructed to sit comfortably on a cushioned seat in front of the radar sensor. The sensor was mounted on a custom stand, facing the participants at a distance of 75 cm. At this distance, the radar beam focused on the torso area, considering −3 dB beamwidths of the antenna (25° and 44°). The participants were told to breathe as they normally would in a relaxed state, without extremely deep and excessive breaths. Additionally, they were asked to refrain from excessive body and hand movements, since the rapid movements could mask the small chest wall movements that come from heartbeats and hence affect the detection. The radar signals obtained for one participant are shown in Figure 2. Three-hundred seconds of data were acquired for each subject.

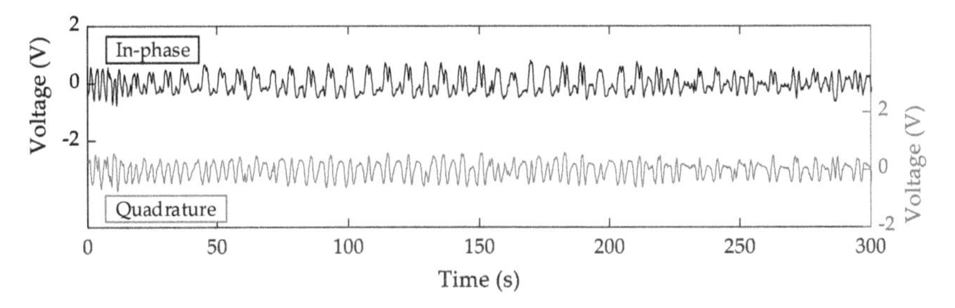

Figure 2. In-phase and quadrature Doppler radar signals during the measurements. The participant was breathing normally in front of the radar sensor.

Figure 3a shows a fragment of recorded time-aligned ECG and radar signals. There was a distinctive signal shape in the radar signals with a time delay in relation to the R wave of the ECG. The heartbeat-related disturbance in the radar signals originated from the mechanical movement of the chest wall during the heartbeat. This disturbance corresponded to the ballistocardiogram J-wave peak [35]. It can be seen that this characteristic signal shape became distorted and was reduced in the presence of breathing and movement.

Many previous works considered that the heartbeat-related displacements could be modeled as a sine wave [16], half-sine pulses [21], Gaussian pulses [22], or as an array of two consecutive pulses [24]. However, the mechanical response of the chest wall has a complex waveform that is difficult to model [33]. Figure 3b shows the heartbeat-related displacement obtained from the radar data shown in Figure 3a, using the extended differentiate and cross-multiply (DACM) demodulation algorithm described in [16]. Before the demodulation, the I/Q imbalance of the radar was measured and compensated offline like in [24]. It can be observed that the heartbeat-related displacement had a complex waveform, which further induced the complex signal shapes of the radar signals. The detector in this paper was trained with the aim to recognize these small distortions in the radar signals, without previous modeling. In order to get a reliable data set for training the detector, the signals were cropped to the period of 200 s of normal breathing (Figure 2, time interval 50–250 s).

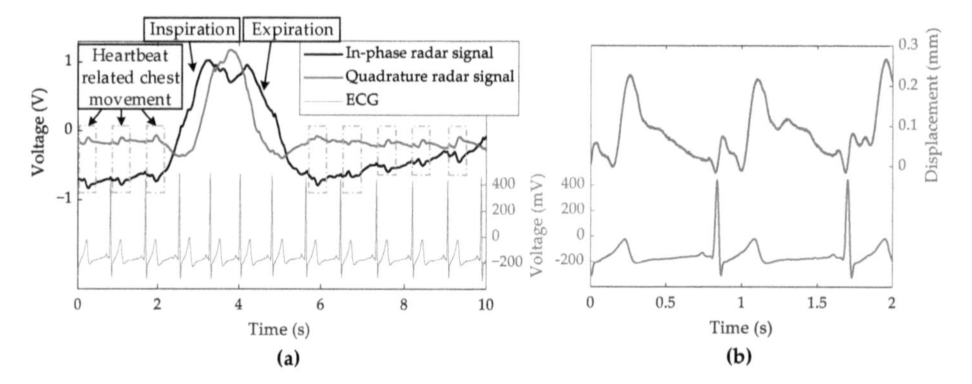

Figure 3. (**a**) A fragment of the recorded ECG and radar signals. It could be noted that the heartbeats acquired by the ECG amplifier are followed with slight movement patterns picked up by the radar. During the inhale–exhale (see 2–5 s) this movement pattern is distorted. (**b**) A heartbeat-related chest wall displacement obtained using the extended differentiate and cross-multiply (DACM) demodulation of the first 2 s of the radar signals.

2.4. Data Preprocessing

The detector was envisioned as a binary classifier detecting the occurrence of each heartbeat, but such implementation required the reshaping of the recorded data into an appropriate format. The continuous nature of the reference ECG signal was not suitable for this approach, so it was instead transformed into a binary on/off signal. R peaks were detected using the Pan-Tompkins algorithm [36]. The surrogate reference binary signal ("binarized target signal") was synthesized in the form of a 400 ms pulse with 200 ms delay after each R wave in order to highlight the temporal relationship between the heart's electrical activity (ECG) and the resulting mechanical displacement (radar signal), as shown in Figure 4. The window of 400 ms width with a latency of 200 ms after the R wave was able to cover most of the mechanical rippling observed within the radar signals. This choice is in line with a study performed on 92 healthy subjects which showed that the duration range of the R–J interval was 203–290 ms [37]. Additionally, the selection of the value of the latency parameter (200 ms) for the binarized target signal was confirmed on a recorded dataset of seven randomly selected subjects from our study. Using the simplest model that was tested in the scope of this paper (a feed-forward artificial neural network with a single hidden layer containing 10 units), a series of training and testing with a range of delays and target widths was performed. The tested delay values ranged between 0 ms to 300 ms, and the target widths were tested in the range between 300 ms and 500 ms. These values were selected based on the visual inspection of the signals, and the values that were chosen as optimal (window of 400 ms width with a latency of 200 ms after the R wave) were those that yielded the best model accuracy in terms of the percentage of detected peaks.

The following stage in the signal preprocessing was the decimation of both the binary reference signal and the radar signal to the same sampling rate, which was a prerequisite related to the ANN employment, as each input state required a corresponding target output. To ensure the fast computation without a loss of information in the physiological range, 100 Hz was selected as the sampling rate during this computation.

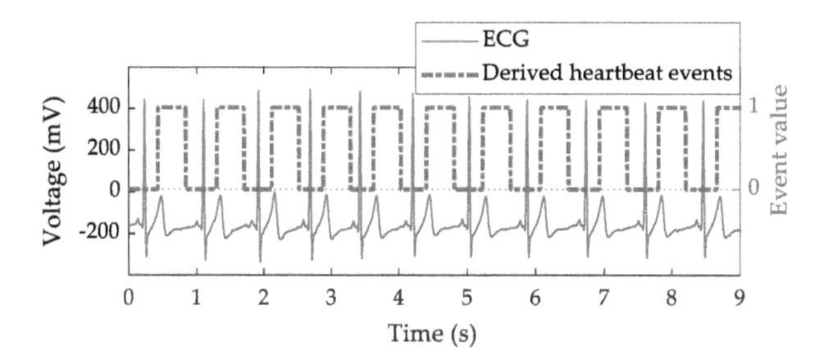

Figure 4. Reference ECG signal (ECG) and the binarized target signal (derived heartbeat events) where the values of "1" represent the presence of the mechanical heart displacement after the R wave, and the values of "0" represent its absence.

The input to the ANN consisted of a 200-sample long vector, corresponding to 1 s of recording, where in-phase and quadrature branches were concatenated so that the first 100 samples in the input corresponded to the in-phase signal and the second 100 samples corresponded to the quadrature signal. Such an input was matched to the value of the reference binary signal as the target output. The choice of 1 s signal memory was based on a small-scale test conducted on a subset of recorded signals in which the memory depth varied between 200 ms and 1 s. The lower bound was sufficient to partially incorporate the mechanical wave observed due to the heartbeats and the upper bound corresponded to the interval between the subsequent heartbeats at a normal heart rate of 60 bpm. These initial test results showed that the increases of memory depth from 200 ms to 500 ms significantly increased ANN performance, while further increases resulted in only incremental gains. Nevertheless, we decided to use the deepest memory as we were not concerned about computational complexity, while using longer intervals was expected to contain more physiologically relevant data that the ANN could learn from, and not overfit to potentially insignificant signal details.

2.5. The Detection Algorithm

Two main approaches were taken in the design of the heartbeat detector: the classical shallow feed-forward neural networks (FF ANN) and the nonlinear autoregressive exogenous model network (NARX) as representative of the feedback-based topologies. The NARX model took as an input the current sample in the radar data stream and the previous 100 samples, together with their calculated output. The NARX topology was tested for a single hidden layer with 10 and 20 neurons, NARX 10 1 and NARX 20 1 respectively. For the classic shallow FF ANN, 4 configurations were tested: (1) a single hidden layer with 10 neurons, FF 10 1, (2) a single hidden layer with 20 neurons, FF 20 1, (3) two hidden layers with 20 and 2 neurons respectively, FF 20 2, and (4) two hidden layers with 40 and 4 neurons respectively, FF 40 4. The output layer in all the cases consisted of a single neuron with a sigmoid activation function. Activation functions for the hidden layers also varied during this topology search, including a hyperbolic tangent and log-sigmoid and linear transfer functions. Furthermore, for the FF ANNs different loss functions were tested: the mean squared normalized error (MSE), the mean squared error with regularization (MSEREG) and the sum squared error (SSE).

As the FF ANN, containing a single hidden layer with 20 neurons, a hyperbolic tangent activation function, trained using Levenberg–Marquardt optimization and MSE loss function, outperformed all the other ANN topologies, the results presented in this paper were focused mainly on this FF 20 1 ANN topology. The whole flowchart, including the preprocessing and the detection algorithm based on the FF 20 1 ANN, is presented in Figure 5. The ANN input consisted of unprocessed in-phase and quadrature components of a Doppler radar signal (discretized and resampled $I(t)$ and $Q(t)$ signals from Equation (3) and Equation (4), respectively). In the FF ANN topology, each neuron in the hidden and output layers calculated a linear combination of its inputs in Equation (5):

$$a_i^j = \sum_{i=1}^{N} w_i^j a_i^{j-1} + b_i^j, \tag{5}$$

where j refers to the current layer, N is the number of inputs to the current layer, w is the weights and b is the corresponding biases. The output of each neuron was passed through the hyperbolic tangent function, with the exception of the output neuron which used a sigmoid function. The weights were calculated through a numerical optimization of the mean squared error loss function in Equation (6):

$$MSE = \frac{1}{M} \sum_{i=1}^{M} (b_i - a_i)^2, \tag{6}$$

where M is the number of data points, b_i is the binarized target signal and a_i is the network output. This was an iterative procedure that was set to run for a maximum of 1000 epochs or to stop early if the solution became sufficiently close to the minimum, that is, if the gradient became smaller than 10^{-7}. The training would also stop if the error on the portion of the data set aside for validation (30%) failed to decrease for 6 consecutive epochs.

To remove fast noisy changes, the sequence of outputs of the ANN calculated for each sample was smoothed. This stage of the detection algorithm was implemented as a moving average filter with a width of 10 consecutive ANN outputs.

The next stage of the algorithm was the peak detection subroutine which marked local maximums of the continuous probabilities output, imposing established constraints on the minimal distance between the consecutive peaks based on the known physiological range in rest 40–120 beats/min [38,39] and their prominence (detection amplitude). When the duration of the detected inter-pulse interval (IPI) was twice as large as the previously detected IPIs (within the established constraints), a beat was interpolated as having occurred at the point in time that was the arithmetic mean between the occurrences of the current and previously detected heartbeats. The detection amplitude was defined empirically for each ANN topology on a small test sample using the error of the number of detected heartbeats as a metric. The same detection amplitude was then used for all the subjects.

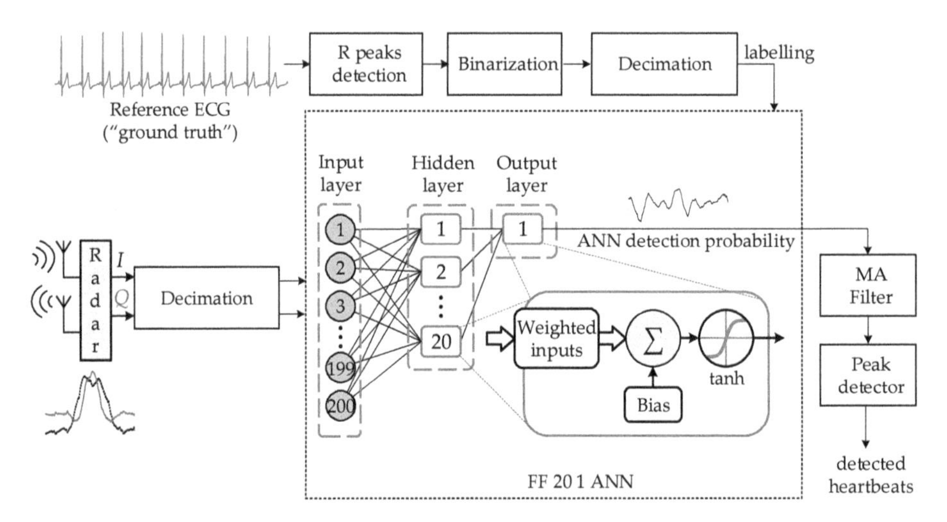

Figure 5. Flowchart for the proposed method for heartbeat detection based on the classical shallow feed-forward neural network with a single hidden layer with 20 neurons (FF 20 1 ANN). The artificial neural network (ANN) input is the 200-sample long vector containing resampled 100 in-phase and 100 quadrature component samples. For the FF 20 1 ANN there are 20 neurons in the single hidden layer and 1 neuron in the output layer. MA Filter—moving average filter.

2.6. Error Estimation and Statistical Analysis

The inter-pulse interval was calculated as the time elapsed between every two adjacent heartbeats. The classification error was determined through the percentage error in the total number of detected heartbeats and the error in the estimation of median IPIs. The similarities were also assessed between the distributions of the ANN-detected IPIs and those extracted from the reference ECG method.

The performance was evaluated using a three-fold cross-validation: the data were split into three equal subject groups (folds), each containing recordings acquired from 7 subjects, out of which two folds were used for training and one for testing. The training and testing were repeated three times for a different fold held out for evaluation (as shown in Figure 6).

The evaluation metrics were calculated over the set of predictions obtained on the three folds used in the test mode. A statistical analysis was performed on the results obtained via radar and those extracted from the reference ECG signal. The number of detected heartbeats and median IPI for all the recordings used as test were compared to those calculated from the reference signal. Apart from group evaluations, a statistical comparison was also performed on the level of the individual heart event detection within each of the 21-subject sets in the test mode. As in all of the statistical tests, one or more samples were found to be significantly non-normal (Lilliefors test with a 0.05 significance level),Wilcoxon signed rank test was used for statistical comparisons.

All processing was done using the Matlab2018b (The MathWorks, Inc., Natick, Massachusetts, United States).

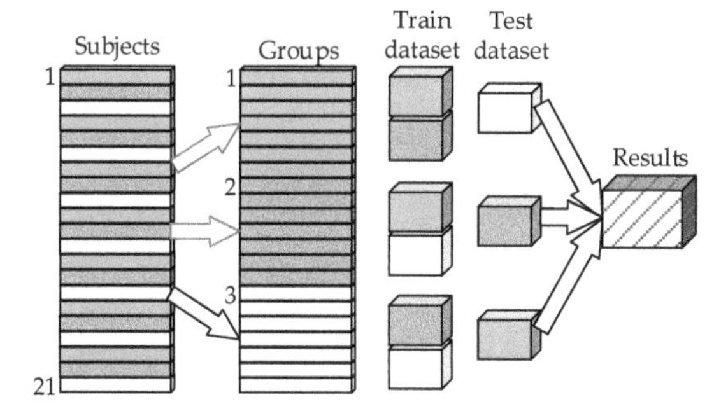

Figure 6. Database organization for the purpose of the ANN training and the cross-validation. The dataset was divided into three equal subsets and two subsets were used for training, while the remaining subset was used for testing. This process was repeated for all the combinations of the subsets in the training set. The three training datasets comprised 3473, 3429 and 3386 heartbeats, while the three testing datasets comprised 1671, 1715 and 1758 heartbeats.

3. Results

The results obtained using six different tested ANN topologies are presented in Table 1.

The smallest error in the number of detected beats (count error—CNT error) was achieved by the FF ANN with 20 neurons in the hidden layer (2%), which was notably better than any other tested topology (12.3% < CNT Error < 34.3%). This topology also had the smallest error when the median IPIs were compared. When comparing at the level of the individual IPIs, the number of subjects whose median was significantly different from the reference was also higher than the rest, but was still relatively low.

Table 1. Error metrics for the six different ANN topologies with reference to the ECG-derived number of heart beats and inter-pulse intervals.

Topology	CNT Error [1] (%)	IPI MRE [2] (%)	No Diff. # [3]
FF [4] 10 1	−22.0	16.4	7
FF 20 1	−2.0	15.3	11
FF 20 2 1	−30.0	15.3	4
FF 40 4 1	−27.6	16.6	7
NARX [5] 10 1	−34.3	16.1	5
NARX 20 1	−12.3	16.4	10

[1] Percentage error of the number of the detected heartbeats out of a total 5144 heartbeats. The negative values correspond to fewer detected heartbeats by ANN compared with the "ground truth"; [2] Inter-pulse interval (IPI) mean relative error; [3] Number of subjects with no significant difference between the medians of the estimated IPI-s and the IPI-s from the ECG reference, out of the total 21 subjects; [4] Feed forward ANN; [5] Nonlinear autoregressive exogenous model; The numbers in the topology descriptions stand for the number of units in each layer.

Figure 7 shows an example of the output of the FF 20 1 ANN, the topology which showed the best performance, smoothed with a moving average window, with the prominent peaks detected and marked. This example shows the typical behavior and errors of the methodology.

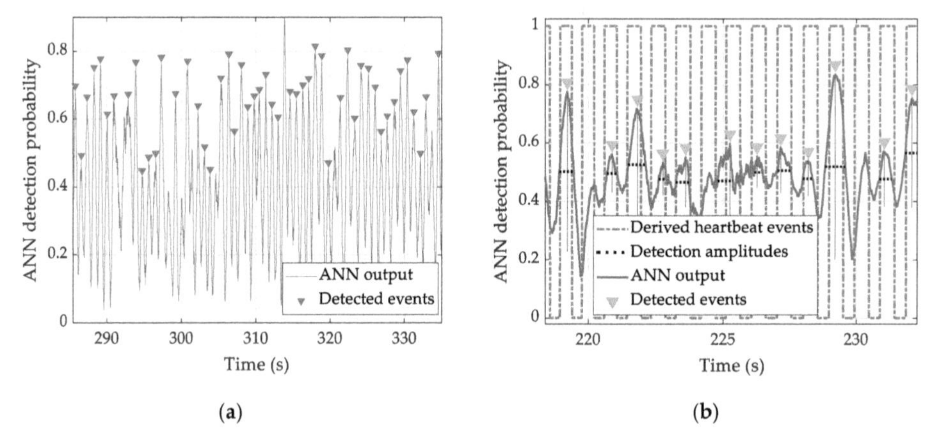

Figure 7. An example of the heartbeat's method detection. Panel (a) shows the output of the FF 20 1 ANN smoothed with a moving average window, with the detected prominent peaks in a 50 s interval. Panel (b) shows the output of the FF 20 1 ANN smoothed with a moving average window on a shorter time scale plotted against the reference ECG (derived heartbeat events) for an easier visualization of the detected heart events.

The error of the total number of detected heart events for the FF 20 1 ANN configuration was −2% (104 undetected beats out of a total of 5144 heartbeats extracted from the ECG). The statistical tests showed that there were no significant differences between these two methods in terms of the number of detected events ($p>0.05$). The difference in the medians of the IPIs calculated using the reference ECG and the FF 20 1 ANN was −2 samples (−20 ms) and was not statistically significant. When it comes to individual heart event detection, for 11 out of the 21 subjects the medians of the ANN detections were not significantly different from the reference.

For the specific ANN that showed the best performance with the recorded database, there were 20 neurons in the first hidden layer, resulting in ~4000 multiply–accumulate operations. In the implementation on an embedded platform (Teensy 4.0 programmed in Arduino IDE) this calculation took 66 μs, which was more than enough for executing the proposed method in real time. As the calculation was done with a 100 Hz rate, there were ~10 ms in between the consecutive heart event estimations.

4. Discussion

The work presented in this paper is intended for the detection of individual heartbeats using a state-of-the-art mm-wave radar sensor. The radar sensor relied on the Doppler shift in the signal reflected from the objects within its field of view to detect any movements, even small ones. The real challenge of such a measurement was separating the influence from the different sources. Furthermore, smaller displacements, such as chest movements due to heart activity could be completely hidden or distorted by other physiological sources, such as breathing, talking or change of posture. Thus, the focus of this research was on the specific radar signal footprint in the time domain resulting from the heartbeats and the method by which to identify such small signal ripples. The computational tool selected for this task was shallow artificial neural networks for their high capacity for generalization, ability to be trained without prior knowledge of the signal properties and being computationally inexpensive, enabling easy implementation in an embedded or a high-level system. For the shallow ANNs that were tested, the dominant part of the computational complexity was related to the first hidden layer which performed multiplications of all the input signal samples with the weights (Figure 5).

With the aim of tracking vital signs and the presence of vehicle drivers in a contactless manner, a database of radar signals, alongside ECG and respiration, was gathered from 21 participants sitting comfortably in a cushioned chair. Up to this date, the number of participants in published papers that presented traditional or machine learning approaches for heartbeat detection was less than 12 [6–32], but considering our goal to obtain the realistic results of using a trained ANN on the unseen data, we acquired a larger dataset than in the previous studies. This objective imposed the strict condition of the testing of the performance on radar signals that were completely new to the ANN. Care was taken to split the data in such a manner that the signals obtained from the same person could all be found either in training or in testing (Figure 6). Although the subjects were instructed to sit quietly in the chair, some of them did substantially move their upper body and head but these recordings were nevertheless included in the database, bearing in mind the potential of neural networks to abstract over a wide variety of inputs and their robustness to noise.

After analyzing the performance of various ANNs it was somewhat surprising to find that a relatively basic ANN outperformed more complex networks. The more complex networks were able to pick up minute details in the signals and use them to model the training set more closely at the cost of loss of generalization, while the reduced single layer network did not have such capacity to overfit. One of the conditions that favored the simpler network architectures could still be the limited database used for the training. The acquisition of yet larger amounts of data in the future could make space for more complex network architectures, such as sequential deep learning models, to further improve the results. This would, however, come at the cost of higher processing requirements.

With respect to the main idea of the ANN-based method, which was the identification of individual heartbeats, the most important metric of the ANN testing was the number of detections. Due to the lack of similar metrics in other scientific publications, the result presented in this paper of 2% undetected heartbeats could not be put into perspective with other approaches.

As another performance evaluation metric, we calculated the time between the consecutive detections. This metric was directly compared with the IPIs from the reference ECG signal and it was shown that the difference between their medians was also not statistically significant, confirming that the method could be used to accurately track averaged heart rates in longer periods. These results were also comparable with the findings presented in [22] where the relative error between the averaged radar and ECG rates was between 0.55% and 1.97%. The method proposed by the authors [24] outperformed all of the previously published methods based on the CW Doppler radar technology regarding the mean relative error (2.07% on the dataset of ten subjects, algorithm latency ~2.5 s). The ANN-based approach presented here brings multiple advantages over other methods presented in relevant papers. The ANN method used raw radar signals without the need for any preprocessing or calibration, which can require the implementation of complex algorithms, nor prior knowledge of the signal properties and process models. The implementation of the shallow ANN within a microprocessor system was quite computationally inexpensive as it comprises only a small number of basic arithmetic operations (multiplications and additions for neural layers, while activation functions can be obtained using look-up tables). In comparison with the FFT-based approaches and the heavy use of digital filters, this feature presented a significant improvement in the computational complexity. This method is also beneficial in applications that require the fast detection of human presence, as the latency was below the width of the processing window (1 s), while to our knowledge, the shortest latency reported in relevant publications is 2.5 s [24]. This estimate of 1 s was made for the worst-case scenario of system power-up which requires the initial filling of the ANN input buffer. Once the system is up and running, with a human entering its field of view, this latency is expected to be significantly lower. Namely, due to the training procedure, the heart detections could occur in the 400 ms window that contains the latest signal samples. The smoothing by moving average filter was done on only 10 ANN outputs, which brought negligible delay. The last step of the detection chain which was peak detection required only a few extra samples to identify a local maximum.

To summarize the above contributions of our work, we listed all the relevant previously published methods for heartbeat extraction in normal breathing conditions based on the CW Doppler radar technology in Table 2. This work is the first approach that used artificial neural networks for the heartbeat detection based on the CW Doppler radar technology. The method was not person-specific (as opposed to the supervised machine learning approach applied in [32]) and we performed a more realistic scenario on 21 subjects where all the results were presented on "unseen data". This was a fast and reliable calibration-free method, with low percentage of failed heartbeat detections and with the latency that outperformed all relevant previously published non-person-specific approaches.

Table 2. Comparison of the methods for heartbeat extraction in normal breathing condition based on the continuous-wave (CW) Doppler radar technology.

Ref.	Radar Freq. (GHz)	N [1]	TC [2]	T [3] (s)	Method [4]	Unseen Data Tested [5]	W [6] (s)	CF [7]	FD [8] (%)	HR/IPIs Avg. Error [9] (%)
[15]	2.4	5	S 80 cm	30	Multiple Signal Classification	NO	8–28	NO	-	~10
[18]	5.8	10	S 50 cm	240	Ensemble Empirical Mode Decomposition	NO	15	NO	-	3.67
[19]	5.8	4	S –	30	Time-window variation	NO	2–5	YES	-	3.3
[20]	5.8	2	S –	60	Wavelet T.	NO	3.5	YES	-	3
[21]	10.225	3	S 1.1 m	90	Discrete Cosine T.	NO	1.5 2 3	NO	-	10.4 7.6 5.1
[22]	2.4	8	S 1.5 m S 75 cm	300	Frequency–Time Phase Regression	NO	10–15	NO	-	2
[24]	24	10	S 75 cm	180	Filter bank and Chirp Z T.	NO	3.5	YES	-	1.54
[31]	24	5	S 80 cm S 30 cm T 30 cm	120	Non-negative factorization matrix	NO	8	NO	-	4.17 3.93 4.22
[32]	5.8	1	- -	600	Gamma filter	YES	15	NO	8.3	3.8
This work	24	**21**	S 75 cm	200	**ANN**	**YES**	**< 1**	**YES**	**2**	15.3

[1] Number of subjects (N); [2] Test conditions (TC) during the measurement (S XX= Sitting at distance XX, T XX = Sitting and typing at distance XX); [3] Total measurement time (T) of normal breathing for each session; [4] Data processing approach; [5] Tested on data that were "unseen" in the training process; [6] Time window (W); [7] Calibration-free (CF) for I/Q imbalance, the offset compensation or usage of any demodulation techniques; [8] Failed detection (FD) of heartbeats; [9] Average Error of estimated heart rate or IPIs (HR/IPIs Avg. Error).

As shown in Table 1, the ANN implemented in this paper showed weakness in the estimation of the IPI with high accuracies. Consequently, the utilization of the methodology for the evaluation of heart rate variability (HRV) parameters is limited. The goal of the present study was to develop a method for the fast detection of individual heartbeats; thus, all of the ANN optimizations were governed by the error of the detected heartbeats count. The HRV-related errors were not included in any of the methodology steps; therefore, it would be unrealistic to expect high accuracies compared with the methods specifically designed for the HRV estimation. In addition, the choice of the targets has a significant influence on the HRV parameters estimation. As aforementioned, the main goal was related to the robust detection of individual heartbeats, so the target window was made relatively wide (400 ms) to enable the identification of any signal waveform that was related to the mechanical displacements due to heartbeats. This wide window, except for providing more room for heartbeat detection, means that if a heartbeat was detected in any part of the target window, it would be considered as successful during the training phase. In an example, a heartbeat could be detected with the highest probability at the beginning of the target window, while the following heartbeat could be detected with the highest probability at the end, without any penalty related to the ANN performance during training. On the other hand, this kind of detection would result in 400 ms error in estimating the beat-to-beat interval.

In future work, the focus will be on increasing the HRV estimation accuracy. To achieve this goal, the error of the R–R interval estimation will be introduced as an additional loss function during the training procedure. This would inherently force the ANN to bind the maximal detection probability with a specific part of the mechanical oscillation. However, this approach would require some topological ANN changes, such as a feedback loop with the previously detected IPI and an extension of the memory depth.

In this study, there were several limitations that should be noted. All the subjects that participated in the study were young and physically fit adults. During the selected time period they were mostly sitting calmly and were instructed to restrain from prominent body movements. In this study, the chair was placed in a room with no moving objects. In a scenario which involves nonstationary objects, or recording within confined environments, such as inside a car, the performance of the system could deteriorate due to clutter and multipath effects. Given the continuous nature of the unmodulated signal, Doppler radar has no exact information on the absolute position of the observed person. Tracking vital signs of multiple people simultaneously would most likely have to be performed using modulated signals that can resolve observed targets in a space.

The usage of a high-carrier frequency provided small dimensions of the radar system, since small antennas were used, unlike for lower frequency radar systems whose antennas usually need to be larger. An additional advantage of the high-frequency radar was its sensitivity to the small chest displacements that come from heartbeats. Since the method in this paper was using in-phase and quadrature radar signals directly, the radar sensitivity was of crucial importance for the heartbeat detection. The usage of even higher carrier frequency could improve the results, since the sensitivity to small displacements would be larger.

The radar used for this study had a relatively broad field of view, which made it susceptible to picking up clutter from the surroundings. Additionally, in applications that would require a larger distance between the radar and a patient, this broad field of view would pick up even more surrounding movements. Using an antenna with a narrower beam could improve the performance of the system in the future. It is expected that the signal-to-noise ratio (SNR) of the received signals would be smaller if the radar was placed at larger distances. This means that further tests need to be done in order to determine the influence of the SNR on the detection accuracy. Future work will also include tests of the ANN performance in cases when subjects perform natural movements, to estimate the reliability of the sensor and the method in an environment saturated with motion originating from different sources.

5. Conclusions

In this paper, we presented a simple and efficient contactless method for detecting individual heartbeats. The method is based on the CW Doppler radar directly coupled with an ANN stage to detect small signal ripples resulting from sub-millimeter chest movements due to heartbeats. The method has lower latency, lower computational complexity and an easier implementation on an embedded platform when compared to the traditional methodologies described in the literature, while still achieving a good heart rate estimation accuracy. With the promising results presented in this paper, we could foresee the application of the system in uses that require real-time operation, such as human detection in an industrial, automotive or clinical environment.

Author Contributions: Conceptualization, N.M., V.P., M.B. and M.J.; methodology, N.M. and M.B.; software, N.M., V.P. and M.B.; formal analysis, N.M. and M.B.; investigation, N.M., V.P., M.B. and M.J.; resources, V.P., V.M. and M.J.; data curation, V.P. and M.J.; writing—original draft preparation, N.M., V.P., M.B. and M.J.; project administration, V.M. and M.J.; funding acquisition, C.A., V.M. and M.J. All authors have read and agreed to the published version of the manuscript.

Funding: This research was funded by the Innovation fund of Serbia, grant number ID50053, the Ministry of Education, Science and Technological Development of the Republic of Serbia, the Promobilia Foundation and Stiftelsen för bistånd åt rörelsehindrade i Skåne. The APC was funded by Lund University Library.

Conflicts of Interest: The authors declare no conflict of interest.

References

1. Webster, J. *Medical Instrumentation: Application and Design*, 4th ed.; John Wiley & Sons: New York, NY, USA, 2009.
2. Gu, C. Short-range noncontact sensors for healthcare and other emerging applications: A review. *Sensors* **2016**, *16*, 1169. [CrossRef]
3. Zhao, M.; Adib, F.; Katabi, D. Emotion recognition using wireless signals. In Proceedings of the 22nd Annual International Conference on Mobile Computing and Networking, New York, NY, USA, 3–7 October 2016.
4. Healey, J.A.; Picard, R.W. Detecting stress during realworld driving tasks using physiological sensors. *IEEE Trans. Intell. Transp. Syst.* **2005**, *6*, 156–166. [CrossRef]
5. Lee, B.G.; Lee, B.L.; Chung, W.Y. Wristband-type driver vigilance monitoring system using smartwatch. *IEEE Sens. J.* **2015**, *15*, 5624–5633. [CrossRef]
6. Droitcour, A.; Lubecke, V.; Lin, J.; Boric-Lubecke, O. A microwave radio for Doppler radar sensing of vital signs. In Proceedings of the 2001 IEEE MTT-S International Microwave Symposium Digest, Phoenix, AZ, USA, 20–24 May 2001.
7. Droitcour, A.D.; Boric-Lubecke, O.; Lubecke, V.M.; Lin, J.; Kovacs, G.T.A. Range correlation and I/Q performance benefits in single-chip silicon Doppler radars for noncontact cardiopulmonary monitoring. *IEEE Trans. Microw. Theory Technol.* **2004**, *52*, 838–848. [CrossRef]
8. Xiao, Y.; Lin, J.; Boric-Lubecke, O.; Lubecke, V.M. Frequency tuning technique for remote detection of heartbeat and respiration using low-power double-sideband transmission in the Ka-band. *IEEE Trans. Microw. Theory Technol.* **2006**, *54*, 2023–2032. [CrossRef]
9. Li, C.; Xiao, Y.; Lin, J. Experiment and spectral analysis of a low-power Ka-band heartbeat detector measuring from four sides of a human body. *IEEE Trans. Microw. Theory Technol.* **2006**, *54*, 4464–4471. [CrossRef]
10. Park, B.-K.; Boric-Lubecke, O.; Lubecke, V.M. Arctangent demodulation with DC offset compensation in quadrature Doppler radar receiver systems. *IEEE Trans. Microw. Theory Technol.* **2007**, *55*, 1073–1079. [CrossRef]
11. Choi, J.H.; Kim, D.K. A remote compact sensor for the real-time monitoring of human heartbeat and respiration rate. *IEEE Trans. Biomed. Circuits Syst.* **2009**, *3*, 181–188. [CrossRef] [PubMed]
12. Massagram, W.; Lubecke, V.; Host-Madsen, A.; Boric-Lubecke, O. Assessment of heart rate variability and respiratory sinus arrhythmia via Doppler radar. *IEEE Trans. Microw. Theory Technol.* **2009**, *57*, 2542–2549. [CrossRef]
13. Li, C.; Ling, J.; Li, J.; Lin, J. Accurate Doppler radar non-contact vital sign detection using the RELAX algorithm. *IEEE Trans. Instrum. Meas.* **2010**, *59*, 687–695. [CrossRef]

14. Bakhtiari, S.; Liao, S.; Elmer, T.W.; Gopalsami, N.; Raptis, A.C. A real-time heart rate analysis for a remote millimeter wave I-Q sensor. *IEEE Trans. Biomed. Eng.* **2011**, *58*, 1839–1845. [CrossRef] [PubMed]
15. Bechet, P.; Mitran, R.; Munteanu, M. A non-contact method based on multiple signal classification algorithm to reduce the measurement time for accurately heart rate detection. *Rev. Sci. Instrum.* **2013**, *84*, 084707. [CrossRef] [PubMed]
16. Wang, J.; Wang, X.; Chen, L.; Huangfu, J.; Li, C.; Ran, L. Noncontact distance and amplitude-independent vibration measurement based on an extended DACM algorithm. *IEEE Trans. Instrum. Meas.* **2014**, *63*, 145–153. [CrossRef]
17. Huang, M.-C.; Liu, J.J.; Xu, W.; Gu, C.; Li, C.; Sarrafzadeh, M. A self-calibrating radar sensor system for measuring vital signs. *IEEE Trans. Biomed. Circuits Syst.* **2015**, *10*, 352–363. [CrossRef] [PubMed]
18. Hu, W.; Zhao, Z.; Wang, Y.; Zhang, H.; Lin, F. Noncontact accurate measurement of cardiopulmonary activity using a compact quadrature Doppler radar sensor. *IEEE Trans. Biomed. Eng.* **2014**, *61*, 725–735. [CrossRef] [PubMed]
19. Tu, J.; Lin, J. Fast acquisition of heart rate in noncontact vital sign radar measurement using time-window-variation technique. *IEEE Trans. Instrum. Meas.* **2016**, *65*, 112–122. [CrossRef]
20. Li, M.; Lin, J. Wavelet-transform-based data-length-variation technique for fast heart rate detection using 5.8-GHz CW Doppler radar. *IEEE Trans. Microw. Theory Technol.* **2018**, *66*, 568–576. [CrossRef]
21. Park, J.; Ham, J.W.; Park, S.; Kim, D.H.; Park, S.J.; Kang, H.; Park, S.O. Polyphase-basis discrete cosine transform for real-time measurement of heart rate with CW Doppler radar. *IEEE Trans. Microw. Theory Technol.* **2018**, *66*, 1644–1659. [CrossRef]
22. Nosrati, M.; Tavassolian, N. High-accuracy heart rate variability monitoring using Doppler radar based on Gaussian pulse train modeling and FTPR algorithm. *IEEE Trans. Microw. Theory Technol.* **2018**, *66*, 556–567. [CrossRef]
23. Yamamoto, K.; Toyoda, K.; Ohtsuki, T. Spectrogram-based non-contact RRI estimation by accurate peak detection algorithm. *IEEE Access* **2018**, *6*, 60369–60379. [CrossRef]
24. Petrović, V.L.; Janković, M.M.; Lupšić, A.V.; Mihajlović, V.R.; Popović-Božović, J.S. High-Accuracy Real-Time Monitoring of Heart Rate Variability Using 24 GHz Continuous-Wave Doppler Radar. *IEEE Access* **2019**, *7*, 74721–74733. [CrossRef]
25. Ahmad, A.; Roh, J.C.; Wang, D.; Dubey, A. Vital signs monitoring of multiple people using a FMCW millimeter-wave sensor. In Proceedings of the 2018 IEEE Radar Conference, Oklahoma City, OK, USA, 23–27 April 2018.
26. Khan, F.; Cho, S.H. A detailed algorithm for vital sign monitoring of a stationary/non-stationary human through IR-UWB radar. *Sensors* **2017**, *17*, 290. [CrossRef] [PubMed]
27. Schires, E.; Georgiou, P.; Lande, T.S. Vital sign monitoring through the back using an UWB impulse radar with body coupled antennas. *IEEE Trans. Biomed. Circuits Syst.* **2018**, *12*, 292–302. [CrossRef] [PubMed]
28. Lee, Y.; Park, J.Y.; Choi, Y.W.; Park, H.K.; Cho, S.H.; Cho, S.H.; Lim, Y.H. A novel non-contact heart rate monitor using impulse-radio ultra-wideband (IR-UWB) radar technology. *Scientific Rep.* **2018**, *8*, 13053. [CrossRef]
29. Wu, S.; Sakamoto, T.; Oishi, K.; Sato, T.; Inoue, K.; Fukuda, T.; Mizutani, K.; Sakai, H. Person-Specific Heart Rate Estimation With Ultra-Wideband Radar Using Convolutional Neural Networks. *IEEE Access* **2019**, *7*, 168484–168494. [CrossRef]
30. Wang, P.; Qi, F.; Liu, M.; Liang, F.; Xue, H.; Zhang, Y.L.V.; Wang, J. Noncontact Heart Rate Measurement Based on an Improved Convolutional Sparse Coding Method Using IR-UWB Radar. *IEEE Access* **2019**, *7*, 158492–158502. [CrossRef]
31. Ye, C.; Toyoda, K.; Ohtsuki, T. Blind source separation on non-contact heartbeat detection by non-negative matrix factorization algorithms. *IEEE Trans. Biomed. Eng.* **2019**, *67*, 482–494. [CrossRef]
32. Saluja, J.J.; Casanova, J.J.; Lin, J. A Supervised Machine Learning Algorithm for Heart-rate Detection Using Doppler Motion-Sensing Radar. *IEEE J. Electromagn. RF Microw. Med. Biol.* **2019**, *4*, 45–51. [CrossRef]
33. Yao, Y.; Sun, G.; Kirimoto, T.; Schiek, M. Extracting cardiac information from medical radar using locally projective adaptive signal separation. *Front. Physiol.* **2019**, *10*, 568. [CrossRef]
34. Coyle, S.; Lau, K.T.; Moyna, N.; O'Gorman, D.; Diamond, D.; Di Francesco, F.; Costanzo, D.; Salvo, P.; Trivella, M.G.; De Rossi, D.E.; et al. BIOTEX—Biosensing Textiles for Personalised Healthcare Management. *IEEE Trans. Inf. Technol. Biomed.* **2010**, *14*, 364–370. [CrossRef]

35. Lindqvist, A.; Pihlajamaki, K.; Jalonen, J.; Laaksonen, V.; Alihanka, J. Static-charge-sensitive bed ballistocardiography in cardiovascular monitoring. *Clin. Physiol.* **1995**, *16*, 23–40. [CrossRef] [PubMed]
36. Pan, J.; Tompkins, W.J. A real-time QRS detection algorithm. *IEEE Trans. Biomed. Eng.* **1985**, *BME-32*, 230–236. [CrossRef] [PubMed]
37. Inan, O.T. Novel Technologies for Cardiovascular Monitoring Using Ballistocardiography and Electrocardiography. Ph.D. Thesis, Stanford University, Stanford, CA, USA, 2009.
38. Umetani, K.; Singer, D.H.; McCraty, R.; Atkinson, M. Twenty-four hour time domain heart rate variability and heart rate: Relations to age and gender over nine decades. *J. Am. Coll. Cardiol.* **1998**, *31*, 593–601. [CrossRef]
39. Jose, A.D.; Collison, D. The normal range and determinants of the intrinsic heart rate in man. *Cardiovasc. Res.* **1970**, *4*, 160–167. [CrossRef]

Article

Uncertainty-Aware Visual Perception System for Outdoor Navigation of the Visually Challenged

George Dimas, Dimitris E. Diamantis, Panagiotis Kalozoumis and Dimitris K. Iakovidis *

Department of Computer Science and Biomedical Informatics, University of Thessaly, 35131 Lamia, Greece; gdimas@uth.gr (G.D.); didiamantis@uth.gr (D.E.D.); pkalozoumis@uth.gr (P.K.)
* Correspondence: dimitris.iakovidis@ieee.org

Received: 23 March 2020; Accepted: 18 April 2020; Published: 22 April 2020

Abstract: Every day, visually challenged people (VCP) face mobility restrictions and accessibility limitations. A short walk to a nearby destination, which for other individuals is taken for granted, becomes a challenge. To tackle this problem, we propose a novel visual perception system for outdoor navigation that can be evolved into an everyday visual aid for VCP. The proposed methodology is integrated in a wearable visual perception system (VPS). The proposed approach efficiently incorporates deep learning, object recognition models, along with an obstacle detection methodology based on human eye fixation prediction using Generative Adversarial Networks. An uncertainty-aware modeling of the obstacle risk assessment and spatial localization has been employed, following a fuzzy logic approach, for robust obstacle detection. The above combination can translate the position and the type of detected obstacles into descriptive linguistic expressions, allowing the users to easily understand their location in the environment and avoid them. The performance and capabilities of the proposed method are investigated in the context of safe navigation of VCP in outdoor environments of cultural interest through obstacle recognition and detection. Additionally, a comparison between the proposed system and relevant state-of-the-art systems for the safe navigation of VCP, focused on design and user-requirements satisfaction, is performed.

Keywords: visually challenged; navigation; image analysis; fuzzy sets; machine learning

---

1. Introduction

According to the World Health Organization (WHO), about 16% of the worldwide population lives with some type of visual impairment [1]. Visually challenged people (VCP) struggle in their everyday life and have major difficulties in participating in sports, cultural, tourist, family, and other types of outdoor activities. The last two decades, a key solution to this problem has been the development of assistive devices able to help, at least partially, the VCP to adjust in the modern way of life and actively participate in different types of activities. Such assistive devices require the cooperation of researchers from different fields, such as medicine, smart electronics, computer science, and engineering. So far, as a result of this interdisciplinary cooperation, several designs and components of wearable camera-enabled systems for VCP have been proposed [2–5]. Such systems incorporate sensors, such as cameras, ultrasonic sensors, laser distance sensors, inertial measurement units, microphones, and GPS, which enable the user identify his/her position in an area of interest (i.e., outdoor environment, hospital, museum, archeological site, etc.), avoid static or moving obstacles and hazards in close proximity, and provide directions not only for navigation support but also for personalized guidance in that area. Moreover, mobile cloud-based applications [6], methodologies for optimal estimation of trajectories using GPS and other sensors accessible from a mobile device [7], and algorithms enabling efficient data coding for video streaming [8] can be considered for enhanced user experience in this context. Users should be able to easily interact with the system through speech in real-time. Moreover, the system should be able to share the navigation experience of the user not strictly as audiovisual information but also through social interaction with remote individuals, including people with locomotor disabilities and the elderly.

During the last two years, several studies and research projects have been initiated, setting higher standards for systems for computer-assisted navigation of VCP. In [9], an Enterprise Edition of a Google glass device was employed to support visually challenged individuals during their movement. Their system comprised a user interface, a computer network platform, and an electronic device to integrate all components into a single assistive device. In [10], a commercial pair of smart glasses (KR-VISION), consisting of an RGB-D sensor (RealSense R200) and a set of bone-conducting earphones, was linked to a portable processor. The RealSense R200 sensor was also employed in [11], together with a low-power millimeter wave (MMW) radar sensor, in order to unify object detection, recognition, and fusion. Another smart assistive navigation system comprised a smart-glass with a Raspberry Pi camera attached on a Raspberry Pi processor, as well as a smart shoe with an IR sensor for obstacle detection attached on an Arduino board [12]. In [13], a binocular vision probe with two charged coupled device (CCD) cameras and a semiconductor laser was employed to capture images in a fixed frequency. A composite head-mounted wearable system with a camera, ultrasonic sensor, IR sensor, button controller, and battery for image recognition was proposed in [14]. Two less complex approaches were proposed in [15,16]. In the first, two ultrasonic sensors, two vibrating motors, two transistors, and an Arduino Pro Mini Chip were attached on a simple pair of glasses. The directions were provided to the user through vibrations. In the second, a Raspberry Pi camera and two ultrasonic sensors attached on a Raspberry Pi processor were placed on a plexiglass frame.

The aforementioned systems incorporate several types of sensors, which increase the computational demands and the energy consumption, the weight of the wearable device, as well as the complexity of the system. In addition, although directions in the form of vibrations are faster perceivable, their expressiveness is limited, and the learning curve required increases with the number of messages needed for user navigation.

This paper presents a novel visual perception system (VPS) for outdoor navigation of the VCP in cultural areas, which copes with these issues in accordance with the respective user requirements [17]. The proposed system differs from others, because it follows a novel uncertainty-aware approach to obstacle detection, incorporating salient regions generated using a Generative Adversarial Network (GAN) trained to estimate saliency maps based on human eye-fixations. The estimated eye-fixation maps, expressing the human perception of saliency in the scene, adds to the intuition of the obstacle detection methodology. Additional novelties of the proposed VPS, include: (a) it can be personalized, based on the user characteristics—the user's height, in order to minimize false alarms that may occur from the obstacle detection methodology and 3D printed to meet the user's preferences; (b) the system implements both obstacle detection and recognition; and (c) the methodologies of obstacle detection and recognition are integrated in the system in a unified way.

The rest of this paper is organized in 6 sections, namely, Section 2, where the current state-of-the-art is presented; Section 3, describing the system architecture; Section 4, analyzing the methodologies used for obstacle detection and recognition; Section 5, examining the performance of the obstacle detection and recognition tasks; Section 6, where the performance of the proposed system with respect to other systems is discussed; and finally Section 7 presenting the conclusions of this work.

2. Related Work

2.1. Assistive Navigation Systems for the VCP

A review on relevant systems proposed until 2008 was presented in [18], where three categories of navigation systems were identified. The first category is based on positioning systems, including the Global Positioning System (GPS) for outdoor positioning, and preinstalled pilots and beacons emitting signals to determine the absolute position of the user in a local structured environment; the second is based on radio frequency identification (RFID) tags with contextual information, such as surrounding landmarks and turning points; and the third concerns vision-based systems that exploit information acquired from digital cameras to perceive the surrounding environment. Moreover, in a survey conducted in 2010 [19], wearable obstacle avoidance electronic travel aids for blind were reviewed and ranked based on their features. A more recent study [20] reviewed the state-of-the-art sensor-based assistive technologies, where it was concluded that most of the current solutions are still at a research stage, only partially solving the problem of either indoor or outdoor navigation. In addition, some guidelines for the development of relevant systems were suggested, including real-time performance (i.e., fast processing of the exchanged information between user and sensors and detection of suddenly appearing objects within a range of 0.5–5 m), wireless connectivity, reliability, simplicity, wearability, and low cost.

In more recent vision-based systems, the main and most critical functionalities include the detection of obstacles, provision of navigational assistance, as well as recognition of objects or scenes in general. A wearable mobility aid solution based on embedded 3D vision was proposed in [21], which enables the user to perceive, be guided by audio messages and tactile feedback, receive information about the surrounding environment, and avoid obstacles along a path. Another relevant system was proposed in [4], where a stereo camera was used to perceive the environment, providing information to the user about obstacles and other objects in the form of intuitive acoustic feedback. A system for joint detection, tracking, and recognition of objects encountered during navigation in outdoor environments was proposed in [3]. In that system, the key principle was the alternation between tracking using motion information and prediction of the position of an object in time based on visual similarity. Another project [2] investigated the development of a smart-glass system consisting of a camera and ultrasonic sensors able to recognize obstacles ahead, and assess their distance in real-time. In [22], a wearable camera system was proposed, capable of identifying walkable spaces, planning a safe motion trajectory in space, recognizing and localizing certain types of objects, as well as providing haptic-feedback to the user through vibrations. A system named Sound of Vision was presented in [5], aiming to provide the users with a 3D representation of the surrounding environment, conveyed by means of hearing and tactile senses. The system comprised an RGB-depth (RGB-D) sensor and an inertial measurement unit (IMU) to track the head/camera orientation. A simple smart-phone-based guiding system was proposed in [23], which incorporated a fast feature recognition module running on a smart-phone for fast processing of visual data. In addition, it included two remotely accessible modules, one for more demanding feature recognition tasks and another for direction and distance estimation. In the context of assisted navigation, an indoor positioning framework was proposed by the authors of [24]. Their positioning framework is based on a panoramic visual odometry for the visually challenged people.

An augmented reality system using predefined markers to identify specific facilities, such as hallways, restrooms, staircases, and offices within indoor environments, was proposed in [25]. In [26], a scene perception system based on a multi-modal fusion-based framework for object detection and classification was proposed. The authors of [27] aimed to the development of a method integrated in a wearable device for the efficient place recognition using multimodal data. In [28], a unifying terrain awareness framework was proposed, extending the basic vision system based on an IR RGB-D sensor proposed in [10] and aiming at achieving efficient semantic understanding of the environment. The above approach, combined with a depth segmentation method, was integrated into a wearable navigation system. Another vision-based navigational aid using an RGB-D sensor was presented in [29], which solely focused on a specific component for road barrier recognition. Even more recently, a live object recognition blind aid system based on convolutional neural network was proposed in [30], which comprised a camera and a computer system. In [9], a system based on a Google Glass device was developed to navigate the user in unfamiliar healthcare environments, such as clinics, hospitals, and urgent cares. A wearable vision assistance system for visually challenged users based on big data and binocular vision sensors was proposed in [13]. Another assistive navigation system proposed in [12] combined two devices, a smart glass and a smart pair of shoes, where various sensors were integrated with Raspberry Pi, and the data from both devices are processed to provide more efficient navigation solutions. In [11], a low-power MMW radar and an RGB-D camera were used to unify obstacle detection, recognition, and fusion methods. The proposed system is not wearable but hangs from the neck of the user at the height of the chest. A navigation and object recognition system presented in [31] consisted of an RGB-D sensor and an IMU attached on a pair of glasses and a smartphone. A simple obstacle detection glass model, incorporating ultrasonic sensors, was proposed in [15]. Another wearable image recognition system, comprising a micro camera, an ultrasonic sensor, an infrared sensor, and a Raspberry Pi as the local processor, was presented in [14]. On the one side of the wearable device were the sensors and the controller and on the other the battery. In [32], a wearable system with three ultrasonic sensors and a camera was developed to recognize texts and detect obstacles and then relay the information to the user via an audio outlet device. A similar but less sophisticated system was presented in [16].

A relevant pre-commercial system, called EyeSynth (Audio-Visual System for the Blind Allowing Visually Impaired to See Through Hearing), promises both obstacle detection and audio-based user communication, and it is developed in the context of a H2020 funding scheme for small medium enterprises (SMEs). It consists of a stereoscopic imaging system mounted on a pair of eyeglasses, and non-verbal and abstract audio signals are communicated to the user. Relevant commercially available solutions include ORCAM MyEye, a device attachable to eyeglasses that discreetly reads printed and digital text aloud from various surfaces and recognizes faces, products, and money notes; eSight Eyewear, which uses a high-speed and high-definition camera that captures whatever the user sees and then displays it on two near-to-eye displays enhancing the vision of partially blind individuals; and the AIRA system, which connects blind or low-vision people with trained, remotely-located human agents who, at the touch of a button, can have access to what the user sees through a wearable camera. The above commercially available solutions do not yet incorporate any intelligent components for automated assistance.

In the proposed system, barebone computer unit (BCU), namely a Raspberry Pi Zero, is employed, since it is easily accessible to everyone and easy to use, contrary to other devices such as Raspberry Pi processors. In contrast to haptic feedback or audio feedback in the form of short sound signals, the proposed method uses linguistic expressions incurring from fuzzy modeling to inform the user about obstacles, their position in space, and scene description. The human eye-fixation saliency used for obstacle detection provides the system with human-like eye-sight characteristics. The proposed method relies on visual cues provided only by a stereo camera system, instead of the various different sensors used in previous systems, thus reducing the computational demands, design complexity, and energy requirements, while enhancing user comfort. Furthermore, the system can be personalized according to the user's height, and the wearable frame is 3D printed, therefore, adjusting to the preferences of each individual user, e.g., head anatomy, and avoiding restrictions imposed by using commercially available glass frames.

2.2. Obstacle Detection

Image-based obstacle detection is a component of major importance for assistive navigation systems for the VCP. A user requirement analysis [17], revealed that the users need a system that aims to real-time performance and mainly detects vertical objects, e.g., trees, humans, stairs, and ground anomalies.

Obstacle detection methodologies consists of two steps: (a) an object detection step and (b) an estimation step of the threat that an object poses to the agent/VCP. The image-based object detection problem has been previously tackled with the deployment of deep learning models. The authors of [33] proposed a Convolutional Neural Network (CNN) model, namely Faster Region-Based CNN, that was used for real-time object detection and tracking [26]. In [3], the authors proposed a joint object detection, tracking and recognition in the context of the DEEP-SEE framework. Regarding wearable navigation aids for VCP, an intelligent smart glass system, which exploits deep learning machine vision techniques and the Robotic Operating System, was proposed in [2]. The system uses three CNN models, namely, the Faster Region-Based CNN [33], You Only Look Once (YOLO) CNN model [34], and Single Shot multi-box Detectors (SSDs) [35]. Nevertheless, the goal of the aforementioned methods was solely to detect objects and not to classify them as obstacles.

In another work, a module of a wearable mobility aid was proposed based on the LeNet model for obstacle detection [21]. However, this machine learning method treats obstacle detection as a 2D problem. A multi-task deep learning model, which estimates the depth of a scene and extracts the obstacles without the need to compute a global map with an application in micro air vehicle flights, has been proposed in [35]. Other, mainly preliminary, studies have approached the obstacle detection problem for the safe navigation of VCP as a 3D problem by using images along with depth information and enhancing the performance by exploiting the capabilities of CNN models [36–38].

Aiming to robust obstacle detection, in this paper we propose a novel, uncertainty-aware personalized method, implemented by our VPS, based on a GAN and fuzzy sets. The GAN is used to detect salient regions within an image, where the detected salient regions are then combined with the 3D spatial information acquired by an RGB-D sensor using fuzzy sets theory. This way, unlike previous approaches, the proposed methodology is able to determine the level of threat posed by the obstacle to the user and its position in the environment with linguistic expressions. In addition, the proposed method takes into consideration the height of the user in order to describe the threat of an obstacle more efficiently. Finally, when compared to other deep learning assisted approaches, our methodology does not require any training regarding the obstacle detection part.

2.3. Object Recognition

Although object detection has a critical role in the safety assurance of VCP, the VPS aims to provide an effective object and scene recognition module, which enables the user to make decisions based on the visual context of the environment. More specifically, object recognition provides the capability to the user to identify what type of object has been detected by the object detection module. Object recognition can be considered as a more complex module compared to object detection, since it requires an intelligent system that can incorporate the additional free parameters required to distinguish between the different detected objects.

In the last decade, object recognition techniques have been drastically improved, mainly due to the appearance of CNN architectures, such as [39]. CNNs are a type of ANNs that consist of multiple convolutional layers with neuron arrangement mimicking the biological visual cortex. This enables CNNs to automatically extract features from the entire image, instead of relying on hand-crafted features, such as color and texture. Multiple CNN architectures have been proposed over the last years, each one contributing some unique characteristics [17]. Although conventional CNN architectures, such as the Visual Geometry Group Network (VGGNet) [40], offer great classification performance, they usually require large, high-end workstations equipped with Graphical Processing Units (GPUs) to execute them. This is mainly due to their large number of free-parameters [40] that increase their computational complexity and inference time, which in some applications, such as the assistance of VCP, is a problem of major importance. Recently, architectures, such as MobileNets [41] and ShuffleNets [42], have been specifically proposed to enable their execution on mobile and embedded devices. More specifically, MobileNets [41] are a series of architectures, which by using depth-wise separable convolutions [43] instead of conventional convolutions, vastly reduce the number of free-parameters of the network, enabling their execution on mobile devices. The authors in [42] proposed the use of the ShuffleNets architecture by using point-wise group convolution and channel shuffling to achieve a low number of free-parameters with high classification accuracy. Both architectures try to balance the trade-off between classification accuracy and computational complexity.

CNNs have also been used for object and scene recognition tasks in the context of assisting VCP. In the work of [21], a mobility aid solution was proposed that uses a LeNet architecture for object categorization in 8 classes. An architecture named "KrNet" was proposed in [29], which relies on a CNN architecture to provide real-time road barrier recognition in the context of navigational assistance of VCP. A terrain awareness framework was proposed in [28] that uses CNN architectures, such as SegNet [44], to provide semantic image segmentation.

In VPS, we make use of a state-of-the-art CNN architecture named Look Behind Fully Convolutional Network light or LB-FCN light [45], which offers high object recognition accuracy, while maintaining low computational complexity. Its architecture is based on the original LB-FCN architecture [46], which offers multi-scale feature extraction and shortcut connections that enhance the overall object recognition capabilities. LB-FCN light replaces the original convolutional layers with depth-wise separable convolutions and improves the overall architecture by extracting features under three different sizes (3×3, 5×5, and 7×7), lowering the number of free parameters of the original architecture. This enables the computationally efficient performance of the trained network while maintaining the recognition robustness, which is important for systems that require fast recognition responses, such as the one proposed in this paper. In addition to the low computational complexity provided by the LB-FCN light architecture, the system is cost-effective, since the obstacle recognition task does not require high-end expensive GPUs. Consequently, multiple conventional low-cost CPUs can be used instead, which enable relatively easy horizontal scaling of the system architecture.

3. System Architecture

The architecture of the cultural navigation module of the proposed VPS, consists of four components; a stereoscopic depth-aware RGB camera, a BCU, a wearable Bluetooth speaker device, and cloud infrastructure. The first three components are mounted on a single smart wearable system, with the shape of sunglasses, capable of performing lightweight tasks, such as risk assessment, while the computationally intense tasks, such as object detection and recognition, are performed on a cloud computing infrastructure. These components are further analyzed in the following Sections 3.1 and 3.2.

3.1. System Components and Infrastructure

As the stereoscopic depth aware RGB camera, the Intel® RealSense™ D435 was chosen, since it provides all the functionalities needed by the proposed system in a single unit. This component is connected via a USB cable to a BCU of the wearable system. The BCU used in the system was a Raspberry Pi Zero. The BCU orchestrates the communication between the user and the external services that handle the computationally expensive deep learning requirements of the system on a remote cloud computing infrastructure. Another role of the BCU is to handle the linguistic interpretation of the detected objects in the scenery and communicate with the Bluetooth component of the system, which handles the playback operation. For the communication of the BCU component with the cloud computing component, we chose to use a low-end mobile phone that connects to the internet using 4G or Wi-Fi when available, effectively acting as a hotspot device.

For the communication between the BCU and the cloud computing component of the system, we chose to use the Hyper Text Transfer Protocol version 2.0 (HTTP/2), which provides a simple communication protocol. As the entry point of the cloud computing component, we used a load balancer HTTP microservice, which implements a REpresentational State Transfer (RESTful) Application Programming Interface (API) that handles the requests coming from the BCU, placing them in a message queue for processing. The queue follows the Advanced Message Queuing Protocol (AMQP), which enables a platform agnostic message distribution. A set of message consumers, equipped with Graphical Processing Units (GPUs), process the messages that are placed in the queue and, based on the result, communicate back to the MPUs using the HTTP protocol. This architecture enables the system to be extensible both in terms of infrastructure, since new works can be added on demand, and in terms of functionality, depending on future needs of the platform.

The VPS component communication is shown in Figure 1. More specifically, the BCU component of the system, receives RGB-D images from the stereoscopic camera at a real-time interval. Each image is then analyzed using fuzzy logic by the object detection component of the system on the BCU itself, performing risk assessment. In parallel, the BCU communicates with the cloud computing component by sending a binary representation of the image to the load balancer, using the VPS RESTful API. A worker then receives the message placed in the queue from the load balancer and performs the object detection task, which involves the computation of the image saliency map from the received images using a GAN. When an object is detected and its boundaries determined, the worker performs the object recognition task using a CNN, the result of which is a class label for each detected object in the image. The worker, using HTTP, informs the MPU about the presence and location of the object in the image along with the detected labels. As a last step, the MPU linguistically translates the object position along with the detected labels provided from the methodology described in Section 4, using the build-in text to speech synthesizer of the BCU. The result is communicated via Bluetooth with the speaker attached to the ear of the user for playback. It is important to mention here that, in case of repeated object detections, the BCU component avoids the playback of the same detected object based on the change of the scenery, which enables the system to prevent unnecessary playbacks. In detail, as users are approaching an obstacle, the system notifies them about the collision risk, which is described using the linguistic expressions low, medium and high and its spatial location and category. To avoid user confusion, the system implements a controlled notification policy, where the frequency of notifications increases as the users are getting closer to the obstacle. The information about the obstacle's spatial location and category are provided only in the first notification of the system. If the users continue moving towards a high-risk obstacle, the system notifies them with a "stop" message.

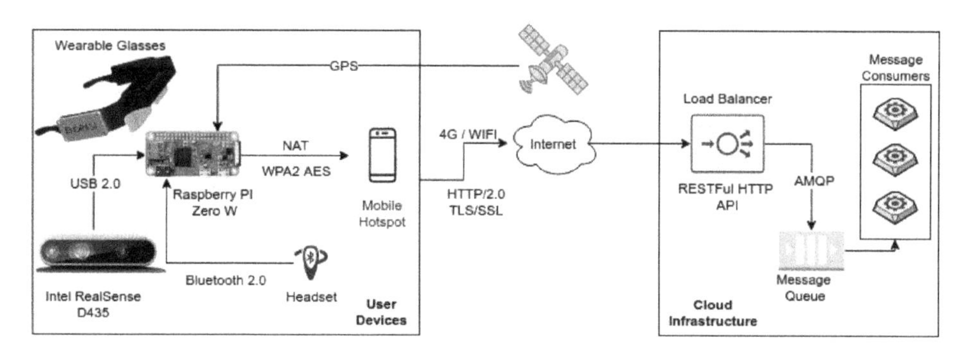

Figure 1. Visual perception system (VPS) architecture overview illustrating the components of the system along with their interconnectivity.

3.2. Smart Glasses Design

The wearable device, in the form of smart glasses, was designed using a CAD software according to the user requirements listed in [17]. The most relevant to the design requirements mentioned that the wearable system should be attractive and elegant, possibly with a selection of different colors, but in a minimalist rather than attention grabbing way. In terms of construction, the system should be robust; last a long time, not requiring maintenance; and be resistant to damage, pressure, knocks and bumps, water, and harsh weather conditions [17].

The design of the model has been parameterized, in terms of its width and length, making it highly adjustable. Therefore, it can be easily customized for each user based on the head dimensions, which makes it more comfortable. The model (Figure 2a,b) comprises two parts, the frame and the glass. In the front portion of the frame, there is a specially designed socket, where the Intel® RealSense™ D435 camera can be placed and secured with a screw at its bottom. In addition, the frame has been designed to incorporate additional equipment if needed, such as Raspberry Pi (covered by the lid with the VPS logo), an ultrasonic sensor, and an IMU. The designed smart-glass model was 3D printed using PLA filament in a Creality CR-10 3D printer. The resulted device is illustrated in Figure 2c.

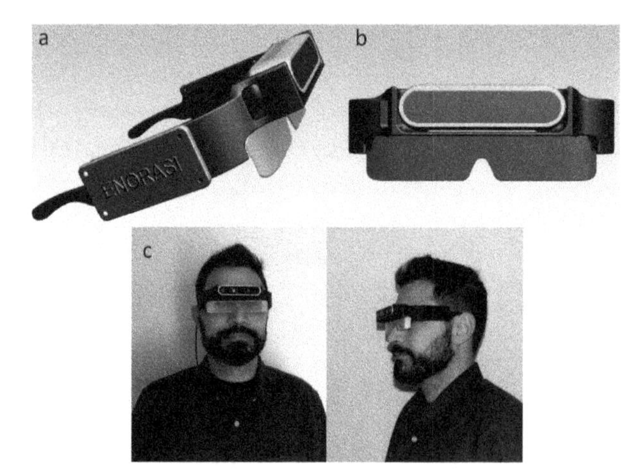

Figure 2. 3D representation of the smart glasses. (**a**) Side view of the glasses; (**b**) front view of the glasses; and (**c**) 3D-printed result with the actual camera sensor. In this preliminary model, the glass-part was printed with transparent PLA filament, which produced a blurry, semi-transparent result. In future versions, the glass-part will be replaced by transparent polymer or glass.

4. Obstacle Detection and Recognition Component

The obstacle detection and recognition component can be described as a two-step process. In the first step, the detection function incorporates a deep learning model and a risk assessment approach using fuzzy sets. The deep learning model is used to predict, eye-human fixations, on images captured during the navigation of the VCP. Then, fuzzy sets are used to assess the risk based on depth values calculated by the RGB-D camera, generating risk maps, expressing different degrees of risk. The risk and saliency maps are then combined using a fuzzy aggregation process through which the probable obstacles are detected. In the second step, the recognition of the probable obstacles takes place. For this purpose, each obstacle region is propagated to a deep learning model, which is trained to infer class labels for objects found in the navigation scenery (Figure 3).

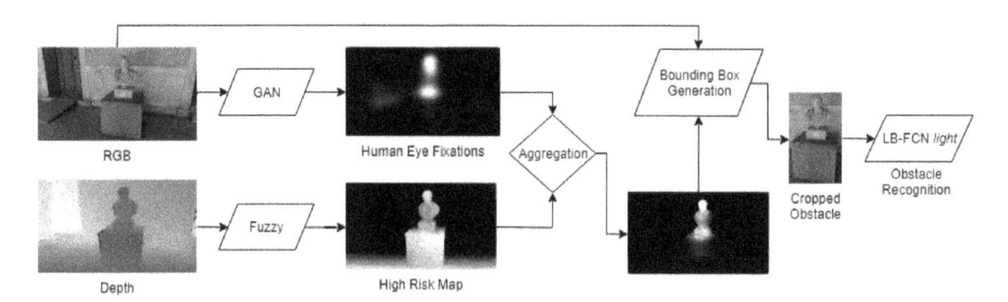

Figure 3. Visualization of the proposed obstacle detection and recognition pipeline.

4.1. Obstacle Detection

The detection-recognition methodology can be summarized as follows:

(a) Eye human fixation estimation model;
(b) Depth-aware fuzzy risk assessment in the form of risk maps;
(c) Obstacle detection and localization via the fuzzy aggregation of saliency maps, produced in Step (a) and the risk maps produced in Step (b);

(d) Obstacle recognition using a deep learning model based on probable obstacle regions obtained in Step (c).

4.1.1. Human Eye Fixation Estimation

The saliency maps used in this work are generated by a GAN [47]. The generated saliency maps derive from human eye fixation points and thus, they make the significance of a region in a scene more instinctual. Such information can be exploited for the obstacle detection procedure, and at the same time, enhance the intuition of the methodology. Additionally, the machine learning aspect enables the extensibility of the methodology, since it can be trained with additional eye fixation data, collected from individuals during their navigation through rough terrains. An example of the saliency maps estimated from a given image can be seen in Figure 4. Since the model is trained on human eye-fixation data, it identifies as salient those regions in the image on which the attention of a human would be focused. As it can be observed in Figure 4, in the first image, the most salient region corresponds to the fire extinguisher cabinet; in the second image, to the people on the left side; and in the last image, to the elevated ground and the tree branch.

(a) (b)

Figure 4. Examples of the generated saliency maps given an RGB image. (**a**) Input RGB images. (**b**) Respective generated saliency maps.

The GAN training utilizes two different CNN models, namely, a discriminator and a generator. During the training, the generator learns to generate imagery related to a task, and the discriminator assists to the optimization of the resemblance to the target images. In our case, the target data are composed of visual saliency maps based on human eye tracking data.

The generator architecture is a VGG-16 [40] encoder-decoder model. The encoder follows an identical architecture to that of VGG-16 unaccompanied by fully connected layers. The encoder is used to create a latent representation of the input image. The encoder weights are initialized by training the model on the ImageNet dataset [48]. During the training, there was no update of the weights of the encoder, with an exception to the last two convolutional blocks.

The decoder has the same architectural structure with the encoder network, with the exception that the layers are placed in reverse order, and the max pooling layers are replaced with up-sampling layers. To generate the saliency map, the decoder has an additional 1×1 convolutional layer in the output, with sigmoidal activation. The decoder weights were initialized randomly. The generator accepts an RGB image I_{RGB} as stimulus and generates a saliency map that resembles the human eye fixation on that I_{RGB}.

The discriminator of the GAN has a simpler architecture. The discriminator model consists of 3×3 convolutional layers, combined with 3 max pooling layers followed by 3 Fully Connected (FC) layers. The Rectified Liner Unit (ReLU) and hyperbolic tangent (tanh) functions are deployed as activation functions for the convolutional and FC layers, respectively. The only exception is the last layer of the FC part, where the sigmoid activation function was used. The architecture of the GAN generator network is illustrated in Figure 5.

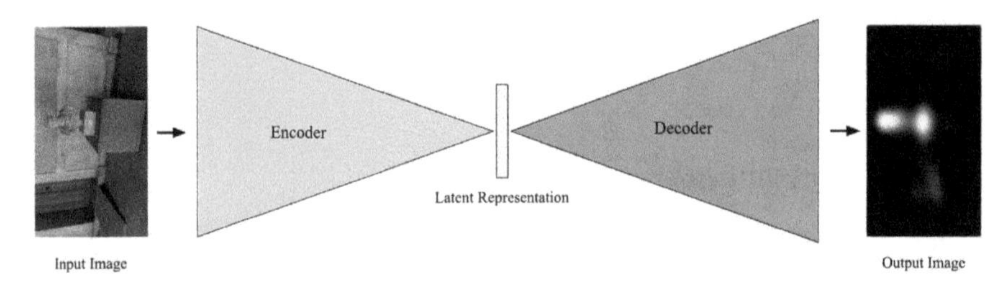

Figure 5. Illustration of the generator architecture. The generator takes as input an RGB image I_{RGB} and outputs a saliency map based on human eye fixation.

4.1.2. Uncertainty-Aware Obstacle Detection

In general, an object that interferes with the safe navigation of a person can be perceived as salient. Considering this, the location of an obstacle is likely to be in regions of a saliency map that indicate high importance, i.e., with high intensities. A saliency map produced by the model described in Section 4.1.1 can be treated as a weighted region of interest, in which an obstacle may be located. High-intensity regions of such a saliency map indicate high probability of the presence of an object of interest. Among all the salient regions in the saliency map, we need to identify these regions that may pose a threat to the person navigating in the scenery depicted in I_{RGB}. Thus, we follow an approach, where both a saliency map and a depth map deriving by an RGB-D sensor are used for the risk assessment. The combination of the saliency and depth maps is achieved with the utilization of Fuzzy Sets [49].

For assessing the risk, it can be easily deduced that objects/areas that are close to the VCP navigating in an area and are salient with regard to the human gaze may pose a certain degree of threat to the VCP. Therefore, as a first step, the regions that are in a certain range from the navigating person need to be extracted, so that they can be determined as threatening. Hence, we consider a set of 3 fuzzy sets, namely, R_1, R_2, and R_3—describing three different risk levels, which can be described with the linguistic values of high, medium, and low risk, respectively. The fuzzy sets R_1, R_2, and R_3 represent a different degree of risk and their universe of discourse is the range of depth values of a depth map. Regarding the fuzzy aspect of these sets and taking into consideration the uncertainty in the risk assessment, there is an overlap between the fuzzy sets describing low and medium and medium and *high* risk. The fuzzy sets R_1, R_2, and R_3 are described by the membership function $r_i(z)$, $i = 1, 2, 3$, where $z \in [0, \infty)$. The membership functions are illustrated in Figure 6c.

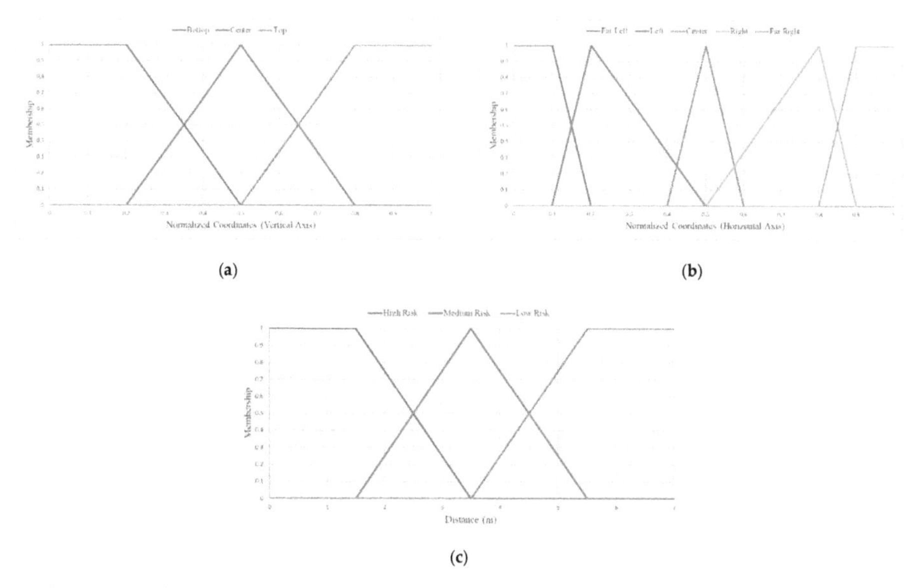

Figure 6. Membership functions of fuzzy sets used for the localization of objects in the 3D space using linguistic variables. (**a**) Membership functions for far left (h_1), left (h_2), central (h_3), right (h_4) and far right (h_5) positions on the horizontal axis. (**b**) Membership functions for up (v_1), central (v_2) and bottom (v_3) positions on the vertical axis. (**c**) Membership functions for low (r_1), medium (r_1), and high risk (r_3) upon the distance of the user from an obstacle.

A major aspect of an obstacle detection methodology is the localization of obstacles and the description of their position in a manner that can be communicated and easily perceived by the user. In our system, the description of the spatial location of an object is performed using linguistic expressions. We propose an approach based on fuzzy logic to interpret the obstacle position using linguistic expressions (linguistic values) represented by fuzzy sets. Spatial localization of an obstacle in an image can be achieved by defining 8 additional fuzzy sets. More specifically, we define 5 fuzzy sets for the localization along the horizontal axis of the image, namely, H_1, H_2, H_3, H_4, and H_5 corresponding to far left, left, central, right, and far right portions of the image. Additionally, to express the location of the obstacle along the vertical axis of the image, we define 3 fuzzy sets, namely, V_1, V_2, and V_3 denoting the upper, central, and bottom portions of the image. The respective membership functions of these fuzzy sets are $h_j(x)$, $j = 1, 2, 3, 4, 5$ and $v_i(y)$, $i = 1, 2, 3$, where $x, y \in [0, 1]$ are normalized image coordinates. An illustration of these membership functions can be seen in Figure 6.

Some obstacles, such as tree branches, may be in close proximity to the individual with respect to the depth but at a certain height that safe passage would not be affected. Thus, a personalization step was introduced to the methodology eliminating false alarms. The personalization aspect and the minimization of false positive obstacle detection instances are implemented through an additional fuzzy set P, addressing the risk an obstacle poses to a person with respect to the height. For the description of this P fuzzy set, we define a two dimensional membership function $p(h_o, h_u)$, where h_o and h_u are the heights of the obstacle and the user, respectively. The personalization methodology is described in Section 4.1.3.

For the risk assessment, since the membership functions describing each fuzzy set were defined, the next step is the creation of 3 risk maps, R_M^i. The risk maps R_M^i, derive from the responses of a membership function, $r_i(z)$, and are formally expressed as:

$$R_M^i(x, y) = r_i(D(x,y)) \tag{1}$$

where D is a depth map that corresponds to an RGB image I_{RGB}. Using all the risk assessment membership functions, namely r_1, r_2, and r_3, 3 different risk maps, R_M^1, R_M^2, and R_M^3, are derived. Each of these risk maps depicts regions that may pose different degrees of risk to the VCP navigating in the area. In detail, risk map R_M^1 represents regions that may pose high degree of risk, R_M^2 medium degree of risk, and finally R_M^3 low degree of risk. A visual representation of these maps can be seen in Figure 7. Figure 7b,c illustrates the risk maps derived from the responses of the r_1, r_2, and r_3 membership functions on the depth map of Figure 7a. Brighter pixel intensities represent higher participation in the respective fuzzy set, while darker pixel intensities represent lower participation.

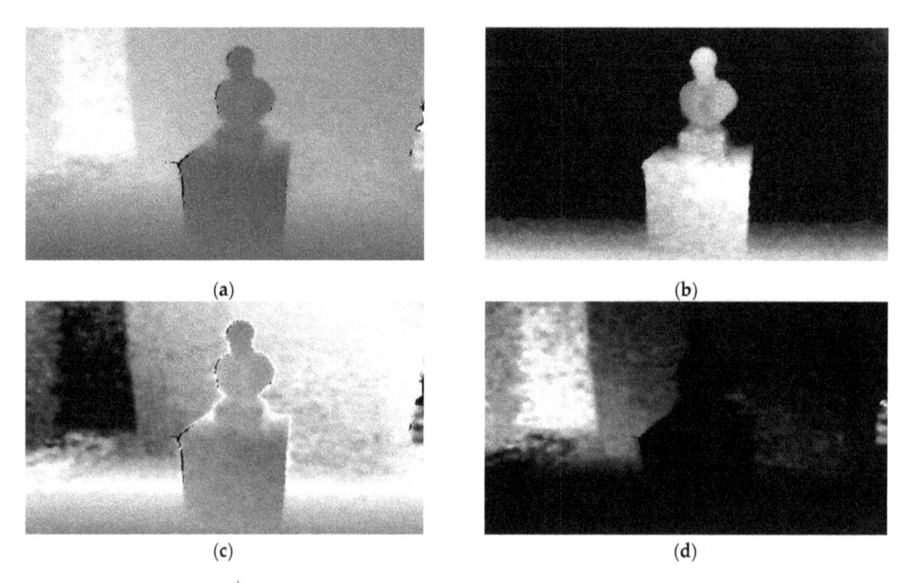

Figure 7. Example of R_M^i creation. (**a**) Depth map D, where lower intensities correspond to closer distances; (**b**) visual representation of R_M^1 representing regions of high risk; (**c**) R_M^2 representing regions of medium risk; (**d**) R_M^3 depicting regions of low risk. Higher intensities in (**b–d**) correspond to lower participation in the respective fuzzy set. All images have been normalized for better visualization.

In the proposed methodology, the obstacle detection is a combination between the risk assessed from the depth maps and the degree of saliency that is obtained from the GAN described in the previous subsection. The saliency map S_M that is produced from a given I_{RGB} is aggregated with each risk map R_M^i, where $i = 1, 2, 3$, using the fuzzy AND ($\wedge$) operator (Godel t-norm) [50], formally expressed as:

$$F_1 \wedge F_2 = \min(F_1(x, y), F_2(x, y)) \tag{2}$$

In Equation (2), F_1 and F_2 denote two generic 2D fuzzy maps with values within the [0, 1] interval, and x, y are the coordinates of each value of the 2D fuzzy map. The risk maps R_M^i are, by definition, fuzzy 2D maps, since they derive from the responses of membership functions r_i on a depth map. The saliency map S_M can be considered as a fuzzy map where its values represent the degree of participation of a given pixel to the salient domain. Therefore, they can be combined with the fuzzy AND operator to produce a new fuzzy 2D map O^i_M as follows:

$$O^i_M = R^i_M \wedge S_M \tag{3}$$

The non-zero values of the 2D fuzzy map O^i_M (obstacle map) at each coordinate (x, y) indicate the location of an obstacle and express the degree of participation in the risk domain of the respective R^i_M. Figure 8d illustrates the respective O^i_M produced using the fuzzy AND operator with the three R^i_M. Higher pixel values of the O^i_M portray higher participation on the respective risk category and the probability of the location of an obstacle.

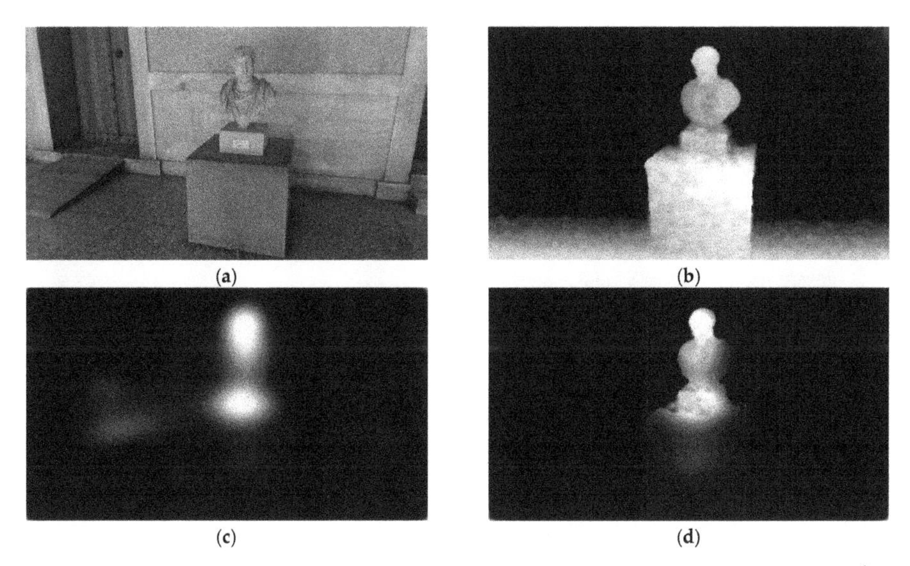

Figure 8. Example of the aggregation process between the saliency map SM and the high-risk map R^1_M. (**a**) Original I_{RGB} used for the generation of the saliency map SM; (**b**) high-risk map R^1_M used in the aggregation; (**c**) saliency map S_M based on the human eye fixation on image (**a**); (**d**) the aggregation product using the fuzzy AND operator between images (**b**) and (**c**).

Theoretically, the O^i_M can be directly used to detect obstacles posing different degrees of risk to the VCP navigating in the area. However, if the orientation of the camera is towards the ground, the ground plane can be often falsely perceived as obstacle. Consequently, a refinement step is needed to optimize the obstacle detection results and reduce the occurrence of false alarm error. Therefore, a simple but effective approach for ground plane extraction is adopted.

The ground plane has a distinctive gradient representation along the Y axis in depth maps, which can be exploited in order to remove it from the O^i_M. As a first step, the gradient of the depth map D is estimated by:

$$\nabla D = \left(\frac{\partial D}{\partial x}, \frac{\partial D}{\partial y} \right) \tag{4}$$

A visual representation of a normalized difference map $\frac{\partial D}{\partial y}$ in the $[0, 255]$ interval can be seen in Figure 9. As it can be seen, the regions corresponding to the ground have smaller differences than the rest of the depth map. In the next step, a basic morphological gradient g [51] is applied on the gradient of D along the y direction $\frac{\partial D}{\partial y}$. A basic morphological gradient is basically the difference between dilation and erosion of the $\frac{\partial D}{\partial y}$ given an all-one kernel $k_{5\times5}$:

$$g\left(\frac{\partial D}{\partial y}\right) = \delta_{k_{5\times5}}\left(\frac{\partial D}{\partial y}\right) - \varepsilon_{k_{5\times5}}\left(\frac{\partial D}{\partial y}\right) \tag{5}$$

where δ and ε denote the operations of dilation and erosion and their subscripts indicate the used kernel. In contrast to the usual gradient of an image, the basic morphological gradient g corresponds to the maximum variation in an elementary neighborhood rather than a local slope. The morphological gradient is followed by consecutive operations of erosion and dilation with a kernel $k_{5\times5}$. As it can be noticed in Figure 9c, the basic morphological filter g gives higher responses on non-ground regions, and thus, the following operations of erosion and dilution are able to eliminate the ground regions quite effectively. The product of these consecutive operations is a ground removal mask G_M, which is then multiplied with O^i_M, setting the values corresponding to the ground, to zero. This ground removal approach has been experimentally proven to be sufficient (Section 5) to eliminate the false identification of the ground as obstacle. A visual representation of the ground mask creation and the ground removal can be seen in Figures 9 and 10, respectively.

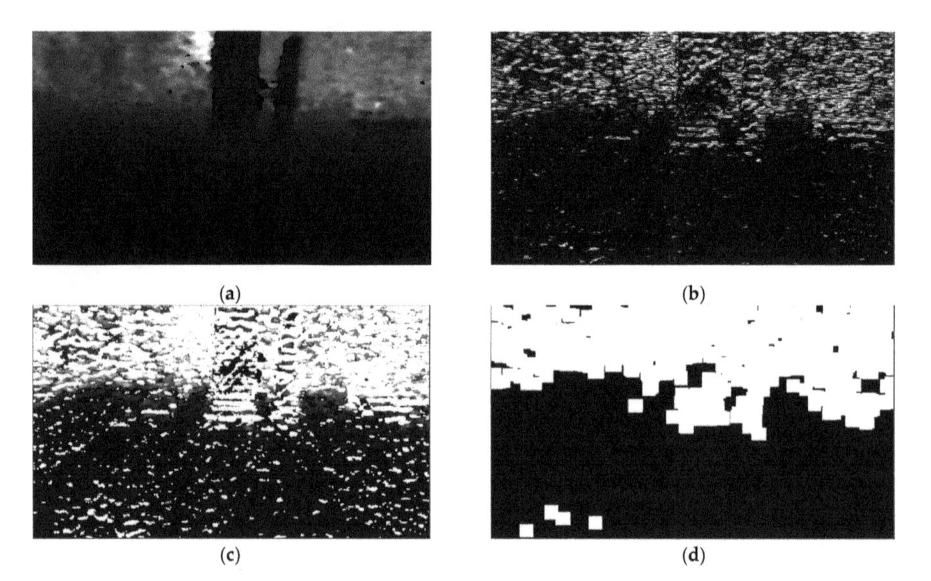

Figure 9. Example of the creation steps of G_M. (**a**) Depth map D, normalized for better visualization; (**b**) visual representation of the difference map Δ_M; (**c**) difference map Δ_M after the application of the basic morphological gradient; and (**d**) the final ground removal mask G_M.

Once the obstacle map of the depicted scene is estimated following the process described above, the next step is the spatial localization of the obstacle in linguistic values. This step is crucial for the communication of the surroundings to a VCP. For this purpose, Fuzzy Sets are utilized in this work. As presented in Section 4.1.1, 5 membership functions are used to determine the location of an obstacle along the horizontal axis (x-axis) and 3 along the vertical axis (y-axis).

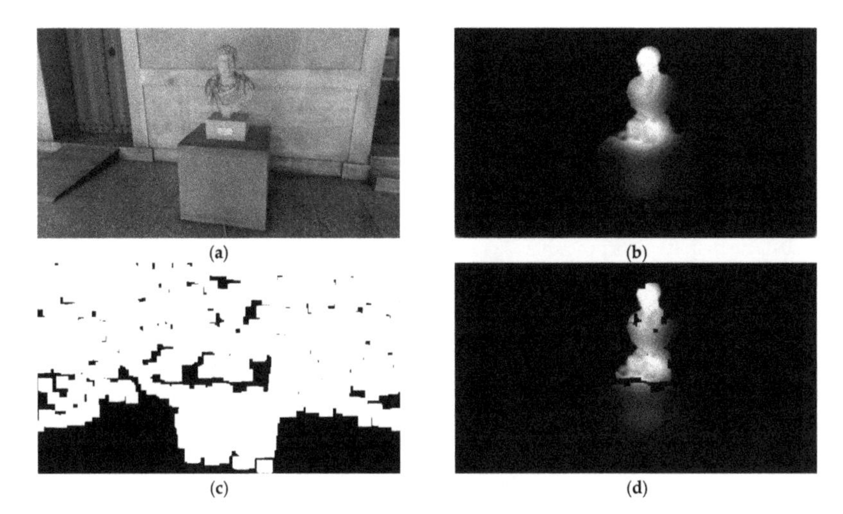

Figure 10. Example of the ground removal procedure. (**a**) Original I_{RGB} image; (**b**) corresponding obstacle map O^1_M; (**c**) respective ground removal mask G_M; (**d**) masked obstacle map O^1_M. In (**d**), the ground has been effectively removed.

Initially, the boundaries of the obstacles depicted in the obstacle maps need to be determined. For the obstacle detection task, the O^1_M obstacle map, through which the high-risk obstacles are represented, is chosen. Then, the boundaries b_l, where $l = 1, 2, 3 \ldots$, of the obstacles are calculated using a border following the methodology presented in [52]. Once the boundaries of each probable obstacle depicted in O^1_M are acquired, their centers $c_l = (c_x, c_y)$, $l = 1, 2, 3, \ldots$ are derived by exploiting the properties of the image moments [53] of boundaries b_l. The centers c_l can be defined using the raw moments m_{00}, m_{10}, and m_{01} of b_l as follows:

$$m_{qk} = \iint_{b_l} x^q y^k I_{RGB}(x, y) dx dy \tag{6}$$

$$c_l = \left(\frac{m_{10}}{m_{00}}, \frac{m_{01}}{m_{00}} \right) \tag{7}$$

where $q = 0, 1, 2, \ldots, k = 0, 1, 2, \ldots$ and x, y denote image coordinates along the x-axis and y-axis respectively. An example of the obstacle boundary detection can be seen in Figure 11, where the boundaries of the obstacles are illustrated with green lines (Figure 11b) and the centers of the obstacles are marked with red circles (Figure 11c).

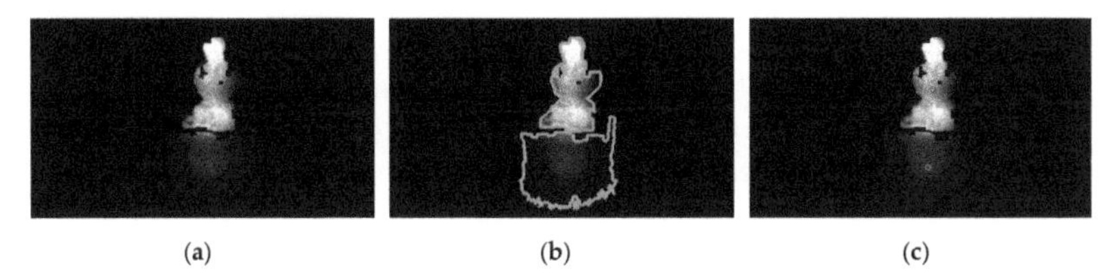

Figure 11. Example of the obstacle boundary extraction and obstacle center calculation. (**a**) O^1_P obstacle map used for the detection of high-risk obstacles; (**b**) boundary (green outline) estimation of the obstacles; (**c**) respective centers of the detected obstacles.

Once the centers have been calculated, their location can be determined and described with linguistic values using the horizontal and vertical membership functions, h_j, where $j = 1, 2, 3, 4, 5$, and

v_i, where $i = 1, 2, 3$. If the response of $h_j(c_x)$ and $v_i(c_y)$ is greater than 0.65, then the respective obstacle with a boundary center of $c_l = (c_x, c_y)$ will be described with the linguistic value that these h_j and v_i represent. Additionally, the distance between object and person is estimated using the depth value of depth map D at the location of $D(c_x, c_y)$. Using this information, the VCP can be warned regarding the location and distance of the obstacle and, as an extension, be assisted to avoid it.

4.1.3. Personalized Obstacle Detection Refinement

The obstacle map depicts probable obstacles that are salient for humans and are within a certain range. However, this can lead to false positive indications, since some obstacles, such as tree branches, can be within a range that can be considered threatening, but at a height greater than that of the user, not affecting his/her navigation. False positive indications of this nature can be avoided using the membership function $p(h_o, h_u)$. To use this membership function, the 3D points of the scene need to be determined by exploiting the intrinsic parameters of the camera and the provided depth map.

To project 2D points on the 3D space in the metric system (meters), we need to know the corresponding depth value z for each 2D point. Based on the pinhole model, which describes the geometric properties of our camera [54], the projection of a 3D point to the 2D image plane is described as follows:

$$\begin{pmatrix} \widetilde{u} \\ \widetilde{v} \end{pmatrix} = \frac{f}{z}\begin{pmatrix} X \\ Y \end{pmatrix} \tag{8}$$

where f is the effective focal length of camera, and $(X, Y, z)^{\mathrm{T}}$ is the 3D point corresponding to a 2D point on the image plane $(\widetilde{u}, \widetilde{v})^T$. Once the projected point $(\widetilde{u}, \widetilde{v})^T$ is acquired, the transition to pixel coordinates $(x, y)^{\mathrm{T}}$ is described by the following equation:

$$\begin{pmatrix} x \\ y \end{pmatrix} = \begin{pmatrix} D_u s_u \widetilde{u} \\ D_v \widetilde{v} \end{pmatrix} + \begin{pmatrix} x_0 \\ y_0 \end{pmatrix} \tag{9}$$

s_u denotes a scale factor; D_u, D_v are coefficients needed for the transition from the metric units to pixels, and $(x_0, y_0)^{\mathrm{T}}$ is the principal point of the camera. With the combination of Equations (8) and (9) the projection which describes the transition from 3D space to the 2D image pixel coordinate system can be expressed as

$$\begin{pmatrix} x \\ y \end{pmatrix} = \begin{pmatrix} \frac{f D_u s_u X}{z} \\ \frac{f D_v Y}{z} \end{pmatrix} + \begin{pmatrix} x_0 \\ y_0 \end{pmatrix} \tag{10}$$

The 3D projection of a 2D point with pixel coordinates (x, y), for which the depth value z is known, can be performed by solving Equation (10) for X, Y formally expressed below [55]:

$$\begin{pmatrix} X \\ Y \end{pmatrix} = z\begin{pmatrix} \frac{x - x_0}{f_x} \\ \frac{y - y_0}{f_y} \end{pmatrix} \tag{11}$$

where $f_x = f D_u s_u$ and $fy = f D_v$. Equation (11) is applied on all the 2D points of I_{RGB} with known depth values z. After the 3D points have been calculated, the Y coordinates are used to create a 2D height map H_M of the scene, where each value is a Y coordinate indicating the height an object at the corresponding pixel coordinate in I_{RBG}. Given the height h_u of the user, we apply the p membership function on the height map H_M to assess the risk with respect to the height of the user. The responses of p on H_M create a 2D fuzzy map P_M as shown below:

$$P_M(x,\ y) = p(H_M(x, y),\ hu) \tag{12}$$

Finally, the fuzzy AND operator is used to combine O^i_M with $\mathrm{P_M}$, resulting in a final personalized obstacle map O^i_P:

$$O^i_P = O^i_M \wedge P_M \tag{13}$$

Non-zero values of O_P^i represent the final location of a probable obstacle with respect to the height of the user and the degree of participation to the respective risk degree, i.e., the fuzzy AND operation between O_P^1 with P_M describes the high-risk obstacles in the scenery.

4.2. Obstacle Recognition

For the object recognition task, the LB-FCN light network architecture [45] was chosen, since it has been proven to work well on obstacle detection-related tasks. A key characteristic of the architecture is the relatively low number of free-parameters compared to both conventional CNN architectures, such as [40], and mobile-oriented architectures, such as [41,42]. The LB-FCN light architecture uses Multi-Scale Depth-wise Separable Convolution modules (Figure 12a) to extract features under three different scales, 3×3, 5×5, and 7×7, which are then concatenated, forming a feature-rich representation of the input volume. Instead of conventional convolution layers, the architecture uses depth-wise separable convolutions [43], which drastically reduce the number of free-parameters in the network.

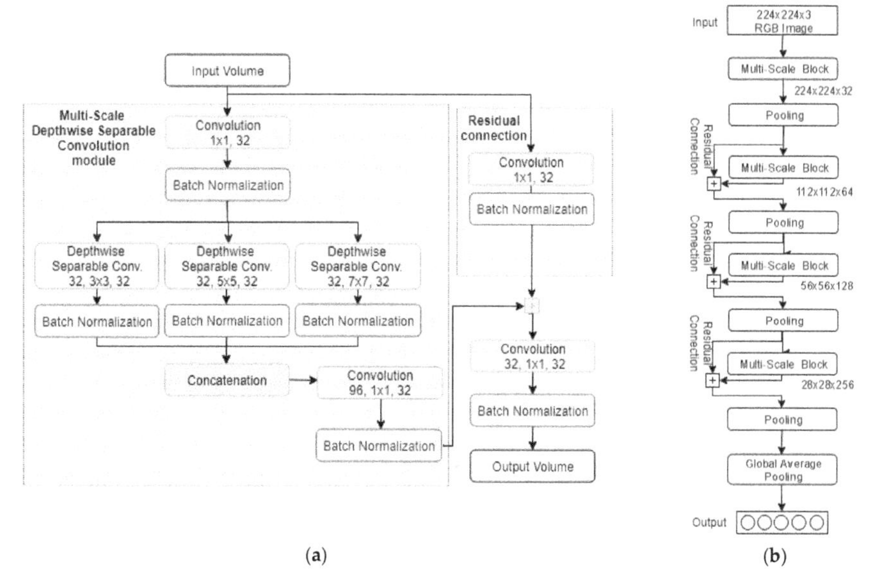

Figure 12. Visualization of (**a**) the multi-scale depthwise separable convolution block and (**b**) the overall Look Behind Fully Convolutional Network (LB-FCN) light network architecture.

The combination of the multi-scale modules and depth-wise separable convolutions enables the reduction of the overall computational complexity of the model without sacrificing significant classification performance. Furthermore, the network uses shortcut connections that connect the input with the output of each multi-scale module, promoting the high-level features to be propagated across the network and encounter the problem of vanishing gradient, which is typical in deep networks. Following the principles established in [56], the architecture is fully convolutional, which simplifies the overall network design and lowers further the number of free-parameters. Throughout the architecture, all convolution layers use ReLU activations and more specifically the capped ReLU activation proposed in [41]. As a regularization technique, batch normalization [57] is applied on the output of each convolution layer, enabling the network to converge faster while reducing the incidence of the overfitting phenomenon during training. It is important to note that compared to the conventional CNN architectures used by other VCP assistance frameworks, such as [21,28,29], the

LB-FCN light architecture offers significantly lower computational complexity with high classification accuracy, making it a better choice for the proposed system.

5. Experimental Framework and Results

To validate the proposed system, a new dataset was constructed consisting of videos captured from an area of cultural interest, namely the Ancient Agora of Athens, Greece. The videos were captured using a RealSense D435 mounted on the smart glasses (Section 3.2) and were divided into two categories. The first category focused on videos of free walk around the area of Ancient Agora and the second category on controlled trajectories towards obstacles found in the same area.

The validation of the system was developed around both obstacle detection and their class recognition. When an obstacle was identified and its boundaries were determined, the area of the obstacle was cropped and propagated to the obstacle recognition network. In the rest of this section, the experimental framework will be further described (Section 5.1) along with results achieved using the proposed methodology (Section 5.2).

5.1. Experimental Framework

The dataset composed for the purposes of this study focuses on vertical obstacles that can be found in sites of cultural interest. The dataset consisted of 15,415 video frames captured by researchers wearing the smart glasses described Section 3.2 (Figure 2). In 5138 video frames the person wearing the camera was walking towards the obstacles but not in a range for the obstacle to be considered threatening. In the rest 10,277 video frames, the person was walking until collision, towards obstacles considered as threatening, which should be detected and recognized. The intervals determining whether an obstacle is considered as threatening or not were set according to the user requirements established by VCP for obstacle detection tasks in [17]. Regarding that, the desired detection distance for the early avoidance of an obstacle according to the VCP user requirements is up to 2 m.

During data collection, the camera captured RGB images, corresponding depth maps, and stereo infrared (IR) images. The D435 sensor is equipped with an IR projector, which is used for the improvement of depth quality through the projection of an IR pattern that enables texture enrichment. The IR projector was used during the data acquisition for a more accurate estimation of the depth. In this study, only the RGB images and the depth maps needed for our methodology were used. The categories of obstacles visible in the dataset were columns, trees, archaeological artifacts, crowds, and stones. An example of types of obstacles included in our dataset can be seen in Figure 13. As previously mentioned, all data were captured in an outdoor environment, in the Ancient Agora of Athens. In addition, it is worth noting that the data collection protocol that was followed excludes any images that include human subjects that could be recognized in any way.

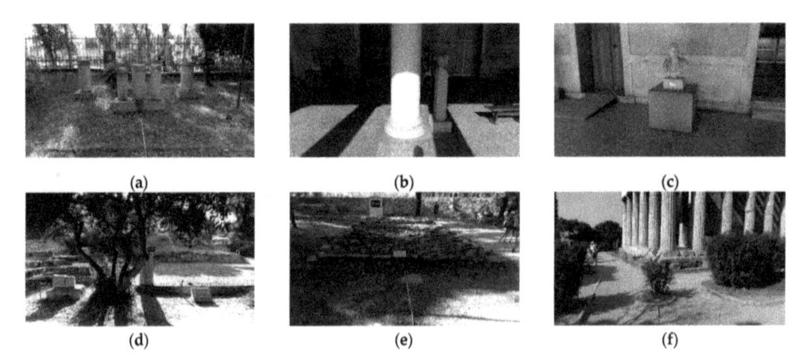

Figure 13. Example of the objects identified as obstacles in our dataset: (**a**–**c**) columns/artifacts; (**d**) tree; (**e**) cultural sight near the ground level; (**f**) small tree/bush.

5.2. Obstacle Detection Results

For the obstacle detection task, only the high-risk map was used, since it depicts objects that pose immediate threat to the VCP navigating the area. The high-risk interval of the membership function r_1 was decided to be at $0 < z < 3.5$ m. By utilizing the fuzzy sets, an immediate threat within the range of $0 < z < 1.5$ m can be identified, since the responses of r_1 in this interval are 1, and then, it degrades until the distance of 3.5 m, where it becomes 0. With this approach, the uncertainty within the interval of $1.5 < z < 3.5$ m is taken into consideration, while at the same time, the requirement regarding the detection up to 2 m is satisfied. The GAN that was used for the estimation of the saliency maps based on the human eye-fixation was trained on the SALICON dataset [58].

The proposed methodology was evaluated on the dataset described in Section 4.1. For the evaluation of the obstacle detection methodology, the sensitivity, specificity, and accuracy metrics were used. The sensitivity and specificity are formally defined as follows:

$$Sensitivity = \frac{TP}{TP + FN} \tag{14}$$

$$Specificity = \frac{TN}{TN + FP} \tag{15}$$

where TP (true positive) are the true positive obstacle detections, e.g., the obstacles that were correctly detected, FP (false positive) are the falsely detected obstacles, TN (true negative) are frames were correctly no obstacles were detected, and FN (false negative) are frames that obstacles were not correctly detected.

Our method resulted in an accuracy of 85.7% on its application of the aforementioned dataset, with a sensitivity and specificity of 85.9% and 85.2%, respectively. A confusion matrix for the proposed method is presented in Table 1. For further evaluation, the proposed method was compared to that proposed in [38], which, on the same dataset, resulted in an accuracy of 72.6% with a sensitivity and specificity of 91.7% and 38.6%, respectively. The method proposed in [38] included neither the ground plane removal in its pipeline nor the personalization aspect. On the other hand, the proposed approach was greatly benefited from these aspects in the minimization of false alarms. As it can be seen in Figure 14, the dataset contains frames where the camera is oriented towards the ground, and without a ground plane removal step, false alarms are inevitable. The obstacles in Figure 14 were not in a range to be identified as a threat to the user; however, in Figure 14a–c, where the ground plane removal has not been applied, the ground has been falsely identified (green boxes) as obstacle. A quantitative comparison between the two methods can be seen in Table 2.

Table 1. Confusion matrix of the proposed methodology. Positive are the frames with obstacles, and negative are the frames with no obstacles.

	Detected	
Actual	**Positive (%)**	**Negative (%)**
Positive (%)	55.1	9.0
Negative (%)	5.3	30.6

Table 2. Results and quantitative comparison between the proposed and state-of-the art methodologies.

Metrics	Proposed (%)	Method [38] (%)	Method [36] (%)
Accuracy	85.7	72.6	63.7
Sensitivity	86.0	91.7	87.3
Specificity	85.2	38.6	21.6

(a) (b) (c)

Figure 14. Qualitative example of false ground detection as obstacle resulting from using the methodology presented in [38]. In all images, the obstacles are not in a threatening distance. (**a**) False positive detection on dirt ground-type. (**b**) False positive detection on rough dirt ground-type. (**c**) False positive detection on tile ground-type.

Qualitative results with respect to the ground detection method can be seen in Figure 15. As it can be observed, the methodology used for the ground plane detection is resilient to different ground types. The ground types that were found in our dataset were grounds with dirt, tiles, marble, and gravels. In addition, using such a method reduces greatly the false alarm rate when the head is oriented towards the ground plane. Even though the masking process is noisy, the obstacle inference procedure is not affected.

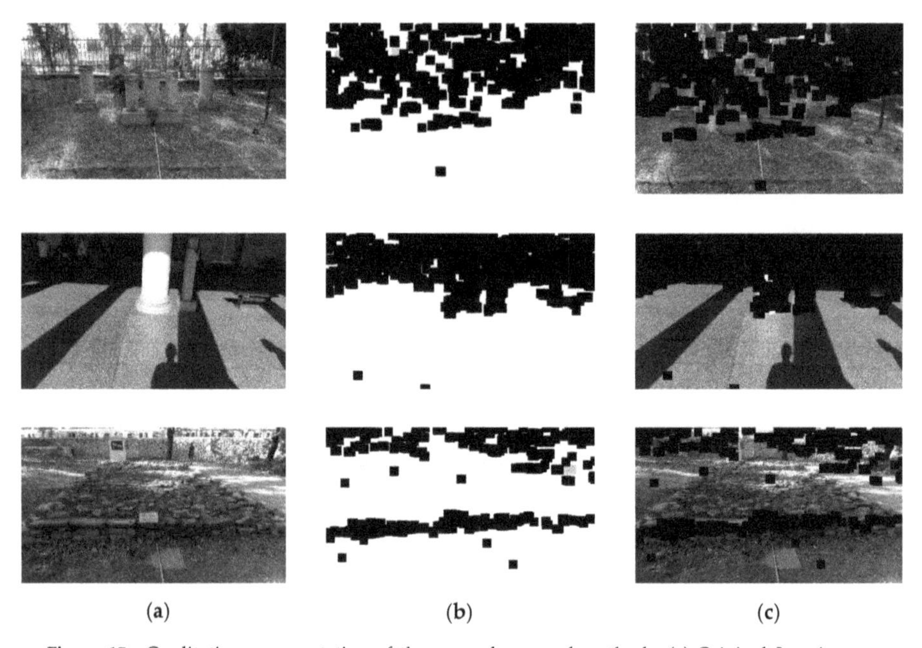

(a) (b) (c)

Figure 15. Qualitative representation of the ground removal method. (**a**) Original I_{RGB} images. (**b**) Ground masks with the white areas indicating the ground plane. (**c**) Images of (**a**) masked with the masks of (**b**).

5.3. Obstacle Recognition Results

The original LB-FCN light architecture was trained on the binary classification problem of staircase detection in outdoor environments. In order to train the network on obstacles that can be found by the VPS, a new dataset named "Flickr Obstacle Recognition" was created (Figure 16) with images, published under the Creative Commons license, found on the popular social media platform "Flickr" [59]. The dataset contains 1646 RGB images of various sizes that contain common obstacles, which can be found in the open space. More specifically, the images are weakly annotated based on their content in 5 obstacle categories: "benches" (427 images), "columns" (229 images), "crowd" (265 images), "stones" (224 images), and "trees" (501 images). It is worth mentioning that the dataset is considered relatively challenging, since the images were obtained by different modalities, under various lighting conditions and different landscapes.

For the implementation of the LB-FCN light architecture, the popular Keras [60] python library with the Tensorflow [61] was used as the backend tensor graph framework. To train the network, the images were downscaled to a size of 224 × 224 pixels and zero-padded where needed to maintain the original aspect ratio. No further pre-processing was applied to the images. For the network training, the Adam [62] optimizer was used with an initial learning rate of alpha = 0.001 and first and second moment estimates exponential decay as rate beta1 = 0.9 and beta2 = 0.999, respectively. The network was trained using a high-end NVIDIA 1080TI GPU equipped with 3584 CUDA cores [63], 11 GB of GDDR5X RAM, and base clock speed of 1480 MHz.

(a) (b) (c) (d) (e)

Figure 16. Sample images from the five obstacle categories: (**a**) "benches", (**b**) "columns", (**c**) "crowd", (**d**) "stones", and (**e**) "trees" from the "Flickr Obstacle Recognition" dataset.

To evaluate the recognition performance of the trained model, the testing images were composed by the detected objects found by the object detection component of the system. More specifically, 212 obstacles of various sizes were detected. The pre-processing of the validation images was similar to that described above for the training set.

For comparison, the state-of-the-art mobile-oriented architecture named "MobileNet-v2" [64] was trained and tested using the same training and testing data. The comparative results, presented in Table 3, demonstrate that the LB-FCN light architecture is able to achieve higher recognition performance, while requiring lower computational complexity, compared to the MobileNet-v2 architecture (Table 4).

Table 3. Comparative classification performance results between the LB-FCN light architecture [45] and the MobileNet-v2 architecture [64].

Metrics	LB-FCN Light [45] (%)	MobileNet-v2 [64] (%)
Accuracy	93.8	91.4
Sensitivity	92.4	90.5
Specificity	91.3	91.1

Table 4. Computational complexity comparison between the LB-FCN light architecture [3] and the MobileNet-v2 architecture [65].

Metrics	LB-FCN Light [45] (%)	MobileNet-v2 [64] (%)
FLOPs × 10^6	0.6	4.7
Trainable free parameters × 10^6	0.3	2.2

6. Discussion

Current imaging, computer vision, speech, and decision-making technologies have the potential to further evolve and be incorporated into effective assistive systems for the navigation and guidance of VCPs. The present study explored novel solutions to the identified challenges, with the aim to deliver an integrated system with enhanced usability and accessibility. Key features in the context of such a system are obstacle detection, recognition, easily interpretable feedback for the effective obstacle avoidance, and a novel system architecture. Some obstacle detection methods such as [21] tackle the problem by incorporating deep learning methods for the obstacle detection tasks and using only the 2D traits of the images. In this work, a novel method was presented, where the 3D information acquired using an RGB-D sensor was exploited for the risk assessment from the depth values of the scenery using fuzzy sets. The human eye fixation was also taken into consideration, estimated by a GAN, in terms of saliency maps. The fuzzy aggregation of the risk estimates and the human eye fixation had as a result the efficient detection of obstacles in the scenery. In contrast to other depth-aware methods, such as the one proposed in [36], the obstacles detected with our approach are described with linguistic values with regard to their opposing risk and spatial location, making them easily interpretable by the VCP. In addition, the proposed method does not only extract obstacles that are an immediate threat to the VCP, e.g., these with non-zero responses from the high-risk membership function r_1, but also obstacles that are of medium and low risk. Therefore, all obstacles are known at any time, even if they are not of immediate high risk. The personalization aspects of the proposed method, alongside with the ground plane detection and removal, provide a significant lower false alarm rate. Furthermore, the method is able to detect and notify the user about partially visible obstacles with the condition that the part of the obstacle is: (a) salient, (b) within a distance that would be considered of high risk and (c) at a height that would be affecting the user. In detail, the overall accuracy of the system based on the proposed method was estimated to be 85.7%, when the methodology proposed in [38] produced an accuracy of 72.6%, based on the dataset described in Section 4.1. Additionally, in contrast to other methodologies such as [2,26,27,31,32], the proposed obstacle detection and recognition system is solely based on visual cues obtained using only an RGB-D sensor, minimizing the computational and energy resources required for the integration, fusion, and synchronization of multiple sensors.

Over the years, there has been a lot of work in the field of deep learning that tempts to increase the classification performance in object recognition tasks. Networks, such as VGGNet [40], GoogLeNet [65], and ResNet [66] provide high classification accuracy but with ever more increasing computational complexity, the result of which limits their usage on high-end devices equipped with expensive GPUs and low inference time [67]. Aiming to decrease the computational complexity and maintain high object recognition performance, this work demonstrated that the LB-FCN light [45] architecture can be used as an effective object recognition solution in the field of obstacle recognition. Furthermore, the comparative results presented in Section 5.2 exhibited that the LB-FCN light architecture is able to achieve higher generalization performance and maintain lower computational complexity compared to the state-of-the-art MobileNet-v2 architecture [64]. It is worth mentioning that single shot detectors, such as YOLO [34] and its variances, have been proved effective in object detection and recognition tasks. However, such detectors are fully supervised, and they need to be trained on a dataset with specific kinds of objects to be able to recognize them. In the current VPS, the obstacle detection task is handled by the described fuzzy-based methodology, which does not require any training on domain-specific data. Therefore, its obstacle detection capabilities are not limited by previous knowledge about the obstacles, and in that sense, it can be considered as a safer option for the VCPs. Using LB-FCN light, which is fully supervised, on top of the results of the fuzzy-based obstacle detection methodology, the system is able to recognize obstacles of predefined categories, without jeopardizing the user's safety. Although the trained model achieved a high overall object recognition accuracy of 93.8%, we believe that by increasing the diversity of the training "Flickr Object Recognition" dataset, the network can achieve an even higher classification performance. This is due to the fact that the original training dataset contains obstacles located in places and terrains that differ a lot from the ones found in the testing dataset.

The human-centered system architecture presented in Section 3.1 orchestrates all the different components of the VPS. The combination of the BCU component with the RGB-D stereoscopic camera and a Bluetooth headset, all mounted on a 3D printed wearable glass frame, enables the user to move freely around the scenery without attracting unwelcome attention. Furthermore, the cloud computing component of the architecture, enables transparent horizontal infrastructure scaling, allowing the system to be expanded based on future needs. Lastly, the communication protocols used by the different components of the system enable transparent component replacement without requiring any redesign of the proposed architecture.

In order to address and integrate the user and design requirements in the different stages of system development, the design process needs to be human-centered. The user requirements for assistive systems, focused on the guidance of VCP, have been extensively reviewed in [17]. Most of the requirements concerned audio-based functions; tactile functions; functions for guidance and description of the surrounding environment; connectivity issues; and design-oriented requirements such as battery life, device size, and device appearance. Relevant wearable systems have embodied, among others, battery and controller [14], 3D cameras with large on-board FPGA processors [68], and inelegant frame design [16], which are contrary to certain user requirements concerning size/weight, aesthetics, and complexity, described in [17]. A major advantage of the proposed configuration is its simplicity, since it includes only the camera and one cable connected to a mobile device. On the contrary, a limitation of the current system is the weight of the camera, which may cause discomfort to the user. Most of this weight is due to the aluminum case. A solution to this issue is to replace the camera with its caseless version, which is commercially available, and make proper adjustments to the designed frame.

7. Conclusions

In this work, we presented a novel methodology to tackle the problem of visually challenged mobility assistance by creating a system that implements:

- A novel uncertainty-aware obstacle detection methodology, exploiting the human eye-fixation saliency estimation and person-specific characteristics;
- Integration of obstacle detection and recognition methodologies in a unified manner;
- A novel system architecture that allows horizontal resource scaling and processing module interchange ability.

More specifically, the proposed VPS incorporates a stereoscopic camera mounted on an adjustable wearable frame, providing efficient real-time personalized object detection and recognition capabilities. Linguistic values can describe the position and type of the detected object, enabling the system to provide an almost natural interpretation of the environment. The 3D printed model of the wearable glasses was designed based on the RealSense D435 camera, providing a discreet and unobtrusive wearable system that should not attract undue or unwelcome attention.

The novel approach followed by the object detection module employs fuzzy sets along with human eye fixation prediction using GANs and enables the system to perform efficient real-time object detection with high accuracy prevailing in current state-of-the-art approaches. This is achieved by incorporating depth-maps along with saliency maps. The module is capable to accurately locate an object that poses a threat to the person navigating the scenery. For the object recognition task, the proposed system incorporates deep learning to recognize the objects obtained from the object detection module. More specifically, we use the state-of-the-art object recognition CNN, named LB-FCN light, which offers high recognition accuracy with relatively low number of free parameters. To train the network, a new dataset was created, named "Flickr Obstacle Recognition" dataset, containing RGB outdoor images from five common obstacle categories.

The novel object detection and recognition modules, combined with the user-friendly and highly adjusTable 3D frame, suggest that the proposed system can be the backbone for the development of a complete, flexible, and effective solution to the problem of visually challenged navigation assistance. The effectiveness of the proposed system was validated for both obstacle detection and recognition using datasets acquired from an outdoor area of interest. As a future work we intend to further validate our system in field tests where VCPs and/or blind-folded subjects will wear the proposed VPS for outdoor navigation. the capacity for further improvements of the background algorithms, structural design, and incorporated equipment provides great potential to the production of a fully autonomous commercial product, available to everyone at low cost. Furthermore, considering that the proposed VPS is developed in the context of a project for assisted navigation in cultural environments, the acquired data can be used also for the 4D reconstruction of places of cultural importance, by exploiting and improving state-of-the-art approaches [69,70]. Such a functionality extension of the system will contribute to further enhancement of cultural experiences for a broader userbase, beyond VCPs, as well as to the creation of digital archives with research material for the investigation of cultural environments over time, via immersive 4D models.

Author Contributions: Conceptualization, G.D., D.E.D., P.K., and D.K.I.; methodology, G.D., D.E.D., P.K., and D.K.I.; software, G.D., D.E.D., and D.K.I.; validation, G.D., D.E.D., and P.K.; formal analysis, G.D., D.E.D.; investigation, G.D., D.E.D., P.K., and D.K.I.; resources, G.D., D.E.D., P.K., and D.K.I.; data curation, G.D., D.E.D.; writing—original draft preparation, D.K.I., G.D., D.E.D., and P.K.; writing—review and editing, D.K.I., G.D., D.E.D., and P.K.; visualization, G.D., D.E.D.; supervision, D.K.I.; project administration, D.K.I.; funding acquisition, D.K.I. All authors have read and agreed to the published version of the manuscript.

Funding: This research has been co-financed by the European Union and Greek national funds through the Operational Program Competitiveness, Entrepreneurship and Innovation, under the call RESEARCH–CREATE–INNOVATE (project code: T1EDK-02070).

Acknowledgments: The Titan X used for this research was donated by the NVIDIA Corporation.

Conflicts of Interest: The authors declare no conflict of interest.

References

1. WHO. *World Health Organization-Blindness and Visual Impairement*; WHO: Geneva, Switzerland, 2018.
2. Suresh, A.; Arora, C.; Laha, D.; Gaba, D.; Bhambri, S. Intelligent Smart Glass for Visually Impaired Using Deep Learning Machine Vision Techniques and Robot Operating System (ROS). In Proceedings of the International Conference on Robot Intelligence Technology and Applications, Daejeon, Korea, 14–15 December 2017; pp. 99–112.
3. Tapu, R.; Mocanu, B.; Zaharia, T. DEEP-SEE: Joint Object Detection, Tracking and Recognition with Application to Visually Impaired Navigational Assistance. *Sensors* **2017**, *17*, 2473. [CrossRef] [PubMed]
4. Schwarze, T.; Lauer, M.; Schwaab, M.; Romanovas, M.; Böhm, S.; Jürgensohn, T. A camera-Based mobility aid for visually impaired people. *KI Künstliche Intell.* **2016**, *30*, 29–36. [CrossRef]
5. Caraiman, S.; Morar, A.; Owczarek, M.; Burlacu, A.; Rzeszotarski, D.; Botezatu, N.; Herghelegiu, P.; Moldoveanu, F.; Strumillo, P.; Moldoveanu, A. Computer Vision for the Visually Impaired: The Sound of Vision System. In Proceedings of the IEEE International Conference on Computer Vision Workshops, Venice, Italy, 22–29 October 2017; pp. 1480–1489.
6. Mahmood, Z.; Bibi, N.; Usman, M.; Khan, U.; Muhammad, N. Mobile cloud based-Framework for sports applications. *Multidimens. Syst. Signal Process.* **2019**, *30*, 1991–2019. [CrossRef]
7. Ahmed, H.; Ullah, I.; Khan, U.; Qureshi, M.B.; Manzoor, S.; Muhammad, N.; Khan, S.; Usman, M.; Nawaz, R. Adaptive Filtering on GPS-Aided MEMS-IMU for Optimal Estimation of Ground Vehicle Trajectory. *Sensors* **2019**, *19*, 5357. [CrossRef]
8. Khan, S.N.; Muhammad, N.; Farwa, S.; Saba, T.; Khattak, S.; Mahmood, Z. Early Cu depth decision and reference picture selection for low complexity Mv-Hevc. *Symmetry* **2019**, *11*, 454. [CrossRef]
9. Bashiri, F.S.; LaRose, E.; Badger, J.C.; D'Souza, R.M.; Yu, Z.; Peissig, P. *Object Detection to Assist Visually Impaired People: A Deep Neural Network Adventure*; Springer International Publishing: Cham, Switzerland, 2018; pp. 500–510.

10. Yang, K.; Wang, K.; Zhao, X.; Cheng, R.; Bai, J.; Yang, Y.; Liu, D. IR stereo realsense: Decreasing minimum range of navigational assistance for visually impaired individuals. *J. Ambient Intell. Smart Environ.* **2017**, *9*, 743–755. [CrossRef]
11. Long, N.; Wang, K.; Cheng, R.; Hu, W.; Yang, K. Unifying obstacle detection, recognition, and fusion based on millimeter wave radar and RGB-Depth sensors for the visually impaired. *Rev. Sci. Instrum.* **2019**, *90*, 044102. [CrossRef]
12. Pardasani, A.; Indi, P.N.; Banerjee, S.; Kamal, A.; Garg, V. Smart Assistive Navigation Devices for Visually Impaired People. In Proceedings of the IEEE 4th International Conference on Computer and Communication Systems (ICCCS), Singapore, 23–25 February 2019; pp. 725–729.
13. Jiang, B.; Yang, J.; Lv, Z.; Song, H. Wearable vision assistance system based on binocular sensors for visually impaired users. *IEEE Internet Things J.* **2019**, *6*, 1375–1383. [CrossRef]
14. Chen, S.; Yao, D.; Cao, H.; Shen, C. A Novel Approach to Wearable Image Recognition Systems to Aid Visually Impaired People. *Appl. Sci.* **2019**, *9*, 3350. [CrossRef]
15. Adegoke, A.O.; Oyeleke, O.D.; Mahmud, B.; Ajoje, J.O.; Thomase, S. Design and Construction of an Obstacle-Detecting Glasses for the Visually Impaired. *Int. J. Eng. Manuf.* **2019**, *9*, 57–66.
16. Islam, M.T.; Ahmad, M.; Bappy, A.S. Microprocessor-Based Smart Blind Glass System for Visually Impaired People. In Proceedings of the International Joint Conference on Computational Intelligence, Seville, Spain, 18–20 September 2018; pp. 151–161.
17. Iakovidis, D.K.; Diamantis, D.; Dimas, G.; Ntakolia, C.; Spyrou, E. Digital Enhancement of Cultural Experience and Accessibility for the Visually Impaired. In *Digital Enhancement of Cultural Experience and Accessibility for the Visually Impaired*; Springer: Cham, Switzerland, 2020; pp. 237–271.
18. Zhang, J.; Ong, S.; Nee, A. Navigation systems for individuals with visual impairment: A survey. In Proceedings of the 2nd International Convention on Rehabilitation Engineering & Assistive Technology, Bangkok, Thailand, 13–18 May 2008; pp. 159–162.
19. Dakopoulos, D.; Bourbakis, N.G. Wearable obstacle avoidance electronic travel aids for blind: A survey. *IEEE Trans. Syst. Man Cybern. Part C* **2009**, *40*, 25–35. [CrossRef]
20. Elmannai, W.; Elleithy, K. Sensor-Based assistive devices for visually-Impaired people: Current status, challenges, and future directions. *Sensors* **2017**, *17*, 565. [CrossRef] [PubMed]
21. Poggi, M.; Mattoccia, S. A wearable mobility aid for the visually impaired based on embedded 3D vision and deep learning. In Proceedings of the 2016 IEEE Symposium on Computers and Communication (ISCC), Messina, Italy, 27–30 June 2016; pp. 208–213.
22. Wang, H.-C.; Katzschmann, R.K.; Teng, S.; Araki, B.; Giarré, L.; Rus, D. Enabling independent navigation for visually impaired people through a wearable vision-Based feedback system. In Proceedings of the 2017 IEEE International Conference on Robotics and Automation (ICRA), Marina Bay Sands, Singapore, 29 May–3 June 2017; pp. 6533–6540.
23. Lin, B.-S.; Lee, C.-C.; Chiang, P.-Y. Simple smartphone-based guiding system for visually impaired people. *Sensors* **2017**, *17*, 1371. [CrossRef] [PubMed]
24. Hu, W.; Wang, K.; Chen, H.; Cheng, R.; Yang, K. An indoor positioning framework based on panoramic visual odometry for visually impaired people. *Meas. Sci. Technol.* **2019**, *31*, 014006. [CrossRef]
25. Yu, X.; Yang, G.; Jones, S.; Saniie, J. AR Marker Aided Obstacle Localization System for Assisting Visually Impaired. In Proceedings of the 2018 IEEE International Conference on Electro/Information Technology (EIT), Rochester, MA, USA, 3–5 May 2018; pp. 271–276.
26. Kaur, B.; Bhattacharya, J. A scene perception system for visually impaired based on object detection and classification using multi-Modal DCNN. *arXiv* **2018**, arXiv:1805.08798.
27. Cheng, R.; Wang, K.; Bai, J.; Xu, Z. OpenMPR: Recognize places using multimodal data for people with visual impairments. *Meas. Sci. Technol.* **2019**, *30*, 124004. [CrossRef]
28. Yang, K.; Wang, K.; Bergasa, L.M.; Romera, E.; Hu, W.; Sun, D.; Sun, J.; Cheng, R.; Chen, T.; López, E. Unifying terrain awareness for the visually impaired through real-Time semantic segmentation. *Sensors* **2018**, *18*, 1506. [CrossRef]
29. Lin, S.; Wang, K.; Yang, K.; Cheng, R. KrNet: A kinetic real-time convolutional neural network for navigational assistance. In *International Conference on Computers Helping People with Special Needs*; Springer: Cham, Germany, 2018; pp. 55–62.

30. Potdar, K.; Pai, C.D.; Akolkar, S. A Convolutional Neural Network based Live Object Recognition System as Blind Aid. *arXiv* **2018**, arXiv:1811.10399.
31. Bai, J.; Liu, Z.; Lin, Y.; Li, Y.; Lian, S.; Liu, D. Wearable Travel Aid for Environment Perception and Navigation of Visually Impaired People. *Electronics* **2019**, *8*, 697. [CrossRef]
32. Maadhuree, A.N.; Mathews, R.S.; Robin, C.R.R. Le Vision: An Assistive Wearable Device for the Visually Challenged. In Proceedings of the International Conference on Intelligent Systems Design and Applications, Vellore, India, 6–8 December 2018; pp. 353–361.
33. Ren, S.; He, K.; Girshick, R.; Sun, J. Faster r-CNN: Towards real-time object detection with region proposal networks. In Proceedings of the Neural Information Processing Systems, Montreal, QC, Canada, 7–9 December 2015; pp. 91–99.
34. Redmon, J.; Farhadi, A. YOLO9000: Better, faster, stronger. *arXiv* 2017.
35. Liu, W.; Anguelov, D.; Erhan, D.; Szegedy, C.; Reed, S.; Fu, C.-Y.; Berg, A.C. SSD: Single shot multibox detector. In Proceedings of the European Conference on Computer Vision, Amsterdam, The Netherlands, 8–16 October 2016; pp. 21–37.
36. Lee, C.-H.; Su, Y.-C.; Chen, L.-G. An intelligent depth-Based obstacle detection system for visually-Impaired aid applications. In Proceedings of the 2012 13th International Workshop on Image Analysis for Multimedia Interactive Services, Dublin, Ireland, 23–25 May 2012; pp. 1–4.
37. Mancini, M.; Costante, G.; Valigi, P.; Ciarfuglia, T.A. J-MOD 2: Joint monocular obstacle detection and depth estimation. *IEEE Robot. Autom. Lett.* **2018**, *3*, 1490–1497. [CrossRef]
38. Dimas, G.; Ntakolia, C.; Iakovidis, D.K. Obstacle Detection Based on Generative Adversarial Networks and Fuzzy Sets for Computer-Assisted Navigation. In Proceedings of the International Conference on Engineering Applications of Neural Networks, Crete, Greece, 24–26 May 2019; pp. 533–544.
39. Krizhevsky, A.; Sutskever, I.; Hinton, G.E. Imagenet classification with deep convolutional neural networks. In Proceedings of the Neural Information Processing Systems, Lake Tahoe, NV, USA, 3–6 December 2012; pp. 1097–1105.
40. Simonyan, K.; Zisserman, A. Very deep convolutional networks for large-scale image recognition. *arXiv* **2014**, arXiv:1409.1556.
41. Howard, A.G.; Zhu, M.; Chen, B.; Kalenichenko, D.; Wang, W.; Weyand, T.; Andreetto, M.; Adam, H. Mobilenets: Efficient convolutional neural networks for mobile vision applications. *arXiv* **2017**, arXiv:1704.04861.
42. Zhang, X.; Zhou, X.; Lin, M.; Sun, J. ShuffleNet: An Extremely Efficient Convolutional Neural Network for Mobile Devices. In Proceedings of the IEEE Conference on Computer Vision and Pattern Recognition, San Juan, PR, USA, 17–19 June 1997; pp. 6848–6856.
43. Chollet, F. Xception: Deep learning with depthwise separable convolutions. In Proceedings of the IEEE Conference on Computer Vision and Pattern Recognition, Honolulu, HI, USA, 22–25 July 2017; pp. 1251–1258.
44. Badrinarayanan, V.; Kendall, A.; Cipolla, R. Segnet: A deep convolutional encoder-Decoder architecture for image segmentation. *IEEE Trans. Pattern Anal. Mach. Intell.* **2017**, *39*, 2481–2495. [CrossRef] [PubMed]
45. Diamantis, D.E.; Koutsiou, D.-C.C.; Iakovidis, D.K. Staircase Detection Using a Lightweight Look-Behind Fully Convolutional Neural Network. In Proceedings of the International Conference on Engineering Applications of Neural Networks, Crete, Greece, 24–26 May 2019; pp. 522–532.
46. Diamantis, D.E.; Iakovidis, D.K.; Koulaouzidis, A. Look-Behind fully convolutional neural network for computer-Aided endoscopy. *Biomed. Signal Process. Control.* **2019**, *49*, 192–201. [CrossRef]
47. Goodfellow, I.; Pouget-Abadie, J.; Mirza, M.; Xu, B.; Warde-Farley, D.; Ozair, S.; Courville, A.; Bengio, Y. Generative adversarial nets. In Proceedings of the Neural Information Processing Systems, Montreal, QC, Canada, 8–13 December 2014; pp. 2672–2680.
48. Deng, J.; Dong, W.; Socher, R.; Li, L.-J.; Li, K.; Li, F.-F. Imagenet: A large-scale hierarchical image database. In Proceedings of the 2009 IEEE Conference on Computer Vision and Pattern Recognition, Miami, FL, USA, 20–25 June 2009; pp. 248–255.
49. Nguyen, H.T.; Walker, C.L.; Walker, E.A. *A First Course in Fuzzy Logic*; CRC Press: Boca Raton, FL, USA, 2018.
50. Feferman, S.; Dawson, J.W.; Kleene, S.C.; Moore, G.H.; Solovay, R.M. *Kurt Gödel: Collected Works*; Oxford University Press: Oxford, UK, 1998; pp. 1929–1936.
51. Rivest, J.-F.; Soille, P.; Beucher, S. Morphological gradients. *J. Electron. Imaging* **1993**, *2*, 326.

52. Suzuki, S.; Abe, K. Topological structural analysis of digitized binary images by border following. *Comput. Vision Graph. Image Process.* **1985**, *29*, 396. [CrossRef]
53. Kotoulas, L.; Andreadis, I. Image analysis using moments. In Proceedings of the 5th International. Conference on Technology and Automation, Thessaloniki, Greece, 15–16 October 2005.
54. Heikkilä, J.; Silven, O. A four-step camera calibration procedure with implicit image correction. In Proceedings of the IEEE Computer Society Conference on Computer Vision and Pattern Recognition, San Juan, PR, USA, 17–19 June 1997; pp. 1106–1112.
55. Iakovidis, D.K.; Dimas, G.; Karargyris, A.; Bianchi, F.; Ciuti, G.; Koulaouzidis, A. Deep endoscopic visual measurements. *IEEE J. Biomed. Heal. Informatics* **2019**, *23*, 2211–2219. [CrossRef]
56. Springenberg, J.T.; Dosovitskiy, A.; Brox, T.; Riedmiller, M. Striving for simplicity: The all convolutional net. *arXiv* **2014**, arXiv:1412.6806.
57. Ioffe, S.; Szegedy, C. Batch normalization: Accelerating deep network training by reducing internal covariate shift. *arXiv* **2015**, arXiv:1502.03167.
58. Jiang, M.; Huang, S.; Duan, J.; Zhao, Q. Salicon: Saliency in context. In Proceedings of the IEEE Conference on Computer Vision and Pattern Recognition, Boston, MA, USA, 7–12 June 2015; pp. 1072–1080.
59. Flickr Inc. Find your inspiration. Available online: www.flickr.com/ (accessed on 21 April 2020).
60. Keras. The Python Deep Learning library. Available online: www.keras.io/ (accessed on 21 April 2020).
61. Abadi, M.; Barham, P.; Chen, J.; Chen, Z.; Davis, A.; Dean, J.; Devin, M.; Ghemawat, S.; Irving, G.; Isard, M.; et al. Tensorflow: A system for large-Scale machine learning. In Proceedings of the 12th USENIX Symposium on Operating Systems Design and Implementation, Savannah, GA, USA, 2–4 November 2016; pp. 265–283.
62. Kingma, D.P.; Ba, J. Adam: A method for stochastic optimization. *arXiv* **2014**, arXiv:1412.6980.
63. Sanders, J.; Kandrot, E. *CUDA by Example: An Introduction to General-Purpose GPU Programming*; Addison-Wesley Professional: Boston, MA, USA, 2010.
64. Sandler, M.; Howard, A.; Zhu, M.; Zhmoginov, A.; Chen, L.-C. Mobilenetv2: Inverted residuals and linear bottlenecks. In Proceedings of the IEEE Conference on Computer Vision and Pattern Recognition, Salt Lake City, UT, USA, 18–23 June 2018; pp. 4510–4520.
65. Szegedy, C.; Liu, W.; Jia, Y.; Sermanet, P.; Reed, S.; Anguelov, D.; Erhan, D.; Vanhoucke, V.; Rabinovich, A. Going deeper with convolutions. In Proceedings of the IEEE Conference on Computer Vision and Pattern Recognition, Boston, MA, USA, 7–12 June 2015; pp. 1–9.
66. He, K.; Zhang, X.; Ren, S.; Sun, J. Deep residual learning for image recognition. In Proceedings of the IEEE Conference on Computer Vision and Pattern Recognition, Las Vegas, NV, USA, 26 June–1 July 2016; pp. 770–778.
67. Ogden, S.S.; Guo, T. Characterizing the Deep Neural Networks Inference Performance of Mobile Applications. *arXiv* **2019**, arXiv:1909.04783.
68. Mattoccia, S.; Macrı, P. 3D Glasses as Mobility Aid for Visually Impaired People. In Proceedings of the European Conference on Computer Vision, Zurich, Switzerland, 6–12 September 2014; pp. 539–554.
69. Ioannides, M.; Hadjiprocopi, A.; Doulamis, N.; Doulamis, A.; Protopapadakis, E.; Makantasis, K.; Santos, P.; Fellner, D.; Stork, A.; Balet, O.; et al. Online 4D reconstruction using multi-images available under Open Access. *ISPRS Ann. Photogramm. Remote. Sens. Spat. Inf. Sci.* **2013**, *2*, 169–174. [CrossRef]
70. Rodríguez-Gonzálvez, P.; Muñoz-Nieto, A.L.; Del Pozo, S.; Sanchez, L.J.; Micoli, L.; Barsanti, S.G.; Guidi, G.; Mills, J.; Fieber, K.; Haynes, I.; et al. 4D Reconstruction and visualization of Cultural Heritage: Analyzing our legacy through time. *Int. Arch. Photogramm. Remote. Sens. Spat. Inf. Sci.* **2017**, *42*, 609–616. [CrossRef]

Article

An Intelligent and Low-Cost Eye-Tracking System for Motorized Wheelchair Control

Mahmoud Dahmani [1], Muhammad E. H. Chowdhury [2], Amith Khandakar [2], Tawsifur Rahman [3], Khaled Al-Jayyousi [2], Abdalla Hefny [2] and Serkan Kiranyaz [2,*]

1 School of Engineering, University of Maryland, College Park, MD 20742, USA; dahmani@umd.edu
2 Department of Electrical Engineering, College of Engineering, Qatar University, Doha 2713, Qatar; mchowdhury@qu.edu.qa (M.E.H.C.); amitk@qu.edu.qa (A.K.); ka1403027@student.qu.edu.qa (K.A.-J.); ah1406234@student.qu.edu.qa (A.H.)
3 Department of Biomedical Physics and Technology, University of Dhaka, Dhaka 1000, Bangladesh; tawsifurrahman@bmpt.du.ac.bd
* Correspondence: mkiranyaz@qu.edu.qa; Tel.: +974-3063-5600

Received: 18 February 2020; Accepted: 12 March 2020; Published: 15 July 2020

Abstract: In the 34 developed and 156 developing countries, there are ~132 million disabled people who need a wheelchair, constituting 1.86% of the world population. Moreover, there are millions of people suffering from diseases related to motor disabilities, which cause inability to produce controlled movement in any of the limbs or even head. This paper proposes a system to aid people with motor disabilities by restoring their ability to move effectively and effortlessly without having to rely on others utilizing an eye-controlled electric wheelchair. The system input is images of the user's eye that are processed to estimate the gaze direction and the wheelchair was moved accordingly. To accomplish such a feat, four user-specific methods were developed, implemented, and tested; all of which were based on a benchmark database created by the authors. The first three techniques were automatic, employ correlation, and were variants of template matching, whereas the last one uses convolutional neural networks (CNNs). Different metrics to quantitatively evaluate the performance of each algorithm in terms of accuracy and latency were computed and overall comparison is presented. CNN exhibited the best performance (i.e., 99.3% classification accuracy), and thus it was the model of choice for the gaze estimator, which commands the wheelchair motion. The system was evaluated carefully on eight subjects achieving 99% accuracy in changing illumination conditions outdoor and indoor. This required modifying a motorized wheelchair to adapt it to the predictions output by the gaze estimation algorithm. The wheelchair control can bypass any decision made by the gaze estimator and immediately halt its motion with the help of an array of proximity sensors, if the measured distance goes below a well-defined safety margin. This work not only empowers any immobile wheelchair user, but also provides low-cost tools for the organization assisting wheelchair users.

Keywords: convolutional neural networks (CNNs); machine learning; eye tracking; motorized wheelchair; ultrasonic proximity sensors

---

1. Introduction

The human eye is considered to be an intuitive way of interpreting human communication and interaction that can be exploited to process information related to the surrounding observation and respond accordingly. Due to several diseases like complete paralysis, multiple sclerosis, locked-in syndrome, muscular dystrophy, arthritis, Parkinson's, and spinal cord injury, the person's physiological abilities are severely restricted from producing controlled movement in any of the limbs or even the head, noting that there are about 132 million disabled people who need a wheelchair, and only 22% of them have access to one [1]. They cannot even use a technically advanced wheelchair. Thus, it is very important to investigate novel eye detection and tracking methods that can enhance human–computer interaction, and improve the living standard of these disable people.

Research for eye tracking techniques has been progressively implemented in many applications, such as driving fatigue-warning systems [2,3], mental health monitoring [4,5], eye-tracking controlled wheelchair [6,7], and other human–computer interface systems. However, there are several constraints such as reliable real-time performance, high accuracy, availability of components, and having a portable and non-intrusive system [8–10]. It is also crucial to achieve higher system robustness against encountered challenges, such as changing light conditions, physical eye appearance, surrounding eye features, and reflections of eye-glasses. Several related works have proposed eye-controlled wheelchair systems; however, these rarely address the constraints of the system's software performance, physical, and surrounding challenges beyond the system, novelty of algorithms, and user's comfort and safety altogether.

Furthermore, the Convolutional Neural Network (CNNs) is a state-of-the-art and powerful tool that enables solving computationally and data-intensive problems. CNN is pioneering in a wide spectrum of applications in the context of object classification, speech recognition, and natural language processing and even wheelchair control [11–13], a more detailed literature review will be discussed in the later sections. However, the paper lacks high accuracy, real-time application, and does not provide the details of such a design, which could useful for further improvement. All of the above shortcomings are addressed in the current paper.

In this paper, we propose a low-cost and robust real-time eye-controlled wheelchair prototype using novel CNN methods for different surrounding conditions. The proposed system comprises of two subsystems: sensor subsystem and intelligent signal processing, and decision-making and wheelchair control subsystem as illustrated in Figure 1. The sensor subsystem was designed using an eye-tracking device and ultrasound sensors, which were interfaced to the intelligent data processing and decision-making module. The motor control module was already available in the powered wheelchair; only control signals based on the eye-tracker needs to be delivered to the microcontroller of the original wheelchair joystick bypassing the mechanical joystick input. An array of ultrasound sensors was used to stop the wheelchair in case of emergency. The proposed system can steer through a crowded place faster and with fewer mistakes than with current technologies that track eye movements. The safety provision, ensured by arrays of ultrasound sensors, helps the wheelchair to steer through a congested place safely. Accordingly, the proposed system can help most of the disabled people with spinal cord injury. Furthermore, as the proposed system is targeted to use inexpensive hardware and open source software platform, it can even be utilized to modify non-motorized wheelchairs to produce a very economical motorized wheelchair solution for third-world countries.

The rest of this paper is outlined as follows. Section 2 provides background and reviews the relevant literature on state-of-the-art Eye-Tracking methods, existing eye-controlled wheelchair systems, and Convolutional Neural Networks (CNNs) for Eye Tracking. Section 3 is the Methodology section, which discusses the design of the various blocks of the work along with the details of the machine learning algorithm. Section 4 provides the details of the implementation along with the modifications done in the hardware. Section 5 summarizes the results and performance of the implemented system. Finally, the conclusion is stated in Section 6.

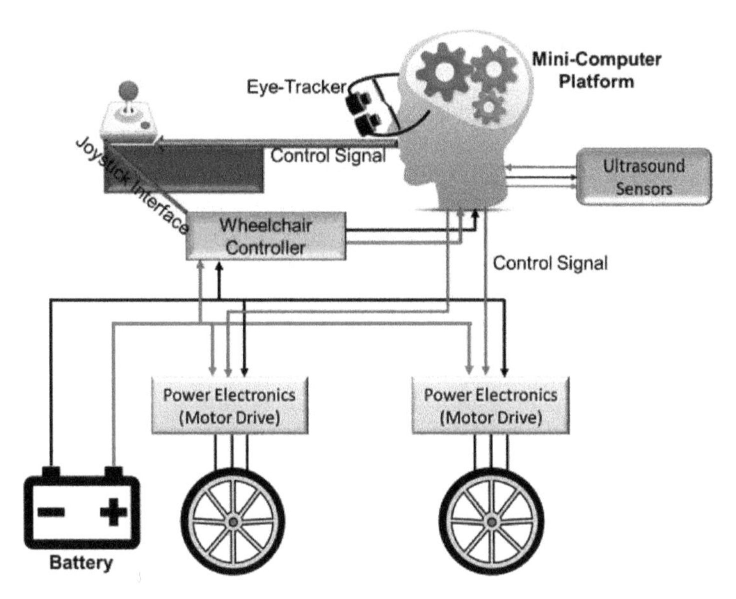

Figure 1. Block representation of the proposed systems for pupil based intelligent eye-tracking motorized wheelchair control.

2. Background and Related Works

Previous studies are explored in this paper within three contexts in order to investigate all major related aspects as well as to cover the relevant literature as much as possible. The aspects are state-of-the-art methods for eye tracking, existing eye-controlled wheelchair systems, and other convolutional neural network (CNN)-based works for eye tracking application.

2.1. State-Of-The-Art Eye Tracking Methods

Generally, there are two different methods investigated widely for eye tracking: Video-based systems and Electrooculography (EOG)-based systems (Figure 2). A video-based system consists mainly of a camera placed at a distance to the user (remote), or attached to the user's head (head-mounted), and a computer for data processing [14,15]. However, the main challenge in remote eye tracking is robust face and eye detection [16,17]. The cameras can be visible-light cameras, referred to as Videooculography (VOG) [14], as examples proposed in [9,18–20], or infrared-illumination cameras such as in [21,22], where infrared (IR) corneal reflection was extracted. Based on near IR illumination, researchers in [17] investigated six state-of-the-art eye detection and tracking algorithms: Ellipse selector (ElSe) [23], Exclusive Curve Selector (ExCuSe) [24], Pupil Labs [25], SET [26], Starburst [27], and Swirski [28], and compared them against each other on large four datasets with an overall 225,569 public labeled eye images of frequently changing sources of noise.

Commercial eye-tracking systems are still very expensive using proprietary tracking algorithms that are not commercially deployed to any powered wheelchair. Note that although pupil tracking is a widely used tool, it is still hard to achieve high-speed tracking with high-quality images, particularly in a binocular system [29].

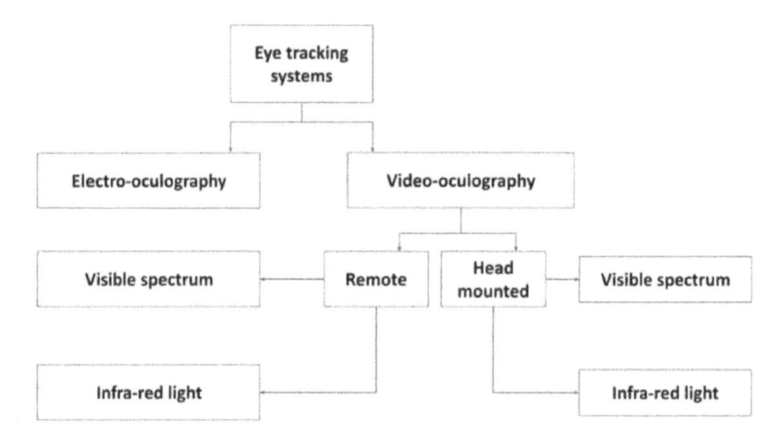

Figure 2. Different eye tracking techniques.

2.2. Existing Eye-Controlled Wheelchair Systems

User's voice [30] and facial expression [31]-based wheelchair control systems were explored by different groups. However, voice control is laborious for the user and sound waves interference or noisy environment distractions can be introduced to the system establishing undesired commands. Facial expression control, on the other hand, is not helpful to all users, especially those who suffer from restrictions in facial expressions due to diseases like facial paralysis. Moreover, classification of facial expressions is more challenging than the eye-controlled system, where only the eye is targeted.

Eye tracking techniques have been previously employed to serve wheelchair systems for the disabled people. An eye-controlled wheelchair prototype was developed in [32] using an infrared (IR) camera fitted with LEDs for automatic illumination adjustment during illumination changes. A flow of image processing techniques was deployed for gaze detection beginning with eye detection, pupil location detection using pupil knowledge, and then converting the pupil location into the user's gaze based on a simple eye model. This was finally converted into a wheelchair command using a threshold angle. Although the algorithm had a fast processing time and robustness against changing circumstances, the used threshold was suitable only to a specific illumination condition enabling automatic illumination adjustment by the camera. This degrades significantly when hit by a strong illumination such as sunlight. In addition, the conceptual key control was quite inconvenient for the user, as the chair stops when the user blinks and when the gaze is deviated from the direction key unless the user looks upwards for free head movement. The user also has to look downwards for forward movement, which is impractical.

In [33], another eye-controlled wheelchair is proposed by processing the images of a head-mounted camera. Gaussian filtering was implemented for removing Gaussian noise from the image. A threshold was then employed for producing a binary image, and erosion followed by dilation was applied for removing white noise. The wheelchair moves in three directions (left, right, and forward) depending on the relative iris position, and starts and stops by blinking for 2 seconds. Although the proposed techniques were simple in implementation, the evaluation parameters of the system were not reported. In addition, the system's performance during pupil transition from one direction to another was not discussed.

Apart from depending on interfaces like joystick control, head control, or sip-puff control [34], the optical-type eye tracking system that controls a powered wheelchair by translating the user's eye movement on the screen positions was used by several researchers, which are reported below. The eye image was divided into nine blocks of three columns and three rows, and depending on the location of the pupil's center, the output of the algorithm was an electrical signal to control the wheelchair's movement to left, right, and straight directions [34]. The system evaluation parameters like response speed, accuracy, and changing illumination conditions were not reported. In addition, safety parameters of the wheelchair's movement, such as ultrasound or IR sensors for obstacles detection, were not discussed.

Another wheelchair control system has been proposed in [35], where positions of the eye pupil were tracked by employing image processing techniques using a Raspberry-Pi board and a motor drive to steer the chair to left, right, or forward directions. The open computer vision (OpenCV) library was used for image processing functions, where the HAAR cascade algorithm was used for face and eye detection, Canny edge was used for edges detection, and Hough Transform methods were used for circle detection to identify the border of the eye's pupil. The eye pupil's center is located depending on the average of two corner points obtained from a corner detection method. The pupil was tracked by measuring the distance between the average point and the eye circle's center point, where the minimum distance indicated that the pupil was at left, and the maximum indicated the eye had moved to the right. If there was no movement, the eye would be in the middle position, and the chair would be moving forward. Eye blinking was needed to start a directional operation, and the system was activated or deactivated when the eye was closed for 3 seconds. Although the visual outputs of the system were provided, the system was not quantitatively assessed, and no rigid evaluation scheme was shown. Therefore, accuracy, response latency, and speed for instance are unknown.

On the other hand, Electrooculography (EOG) is a camera-independent method for gaze tracking, and generally, it requires lower response time and operating power than the video-based methods [10]. In EOG, electrodes are placed on the skin at different positions around the eyes along with a reference electrode, known as ground electrode, placed on the subject's forehead. The eye is modeled as an electric dipole, where the negative pole is at the retina and the positive pole is at the cornea. When the eyes are in their origin state, the electrodes measure a steady electric potential, but when an eye movement occurs, the dipole's orientation changes making a change in the electric corneal–retinal potential that can be measured.

An EOG-based eye controlled wheelchair with an attached on-board microcontroller was proposed in [36] for disabled people. Acquired biosignals were amplified, noise-filtered, and fed to a microcontroller, where the input values for each of the movement (left, right, up, and down) or stationary conditions were given, and the wheelchair movement in the respective direction was performed according to the corresponding voltage values from the EOG signals. The EOG-based system was cost effective, independent from changing light conditions, and containing lightweight signal processing with reasonable functional feasibility [10]; however, the system was not yet fully developed for commercial use because of the electrodes used for signal acquisition. Moreover, it was also restricted for particular horizontal and vertical movements of the eye; it did not respond effectively for oblique rotation of the eye.

Another EOG-guided wheelchair system was proposed in [37] using the Tangent Bug Algorithm. The microcontroller identified the target point direction and distance by calculating the gaze angle that the user was gazing at. Gaze angle and blinks were measured, and used as inputs for the controlling method. The user only asked to look at the desired destination and blink to give the signal to the controlling unit for starting navigation. After that, the wheelchair calculated the desired target position and distance from the measured gaze angle. Finally, the wheelchair moved towards the destination in a straight line and go around obstacles when detected by sensors. Overall, EOG-based systems are largely exposed to signal noise, drifting, and artifacts that affect EOG signal acquisition. This is due to the interference of noise from residential power lines, electrodes, or circuitry [14]. In addition, with placing electrodes at certain distances around the eye, EOG-based systems are considered to be impractical for everyday use.

There are some recent works on using commercially available sensors for eye controlled powered wheelchair for Amyotrophic lateral sclerosis (ALS) patients [38]. A similar work using commercial brain–computer interface and eye tracker devices was done in [6].

2.3. Convolutional Neural Networks (CNNs) for Eye Tracking

CNN is a pioneer in a wide spectrum of applications in the context of object classification; however, only a few previous studies have presented CNNs for the specific task of real-time eye gaze classification to control a wheelchair system as in the current paper. Some examples for CNNs employed for eye gaze classification application are discussed below.

The authors of [39] proposed an eye-tracking algorithm that can be embedded in mobiles and tablets. A large training dataset of almost 2.5 M frames was collected via crowdsourcing of over 1450 subjects, and used for training the designed deep end-to-end CNN. Initially, the face image was used as original image, and the images of the eyes were used as inputs to the model. For real-time practical application, dark knowledge was then applied to reduce the computation time and model complexity by learning a smaller network that achieves a similar performance running at 10–15 frames per second (FPS). The model's performance increases significantly when calibration was done, and when there was variability in the collected data with higher number of subjects rather than higher number of images per subject, but not neglecting the importance of the second. Although the model has achieved robustness in eye detection, the FPS rate should not be less than 20–25 FPS for reliable real-time performance; a rate of which the error would increase significantly in this case if was not addressed.

The authors of [40] proposed a real-time framework for classifying eye gaze direction applying CNNs, and using low-cost off the shelf webcams. Initially, the face region was localized using a modified version of the Viola–Jones algorithm. The eye region was then obtained using two different methods: the first one was geometrically from the face bounding box, and the second (which showed better performance) was localizing facial landmarks through a facial landmark detector to find the eye corners and other fiducial points. The eye region was then catered to the classification stage, where classes of eye access cues are predicted and classified using CNN into seven classes (center, upright, up left, right, left, downright, and down left), where three classes (left, right, and center) showed higher accuracy rates. The algorithm's evaluation was performed on two equal data 50% subsets of testing and training, where CNNs are trained for left and right eyes separately, but eye accessing cues (EAC) accuracy was improved when combining information from both eyes (98%). The algorithm has achieved an average rate of 24 FPS.

The authors of [12] implemented an eye-controlled wheelchair using eye movement captured using webcam in front of the user and using Keras deep learning pre-trained VGG-16 model. The authors also discuss the benefits of the project working for people with glasses. Details of its real-time implementation, in terms of FPS, was not mentioned in the paper.

3. Methodology

Considering the pros-and-cons of the previous works, we propose an eye-controlled wheelchair system running at real-time (30 FPS) using CNNs for eye gaze classification, and made up of several low-cost, lightweight controller subsystems for wheelchair. The system takes the input images from an IR camera attached to a simple headset, providing comfort and convenient movement control to the user with complete free-eye movement. Ultrasonic sensors were mounted to avoid collisions with any obstacles. The system classifies the targeted direction based on a robust CNN algorithm implemented on Intel NUC (a mini-computer with I7-5600U) at 2.4 GHz (2 central processing units (CPUs)) and 8 GB random access memory (RAM) using C++ in Microsoft Visual Studio 2017 (64 bit). Although it is not a graphics processing unit (GPU)-optimized implementation, multiprocessing with a shared memory was attained by deploying the Intel®OpenMP (Open Multi-Processing) application programming interface (API).

A block diagram for the eye-controlled wheelchair is shown in Figure 3. The diagram shows different steps, starting from capturing a new image, until the appropriate command, which is given to the wheelchair.

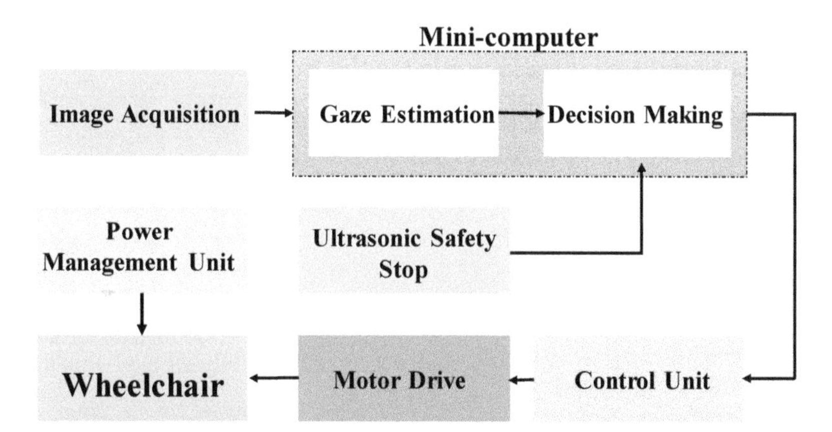

Figure 3. Eye Controlled Wheelchair block diagram.

The wheelchair is primarily controlled through the eye movements that are translated into commands to the chair's motor drives. This is achieved through a gaze detection algorithm implemented on the minicomputer. Yet, a secondary commanding system was added to the design for the means of safety of the user. This safety system is ultrasonic-based that can stop the wheelchair in cases of emergency, suddenly appearing objects, unawareness of the user, etc.

The minicomputer should make the decision of whether the wheelchair should move next mainly according to the gaze direction, or to stop the wheelchair if the safety system is activated. In either case, a command is sent to the control unit, which in turn produces the corresponding command code to send to the wheelchair controller to drive the motor to move the wheelchair in the respective direction.

The Titan x16 power wheelchair was converted to an eye-tracking motorized wheelchair (Figure 4). It is originally equipped with a joystick placed at the right arm of the chair. An eye-tracking controller (image acquisition system and mini-computer) was connected to the electronic control system beside the joystick-based controller. Furthermore, the functionality of the main joystick was not altered; the new control system can be superseded by the original control system.

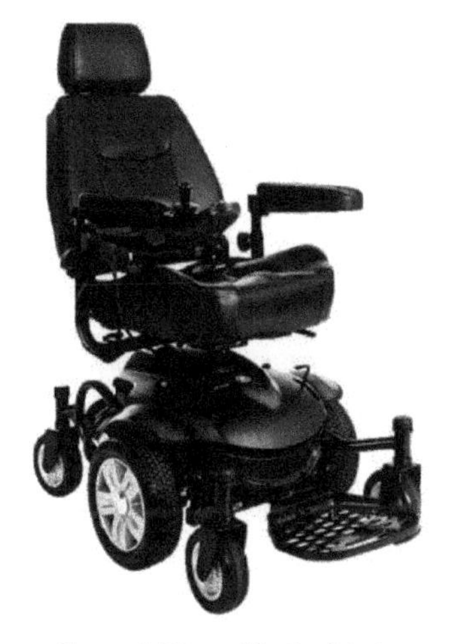

Figure 4. Titan x16 wheelchair.

Based on this overview, the system implementation can be divided into the following steps; design and implementation of an image acquisition mechanism, implementation of the gaze estimation algorithm, design and implementation of the ultrasonic safety system, and finally modifying the joystick controller. Each of these parts is discussed separately as below.

3.1. *Image Acquisition Frame Design*

The concept of the design for an eye-frame was first visualized using Fusion 360 software to validate its user friendliness. Two infrared (IR) [41] cameras were mounted on small platforms supported by flexible arms below eye level. Flexible arms were used to allow position adjustment of the two cameras for different users. The cameras were positioned at a distance that allowed the full eye movement range to be recorded, as seen in Figure 5. The reason for using the IR cameras was primary because (i) the acquired images were almost the same under different lighting conditions and (ii) the user was not irritated by the light emitted from the IR LEDs.

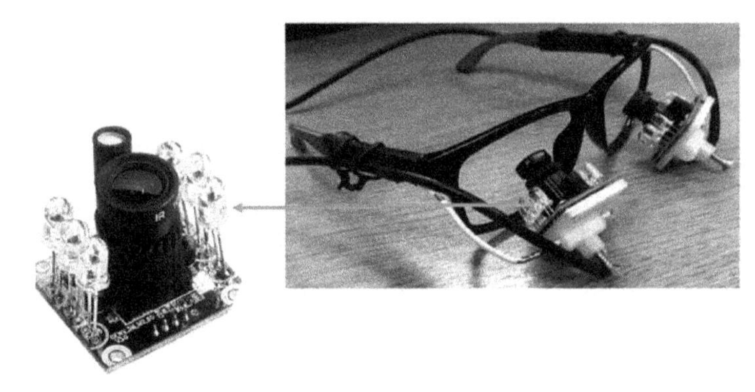

Figure 5. Image acquisition system using IR camera.

3.2. *Gaze Estimation Algorithm*

Accuracy, speed, and robustness to variation in illumination are the key parameters of any gaze estimation algorithm. The work presented a comparison between different algorithms of gaze estimation using these three parameters. Numerous gaze estimation algorithms were implemented by the researchers; nevertheless, the focus of this work is on those algorithms which are user-specific. Figure 6 shows a set of such algorithms. The user specificity means that the algorithms initially trained for a particular user using sample images and then use the trained algorithm to estimate gaze directions. Those four algorithms with blue color in Figure 6 are those tested in this work.

To facilitate the disabled user to have full mobility for the wheelchair, at least three commands are required: forward, right, and left. Furthermore, more than two commands are needed to start and stop the process of moving the chair. It was decided to use left eye winking to start acquisition of images and moving the chair, and another left eye winking to stop acquisition. Correspondingly, for each eye, four states are required: Right Gazing, Forward Gazing, Left Gazing, and Closing the eye.

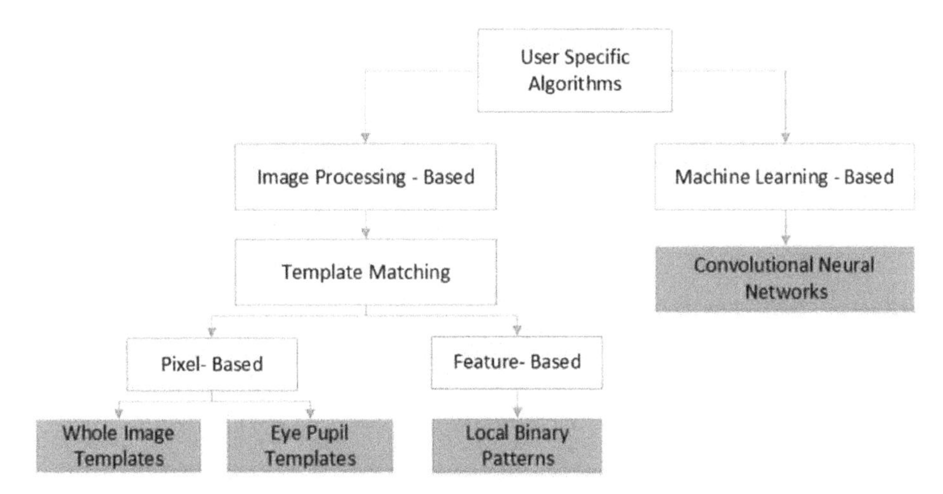

Figure 6. Classification of algorithms for gaze estimation.

The following subsection discusses different algorithms tested in this work.

(a) Template Matching

The process of template matching between a given patch image (known as a template T) and a search image S, simply involves finding the degree of similarity between the two images. The template images are usually smaller in size (lower resolution). The simplest way to find the similarity is the Full Search (FS) method [25], where the template image moves (slides) over the search image, and for each a new position of T, the degree of similarity of the pixels of both images is calculated. For each new position of T, a mathematical equation was used to measure the degree of matching, which was a correlation operation as represented in Equation 1, where T stands for template, S for search image, and the prime superscript represents the X or Y coordinates for the template image.

$$R(x,y) = \sum_{x'y'} T(x'.y').S(x+x'.y+y') \quad (1)$$

Template Matching Using Whole Image Templates

To use template matching for gaze estimation, a set of search images should be available for each user (this set of images were collected in a calibration phase). Four search images were used: three for the three different gaze positions and one for the closed eye. The template image was the image with unknown gaze direction (but certainly, it was one of the four classes). For each new template image, the correlation is done four times (one for each class).

It is worth mentioning that the search and template images should be acquired with the same resolution. The template matching was not sliding, rather it was using a single correlation. The correlation process returned a correlation coefficient for each search image, and the search image with the highest correlation was selected as the class for the template image. Figure 7 shows how the gaze estimation algorithm using equal-sized images was applied.

The process shown in Figure 7 requires that the template and search image should be acquired in the same illumination condition. However, the search images should be collected only once at the beginning and should not be changed, but the template image was subjected to such changes in the lighting conditions. Accordingly, to manage the changes in illumination, Histogram Equalization (HE) was used. Histogram Equalization changes the probability distribution of the image to a uniform distribution to improve the contrast.

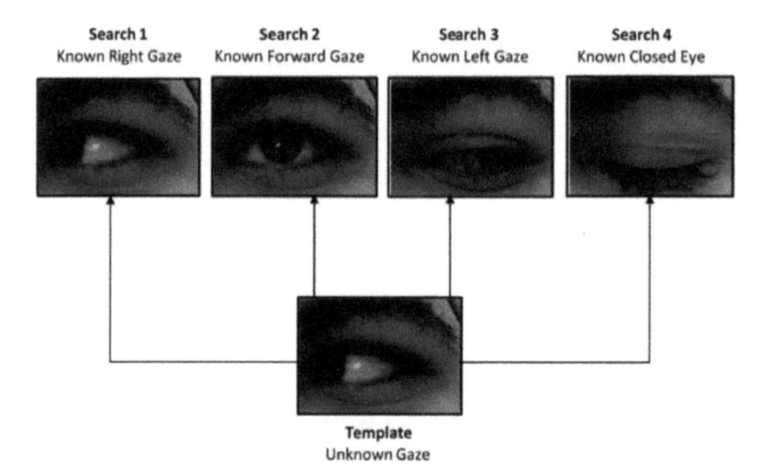

Figure 7. Gaze estimation using template matching.

Template Matching Using Eye Pupil Templates

The second template matching based method uses correlation, but with a slight variation. In this method, the template is an image that contains only the pupil of the user. It was extracted in the calibration phase while the user was gazing forward; this gaze direction allowed for easier and simpler pupil extraction. The iris could have been used as a template; however, the pupil was a better choice because it always appears in all gaze directions (assuming an open eye). This is evident in Figure 7, where the whole iris does not appear in the case of right and left gazes. It is notable that in this template matching approach, only an image of the pupil was required as the template. The newly acquired images (with unknown pupil position) were considered as the search images.

The correlation was applied for each new search image by sliding the pupil on all possible positions of the search image with the template image to locate the position of the pupil on the search image that was maximally matched to the template. This method has two main advantages over the previous method. First, the correlation was done only once for each new image (instead of four times). The second merit was eliminating the constraint of having only few number of known gaze directions. This method can locate the pupil position at any location in the eye, not only forward, right, and left.

Feature-Based Template Matching

Another template matching technique was used to correlate feature values rather than raw pixel intensities. Local Binary Patterns (LBP) represent a statistical approach to analyze image texture [42]. The term "local" is used because texture analysis is done for each pixel with pixels in the neighborhood. The image was divided into cells, 3×3 pixels each. For each cell, the center pixel is surrounded by eight pixels. Simply stated, the value of the center pixel was subtracted from the surrounding pixels. Let x denote the difference, then the output of each operation $s(x)$ was a zero or one, depending on the following thresholding.

$$s(x) = \begin{cases} 1, & x \geq 0 \\ 0, & x < 0 \end{cases} \quad (2)$$

Starting from the upper left pixel, the outputs were concatenated into a string of binary digits. The string was then converted to the corresponding decimal representation.

This decimal number was saved in the position of the center pixel in a new image. Therefore, the LBP operator returned a new image (matrix) that contains the extracted texture features. The template matching was done between the LBP outputs of the source and template images. Although applying this operation requires more time than directly matching the raw pixels, the advantage provided by LBP is the robustness to monotonic variations in pixels intensities due to changes in illumination conditions [43]. Thus, applying the LBP operator eliminates the need of equalizing the histogram.

(b) Networks Architecture and Parameters

In compliance with the notion of a user-specific approach, Convolutional Neural Networks (CNNs) were proposed as another alternative to classify gaze directions in a fast and accurate manner. Like other template matching-based methods, a calibration process should to be carried out to generate a labeled dataset that is required to train the supervised network. This is required to be performed only once for the first time the user utilizes the wheelchair. As a result, relatively small user-specific data, which can be acquired in less than 5 minutes at 30 frames/sec, were employed to train a dedicated CNN for each user. Therefore, the training time will be short and the trained classifier was well suited for real-time eye tracking by predicting the probabilities of the input image (i.e., the user's eye) being one of four classes: right, forward, left, and closed.

This proposed solution was data-driven approach and comprises the following stages.

1. Network Architecture Selection
2. Data Preprocessing
3. Loss function
4. Training algorithm selection
5. Hyperparameters setting

Networks Architecture Selection

The convolutional neural network (CNN) was selected as a network model because it produces state-of-the-art recognition results according to the literature. The main merit of CNNs is the fact that they combine the two major blocks of feature extraction and classification into a unified learner, as shown in Figure 8.

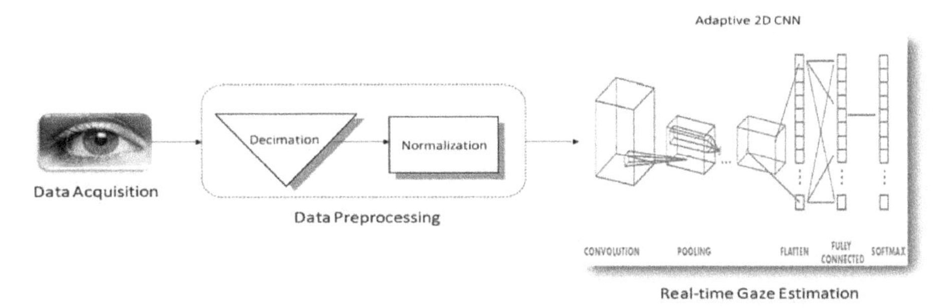

Figure 8. Block diagram of convolutional neural network (CNN).

In fact, this advantage is very convenient to this application for two reasons: First, CNNs learn directly on raw data (i.e., pixel intensities of the eye images), and thus eliminate the need for manual feature extraction. More importantly, the automation of feature extraction by means of training goes hand in hand with the philosophy of user-specificity because features that best characterize the eye pupil are learned for each individual user. By contrast, the previously discussed template-matching approach is not truly user-specific, as it operates on features that are fixed and handcrafted for all users, which does not necessarily yield the optimal representation of the input image.

Architecture wise, CNNs are simply feedforward artificial neural networks (ANNs) with two constraints:

1. Neurons in the same filter are only connected to local patches of the image to preserve spatial structure.
2. Their weights are shared to reduce the total number of the model's parameters.

A CNN consists of three building blocks:

1. Convolution layer to learn features.
2. Pooling (subsampling) layer to reduce the dimensionality the activation maps.
3. Fully-connected layer to equip the network with classification capabilities. The architecture overview is illustrated in Figure 9.

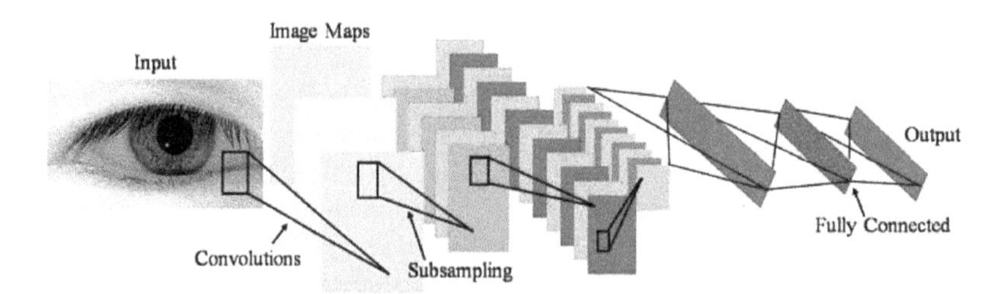

Figure 9. Convolutional Neural Network (CNN) architecture.

Data Preprocessing

Before feeding input images to the network, they were decimated to 64 × 64 images regardless of their original sizes. This reduced the time for the forward propagation of a single image when the CNN is deployed for real-time classification. Then, they were zero-centered by subtracting the mean image and normalized to unit variance by dividing over the standard deviation:

$$I_N(x,y) = \frac{I(x,y) - I_{mean}}{\sigma_I} \tag{3}$$

where I is the original image; the I_N is the preprocessed one; and I_{mean} and σ_I are the mean and standard deviation of the original image, respectively.

The preprocessing stage will help reduce the convergence time of the training algorithm.

Loss Function

The mean squared error (MSE) is the error function chosen to quantify how good the classifier is by measuring its fitness to the data:

$$E(weights, biases) = \sum_{i=1}^{N_c} (y_i - t_i)^2 \tag{4}$$

where E is the loss function; y_i and t_i are the predicted and the actual scores of the ith class, respectively; and N_c denotes the number of classes which is 4 in this classification problem.

Training Algorithm Selection

Learning is actually an optimization problem where the loss function is the objective function and the output of the training algorithm is the network's parameters (i.e., weights and biases) that minimize the objective.

Stochastic gradient descent with adaptive learning rate (adaptive backpropagation) is the employed optimization algorithm to train the CNN. Backpropagation involves iteratively computing the gradient of the loss function (the vector of the derivatives of the loss function with respect to each weight and bias) to use it for updating all of the model's parameters:

$$w_{ki}^{l}(t+1) = w_{ki}^{l}(t) - \varepsilon \frac{\delta E}{\delta w_{ki}^{l}} \tag{5}$$

$$b_{ki}^{l}(t+1) = b_{ki}^{l}(t) - \varepsilon \frac{\delta E}{\delta b_{ki}^{l}} \quad (6)$$

where w_{ki}^{l} is the weight of the connection between the *kth* neuron in the *lth* layer and the *ith* neuron in the *(l-1)th* layer, b_{ki}^{l} is the bias of the *kth* neuron in the lth layer, and ε is the learning rate.

Hyperparameters Setting

Five-fold cross-validation is employed to provide more confidence in the decision-making on the model structure to prevent overfitting.

First, the whole dataset is split into a test set on which the model performance will be evaluated, and a development set utilized to tune the CNN hyperparameters. The latter set is divided into five folds. Second, each fold is treated in turn as the validation set. In other words, the network is trained on the other four folds, and tested on the validation fold. Finally, the five validation losses are averaged to produce a more accurate error estimate since the model is tested on the full development set. The hyperparameters are then chosen such that they minimize the cross-validation error.

To test and compare the performance of these four gaze estimation techniques, a testing dataset was needed.

3.3. Building a Database for Training and Testing

All the discussed algorithms for gaze tracking require a database for the purpose of evaluation of the algorithms. Although three algorithms (template matching, LBP, and CNN) require a labeled dataset, the template matching algorithm that uses the eye pupil as a template does not require having such labels. There is more than one available database that could be used in testing the algorithms. Columbia Gaze Data Set [44] has 5880 images for 56 people in different head poses and eye gaze directions. Also, Gi4E [45] is another public dataset of iris center detection, it contains 1339 images that are collected using a standard webcam. UBIRIS [46] is another 1877-image database that is specified for iris recognition.

Nevertheless, neither of these databases was used because each one has one or more of the following drawbacks (with respect to our approach of testing).

1. The database may have changes in face position, which requires applying more stages to localize the eyes' area. Besides, the set-up for this project is based on having only one head pose, disregarding the gaze direction.
2. The database comprises only one gaze direction, which eliminates the possibility of using the dataset for testing for gaze tracking.
3. The dataset is not labeled, and the time and effort needed to label it is comparatively higher than building a similar new dataset.
4. One important feature that is missing in all of the available datasets is the transition between one gaze direction and another. This time should be known a priori, and be compared with the time needed by all the proposed algorithms.
5. Furthermore, all these datasets lack variations in lighting conditions.

Accordingly, a database was created for eight different users. Videos of eight users gazing in three directions (right, left, and forward) and closing their eyes were captured under two different lighting conditions. The dataset consists of the four selected classes, namely, Right, Forward, Left gazing, and one more class for closed eyes. For each user, a set of 2000 images were collected (500 for each class). The collected dataset includes images for indoors and outdoors lighting conditions. Besides, a dataset was collected for different ten users, yet, only for indoors lighting condition. Then, 500 frames per each class for each user were taken to constitute a benchmark dataset of 16,000 frames. Next, the dataset was partitioned into training set and testing set, where 80% of the data were used for training. Consequently, the algorithm learnt from 12,800 frames and were tested on 3200 unseen frames.

3.4. Safety System—Ultrasonic Sensors

For safety purposes, an automatic stopping mechanism installed on the wheelchair should disconnect the gazing-based controller, and then immediately stop the chair in case that the wheelchair becomes close to any object in the surroundings. Throughout this paper, the word "object" refers to any type of obstacle, which can be as small as a brick, or a chair, or even a human being.

A proximity sensor qualitatively measures how close an object is to the sensor. It has a specific range, and it raises a flag when any object enters this threshold area. There are two main technologies upon which proximity sensors operate: ultrasonic technology or infrared (IR) technology proximity sensors. In fact, the same physical concept applies for both, namely, wave reflections. The sensor transmits electromagnetic waves and receives them after they are reflected from surrounding objects. The process is very analogous to the operation of radar. The difference between ultrasonic-based and IR-based proximity sensors is obviously the type of the electromagnetic radiation.

The different types of proximity sensors can be used in different applications. As infrared sensors operate in the IR spectrum, they cannot be used in outdoors applications where sunlight interferes with their operation [47]. Besides, it is difficult to use IR sensors in dark areas [47]. On the other hand, ultrasonic waves are high-frequency sound waves that should not face any sort of interference. Furthermore, ultrasonic reflections are insensitive to hindrance factors as light, dust, and smoke. This makes ultrasonic sensors advantageous over IR sensors for the case of the wheelchair. A comparison between the two technologies [47] suggests that combining the two technologies together gives more reliable results for certain types of obstacles like rubber and cardboard. However, for this paper, there is no need to use the two sensors together; ultrasonic ones should be enough to accomplish the task of object detection.

Usually, an ultrasonic sensor is used along with a microcontroller (MCU) to measure the distance. As seen in Figure 10, the sensor has an ultrasonic transmitter (Tx) and a receiver (Rx). The microcontroller sends a trigger signal to the sensor, and this triggers the "Tx" to transmit ultrasonic waves. The ultrasonic waves then reflect from the object and are received by the Rx port on the sensor. The sensor accordingly outputs an echo signal (digital signal) whose length is equal to the time taken by the ultrasonic waves to travel (the double-way distance).

$$\mathbf{distance(m)} = \frac{\mathbf{travelling\ time(seconds)}}{2} * \mathbf{speed\ of\ ultrasonic\ waves}\left(\frac{\mathbf{meter}}{\mathbf{second}}\right) \quad (7)$$

$$\mathbf{Error\ in\ distance} = \mathbf{total\ delay\ in\ the\ system} * \mathbf{speed\ of\ the\ wheelchair} \quad (8)$$

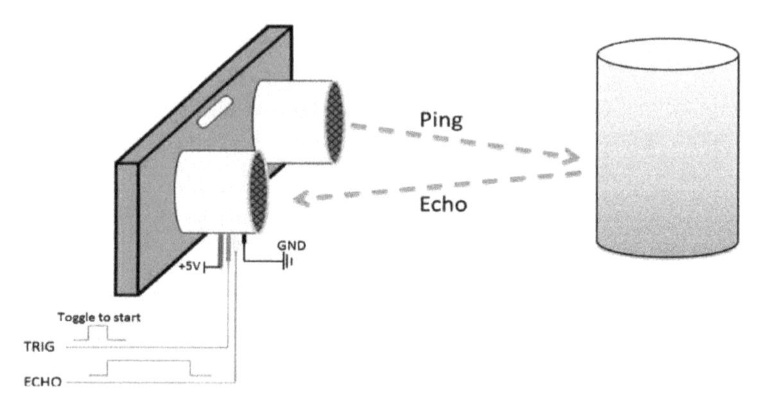

Figure 10. Proximity sensor with microcontroller working.

By knowing the travelling time of the signal, the distance can be calculated using Equation (7). Being analogous to the radars' operation, the time should be divided by two because the waves travel the distance twice. Ultrasonic waves travel in air with a speed of 340 m/sec.

As the wheelchair is moving with a certain speed, this affects the echo-based approach of measuring the distance (Doppler Effect). This introduces some inaccuracy in the calculated distance. However, this error did not affect the performance of the system. The first thing to do is to clarify that any fault in the measurement is completely dependent on the speed of the wheelchair, not the speed of the ultrasonic waves. This makes the error extremely small; as the error is directly proportional to the speed. The faulty distance can be calculated using Equation (8). The wheelchair moves at a maximum speed of 20 km/h (5.56 m/s). The delay is not a fixed parameter; thus, for design purposes, the maximum delay (worst-case scenario) was used.

The total delay in the system is the sum of delay introduced by the MCU and that of the sensor. The delay of the sensor strictly cannot exceed 1 millisecond because the average distance the sensor can measure is ~1.75 m. The average time of travelling of the ultrasonic wave can be calculated using Equation (7). The delay that may occur due to the MCU processing, it is in microseconds range. When applying this delay to Equation (8), the maximum inaccuracy in the measured distance is 0.011 m (1.1 cm).

To cover a wider range for the proximity sensors, a number of sensors was used to cover at 1200 as the chair is designed to move only in the forward direction (of course with right and left steering), but not towards back, refer Figure 11. A sketch program (Fusion 360) was used to visualize the scope of the proximity sensors and specify the number of sensors to use.

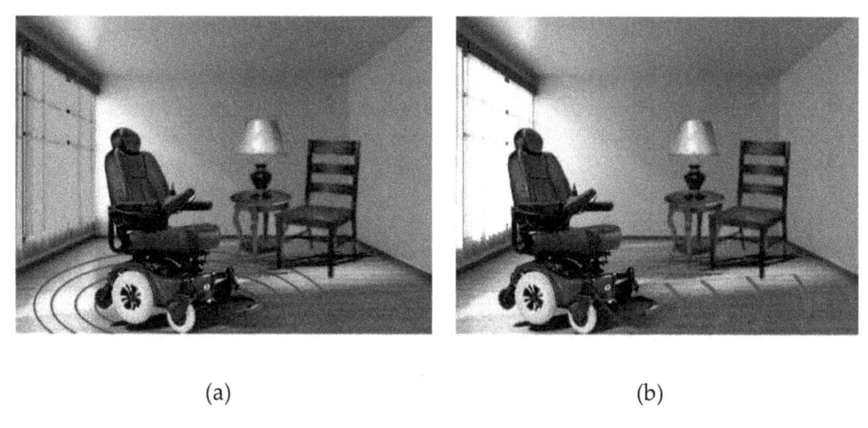

(a) (b)

Figure 11. Proximity sensor covering range: (**a**) 360 degrees and (**b**) 120 degrees.

3.5. *Modifying the Wheelchair Controller*

The electric wheelchair Titan X16 has a joystick-based controller (DK-REMD11B). The objective is to adapt it for gaze direction control signals while maintaining the original joystick's functionality.

Joystick Control Mechanism

To tackle this problem, one must first uncover the underlying principle of operation of the built-in controller. After carrying out rigorous experimentation on the disassembled control unit, it was deduced that it operates according to the principle of electromagnetic induction.

As illustrated in Figure 12, the joystick handle is connected to a coil excited by a sinusoidal voltage and acts like a primary side of a transformer. Underneath the lever lies four coils in each direction (right, left, forward, and backward), which, although electrically isolated from the primary coil, are nevertheless magnetically coupled to it and they behave as the secondary side of a transformer.

Consequently, a voltage was induced across each of the four secondary coils based on the rotation of the Joystick.

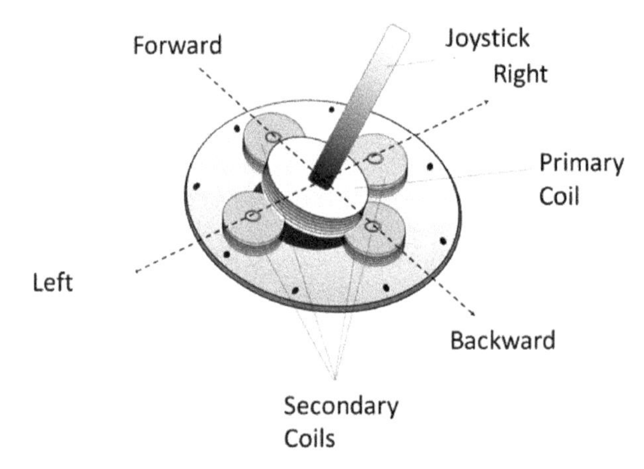

Figure 12. Working principle of existing joystick controller.

Furthermore, moving the joystick in a certain direction changes the coupling (i.e., mutual inductance) between them and the primary coil, thus altering the signals developed on each secondary coil. These signals were then processed to produce movement in the corresponding direction.

To preserve the joystick control capability, we devised a scheme that does not interfere with the built-in controller circuitry and was implemented on the existing joystick controller depicted in Figure 12. The designed system necessitates only three directions (right, left, and forward). There was mechanical arrangement was designed and implemented to control the existing joystick without modifying its functionality. The scheme for controlling the joystick is explained in details in the implementation section.

4. Implementation

This section describes the design methodology discussed in the previous section and implementation details of the final prototype. Implementation details cover eye frame design for image acquisition, Gaze estimation algorithm along with the dataset acquisition and training based on the dataset for the CNN, modified joystick controller, safety system implementation, and the final prototype.

4.1. Frame Implementation

The final frame is shown in Figure 5. The sunglasses were bought from a local grocery store, where the lenses were removed to not alter the wearers' vision. The white platforms are 3D printed to conveniently hold the IR cameras below eye level (Figure 5). The metallic arms holding the platforms were formed using clothing hangers, where the material provides enough flexibility to easily alter its position. The camera cables are tied to the sides of the sunglasses using zip-ties to make the overall unit more robust. The camera is equipped with a light-dependent resistor (LDR) to detect the amount of light present, which will automatically control the brightness of six on-board IR LEDs to illuminate accordingly. The system was checked under two extreme cases where there could be the possibility of deferring of the image quality due to absence and presence of strong sunlight. In small light variations, the IR LEDs brightness is automatically controlled by on-board LDR, and there were no significant image quality variations observed. The camera has a resolution of 2 megapixels and an adjustable lens focus. Also, it supports filming 1920 × 1080 at 30 frames per second, 1280 × 720 at 60 frames per second, and other filming configurations. Furthermore, the camera has a USB connection for computer interfacing.

4.2. Gaze Estimation Algorithm

In this paper, different solutions are proposed to estimate gaze directions: Template matching-based classifiers, LBP, and the CNN. Although the former alternative yielded a very satisfactory performance (i.e., average accuracy of 95%) for all indoors lighting condition, it failed to preserve this accuracy when images were collected outdoors because of the simplicity of the underlying algorithm. To clarify, this depreciation in performance is not because of the change of illumination, however, it is because of the physical difference of the shape of the eye between indoors and outdoors conditions. Two major changes occur when the user is exposed to direct sunlight: the iris shrinks when exposed to high intensity light and the eyelids tend to close.

One way to tackle this issue is to take distinct templates for the various lighting conditions that can be discerned by a light sensor, and then match the input with the template of the corresponding lighting condition. Conversely, the CNN is inherently complex enough to be able to generalize to all lighting conditions without the need of any additional hardware. Template matching technique was used as an aiding tool to the CNN to achieve better performance.

4.2.1. Collecting Training Dataset for the CNN—Calibration Phase

To make use of the accurate and fast CNNs, a training dataset should be available to train the model. The performance of the CNN is dependent on the selection of the training dataset; if the training dataset is not properly selected, the accuracy of the gaze estimation system dramatically decreases. Thus, a calibration phase is required for any new user of the wheelchair. The output of this calibration phase is the trained CNN model that can be used later for gaze estimation. To prepare a nice training dataset, a template matching technique was used. Template matching does not require any prior knowledge of the dataset. Note that the calibration phase faces two problems: the first problem arises from the fact that the user blinks during the collection of the training data. This leads to a flawed model prediction. The second issue is that the user may not respond immediately to the given instructions, i.e., the user may be asked to gaze right for couple of seconds, but he or she responds one or two seconds later.

Fortunately, template matching, if smartly used, can overcome these two problems, and can be used to make sure that the training dataset is 100% correctly labeled. Moreover, the training dataset should be diverse enough to account for different lighting conditions, and different placement of glasses. The cameras could zoom the user's eyes, thus if the user slightly changes the position of glasses, noticeable differences occur between the images.

Figure 13 shows a flowchart of the calibration phase. For simplicity, illustration is discussed on only one eye; however, the same applies for the second eye. First, to have a wide-ranging dataset, the calibration stage was repeated for different conditions, each condition is named as a Scenario.

For each lighting condition, the user was asked to change the position of glasses (slightly slide the glasses either down or up). Thus, four different scenarios were needed. However, that is not enough to ensure a diverse dataset. Therefore, for each scenario, images were acquired for the four classes in three nonconsecutive attempts to capture all the possible patterns of the user's gaze in a particular direction. Every time, the user, indeed, gazed in a different manner. Thus, the importance of collecting images three times can be seen. Each time, 200 frames were collected (a net of 600 frames per class). These frames (that are contaminated with blinks) constitute a temporary training dataset. As previously mentioned, template matching was used to remove these blinks; a random frame was selected from each class to be used as a template.

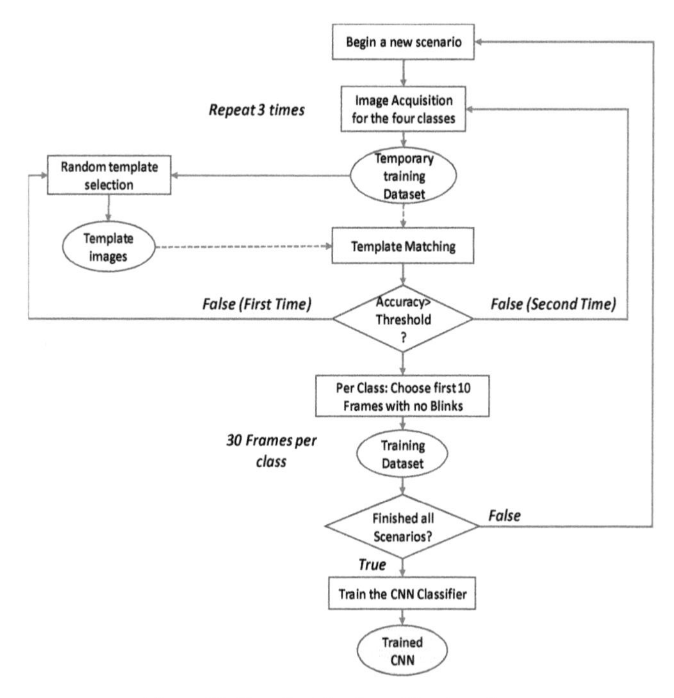

Figure 13. Flowchart for the calibration phase.

There are numerous scenarios that may happen: the first and most probable scenario is that the user has followed the instructions and the template selection was successful, that is, the template image is not a blinking frame. In such a case, and from the testing on the benchmark dataset, the accuracy of template matching was higher than 80%. This accuracy had been set as the threshold; if this accuracy was attained, this means that the image acquisition was successful. The second possibility was that while running the template-matching algorithm, a low accuracy was returned. The cause of this low accuracy can be wrong acquisition or wrong template selection. Thus, it was better to test first for a wrong template selection; thus, a different template was randomly selected, and the template matching was carried out again. If the accuracy increases to reach the threshold, the issue was with the template; i.e., it might have been an irrelevant blinking image. However, if the low accuracy persists, there is a high probability that the user has gazed in a wrong direction during the image acquisition phase, and therefore the image acquisition was repeated.

With the threshold accuracy obtained, it was certainly known that the template image is correctly selected. The next stage was to clean the 600 images of each class from those blinking frames and to keep 500 useful images for each class. Then, 400 images were randomly selected for training while 100 images tested the trained model. After finishing all the four scenarios, the final training dataset was fed to the CNN to get the trained CNN model. It was worth mentioning that even if this calibration requires few minutes to accomplish, it is required to execute once only.

By having the trained CNN, the system was ready for real-time gaze estimation to steer the wheelchair. Between the real-time gaze estimation and moving the wheelchair, there are two points that should be taken into consideration. The first is the safety of the user; whether there is an obstacle in his/her way. The second point is that the real-time estimation for the right eye may give a different classification from the left eye (because of a wrong classification or because of different gazing at the instance the images are collected).

Figure 14 shows a flowchart for the real-time classification, where these two points are tackled. As discussed earlier, proximity sensors were used to scan the region in the wheelchair way. In the decision-making part, the minicomputer first measured the distance for the closest obstacle. If there was no close obstacle, then the wheelchair moved in the predicted direction. Nevertheless, if there was an object in the danger area (threshold), the wheelchair stopped immediately.

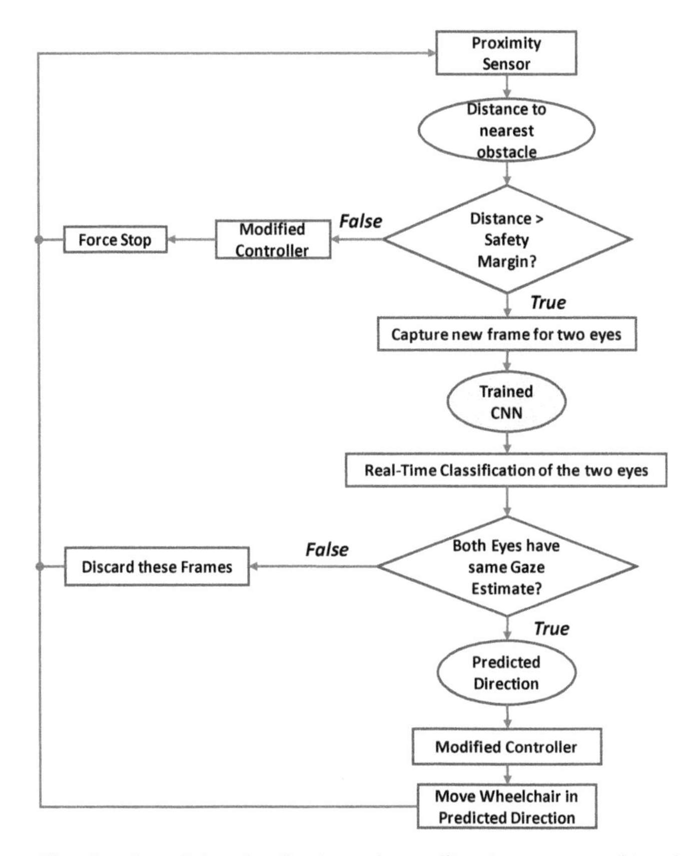

Figure 14. Flowchart for real-time classification and controlling the movement of the wheelchair.

If there was no close obstacle, when the real-time classification runs, two classification results are returned, one for each eye. In fact, it is very important to make use of this redundancy of results; otherwise, there is no need of having two cameras. If the two eyes returned the same class, and considering the high classification performance of the CNN, this predicted class was used to move the wheelchair in the determined direction. However, if the two eyes gave different classifications, no action will be taken, as the safety of the user had the priority.

4.2.2. Training the CNN

The 2-D CNN-based gaze estimator had a relatively shallow and compact structure with only two CNN layers, two subsampling layers, and two fully connected (FC) layers, as illustrated in Table 1. This boosted the system's computational efficiency for training, and most importantly for real-time classification. In this configuration, the subsampling factor (in both dimensions) of the last subsampling layer was adaptively set to 13 to ensure that its output was a stack of 1×1 feature maps (i.e., scalar features).

Table 1. CNN hyperparameters.

CNN Layers				**Subsampling Layers**			**FC Layers**	**Activation Function**
Number	Filter Size	Stride	Zero Padding	Number	Pooling Type	Subsampling Factor (x,y)	Number	
2	3	1	0	2	max	4	2	tanh

The used network had 16 and 12 filters (i.e., hidden neurons) in the two hidden CNN layers, respectively, and 16 hidden neurons in the hidden FC layer. The output layer size was 4, which corresponds to the number of classes. RGB images were fed to the 2-D CNN, and thus the input depth was 3 channels, each of which was a 64×64 frame. Training was conducted by means of backpropagation (BP) with 3 stopping criteria: the minimum train classification error was 1%, the maximum number of BP iterations was 100, or the minimum loss gradient was 0.001 (optimizer converges).

The learning rate ε was initialized to 0.001 and then global adaptation was carried out for each BP iteration: ε was increased by 5% in the next iteration if the training loss decrease, and it was reduced by 30% otherwise.

4.3. Modifying the Joystick Controller

The original controller was kept intact to maintain its original functions, while external modifications were done to the joystick of the controller as shown in Figure 15. The gaze estimation algorithm produces decision on every frame; however, the average of each 10 frames was used to send command to the joystick. This step reduces any potential random error of the control system and reduces ambiguous command. An Arduino Uno microcontroller board was interfaced to Mini-computer using USB interface, which received the commands from gaze algorithm and controlled two servo motors accordingly. Figure 16 shows how two servo motors can produce stop, forward, left, and right directions using two slider with the help of existing joystick controller.

Figure 15. Two servo motors along with two moving arms were attached to original joystick with 3D printed support structure for motion control.

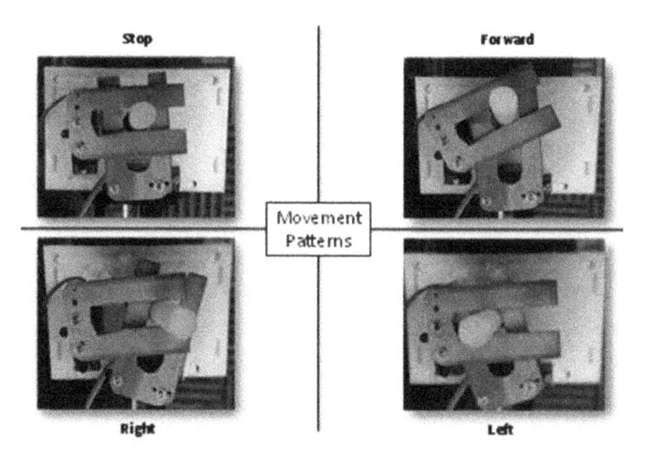

Figure 16. Four different positions of the servo motors to control four different classes of the system.

4.4. Safety System Implementation

As discussed in the methodology section, an array of ultrasonic sensors was needed to cover the all mobility range of the wheelchair. Two assumptions were made when choosing the number and positions of proximity sensors: the stopping distance is 1 m away from detected objects and that the electric wheelchair stops instantaneously. The electric wheelchair moves only forwards and rotates. Thus, the sensors were placed in front of the wheelchair to detect outward objects. An aluminum frame was built and mounted on the wheelchair to hold the sensors.

The three outward sensors, as shown in Figure 17, provided a horizontal detection range of 1 meter, with gaps of 8 centimeters between the covered regions at a 1 m distance from the aluminum frame. Moreover, the three sensors provided a 43 degrees detection angle. Two sensors were placed in between the three previous sensors, but slanted downwards. These sensors detected obstacles shorter than the aluminum frame.

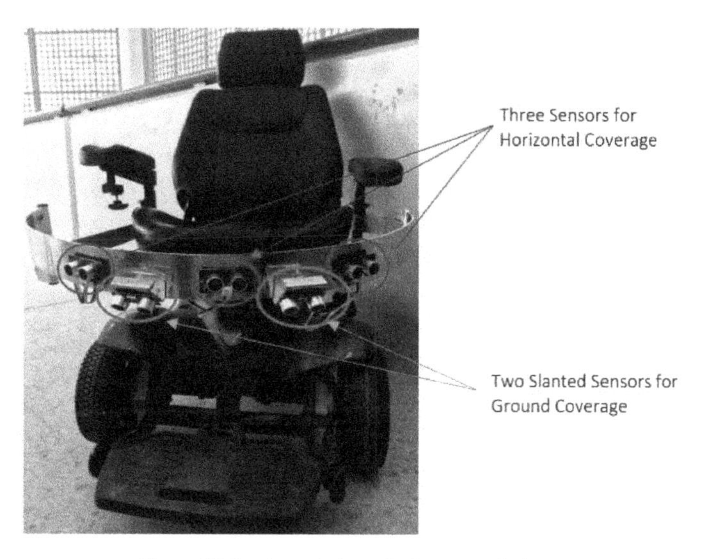

Figure 17. Implemented proximity sensor design.

5. Results and Discussion

The complete system prototype is presented in Figure 17. This section provides the details of the metrics which can show the performance of real time wheel chair movement for disabled people using eye gazing.

5.1. Computation Complexity Analysis

The average time for one backpropagation (BP) iteration per frame is ~19.7 ms with the aforementioned computer implementation. Considering a full BP run with the maximum of 100 iterations over the train set of 60 (for examples), the maximum training time is $60 \times 100 \times 19.7 = 1.97$ min. The total time for a forward pass of a single image to obtain the score vector, which corresponds to the system latency is about 1.57 ms. A system with a latency of 50 ms or less, is considered a real-time system. However, the proposed system operates at a speed that is higher than 30x the real-time speed. The average frame rate that was practically achieved was ~99 frames per sec (fps). This is a significant increase in the frame rate comparing other works reported in the literature.

5.2. Real-time Performance

A Softmax layer was added after the CNN in the real-time gaze estimation. As this is a four-class classification problem, the Softmax block mapped the 4-D class scores vector, output by the CNN, to an equivalent 4-D class probabilities vector. The probabilities in this vector quantify the model's confidence level in thinking that the input frame is of a particular class. Ultimately, the one with the highest probability value was the predicted class. A sample of the results is shown in Figure 18.

```
D:\Users\dahmani\Desktop\QU\DAHH\ML\CNN\CNN F
41% 13% 33% 13% Right
42% 13% 33% 13% Right
41% 13% 34% 13% Right
41% 13% 33% 12% Right
41% 13% 33% 12% Right
42% 13% 32% 13% Right
41% 14% 32% 13% Right
42% 13% 33% 13% Right
41% 13% 34% 12% Right
42% 13% 33% 12% Right
43% 13% 32% 12% Right
42% 13% 32% 13% Right
43% 12% 33% 12% Right
41% 13% 34% 12% Right
39% 12% 38% 11% Right
36% 14% 37% 13% Left
36% 14% 37% 13% Left
34% 15% 38% 14% Left
34% 15% 37% 14% Left
33% 15% 38% 14% Left
33% 15% 38% 14% Left
33% 15% 38% 14% Left
33% 15% 38% 15% Left
32% 15% 38% 14% Left
33% 14% 38% 14% Left
33% 15% 37% 14% Left
34% 14% 38% 14% Left
36% 13% 39% 12% Left
36% 13% 39% 12% Left
```

Figure 18. Real time prediction results.

5.3. Classification Results

Five-fold cross-validation was performed to obtain precise classification accuracy (i.e., the ratio of the number of correctly classified patterns to the total number of patterns classified). This estimates the model's true ability to generalize and extrapolate to new unseen data. This was accomplished by testing on 20% of the entire benchmark dataset of eight users that contains 3200 images. With 5-fold cross-validation, five CNNs were trained, and therefore five 4x4 confusion matrices (CM) per user were computed. These five CMs were then accumulated to yield a cross-validation confusion matrix per user, out of which the cross-validation classification accuracy for every user was computed and shown in Figure 19. Subsequently, the eight cross-validation CMs were accumulated to produce an overall CM that represents the complete model performance over the whole dataset as illustrated in Table 2. Finally, the overall CNN classification accuracy was computed. It is worth noting from Figure 19 that the lowest cross-validation classification accuracy among all the subjects is 96.875%, which is quite satisfactory for accurately estimating a user's gaze. Moreover, five out eight users had a 100% cross-validation accuracy.

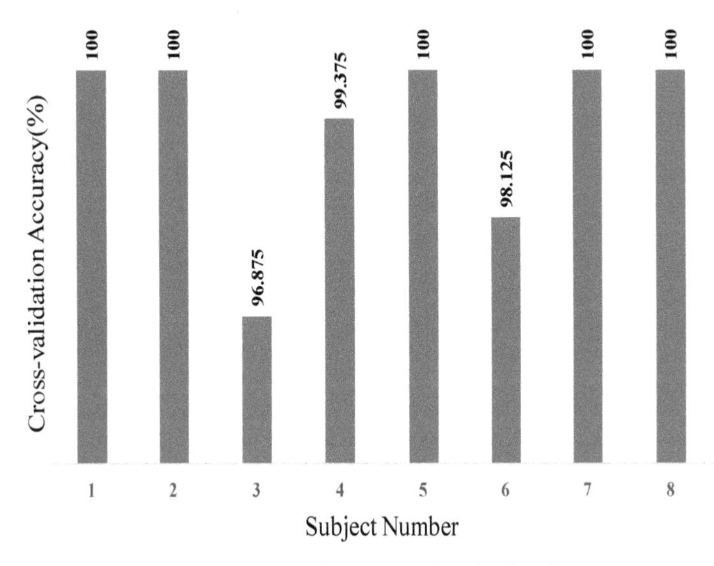

Figure 19. Cross-validation accuracies of eight subjects.

Table 2. Normalized confusion matrix of % of classification results.

		Actual			
		Right	Forward	Left	Closed
Predicted	Right	98.75	0	1.25	0
	Left	1.56	98.44	0	0
	Forward	0	0	100	0
	Closed	0	0	0	100

According to the normalized CM shown in Table 2, the classification result of the designed CNN yields probabilities higher than 98.4% for an accurate gaze estimation for all classes. As the ground truth probabilities (columns) indicate, right and forward classes have extremely low confusion probabilities (1.25% and 1.56%) to left and forward classes, respectively. The overall CNN classification accuracy is 99.3%, which makes the gaze estimation reliability of the designed system practically 100%. The implemented system's performance is summarized in Table 3.

Table 3. Summary of the implemented system's evaluation metrics.

Accuracy (%)	99.3
Frame Rate (frames/sec)	99
Maximum Training Time (min)	1.97

6. Conclusions

This paper aimed to design and implement a motorized wheelchair that can be controlled via eye movements. This is considered as a pivotal solution for people with complete paralysis of four limbs. The starting point for this project was a joystick-controlled wheelchair; all other building blocks were designed from scratch. Hardware-wise, the design stage aimed to build an image acquisition system and to modify the existing controller to move the wheelchair based on commands given from users' eyes instead of the manual joystick. The core and most challenging part of this prototype was to build a gaze estimation algorithm. Two approaches were followed: machine learning-based algorithms and template matching-based ones. Both approaches showed extremely high accuracy for a certain lighting condition. Nevertheless, template matching did not have a robust performance, as the accuracy had dropped dramatically when it was tested in another lighting condition. The Convolutional Neural Networks had superior performance in this regard, and were therefore the chosen technique for gaze estimation. Based on Table 3, the proposed system is very suitable for real time application. The classification accuracy of the gaze estimation is 99.3% which is above 95%, i.e., the norm for such type of applications. The model can process 99 frames per second and can make a prediction in ~1.57 ms, which 5x faster than the real-time requirements and the recent works reported in the literature. The safety of the user was given a paramount concern, and was ensured by incorporating an array of five ultrasonic sensors to cover the whole range of motion of the wheelchair. It can instantly stop if any obstacle is detected in this range. The system can be further developed by opting for a global approach rather than a user specific one, to eliminate the need for a calibration phase. Moreover, the current gaze estimation algorithm can stop the wheelchair or make it move only in three directions, namely, forward, right, and left. However, if a regression was used instead of classification, more degrees of freedom could be provided for movement. The authors are investigating on the joystick controller protocol to replicate it in the microcontroller used for joystick control. In this way, a non-mechanical system of a joystick controller can be developed to control the wheelchair controller directly.

Author Contributions: Experiments were designed by M.E.H.C., A.K. and S.K. Experiments were performed by M.D., T.R., K.A.-J. and A.H. Results were analyzed by M.D., M.E.H.C. and S.K. All authors were involved in the interpretation of data and paper writing. All authors have read and agreed to the published version of the manuscript.

Funding: The publication of this article was funded by the Qatar National Library and Qatar National Research Foundation (QNRF), grant numbers NPRP12S-0227-190164 and UREP22-043-2-015.

Conflicts of Interest: The authors declare no conflicts of interest.

References

1. Analysis of Wheelchair Need/Del-Corazon.org. Available online: http://www.del-corazon.org/analysis-of-wheelchair-need (accessed on 10 March 2017).
2. Zandi, A.S.; Quddus, A.; Prest, L.; Comeau, F.J.E. Non-Intrusive Detection of Drowsy Driving Based on Eye Tracking Data. *Transp. Res. Rec. J. Transp. Res. Board* **2019**, *2673*, 247–257. [CrossRef]
3. Zhang, J.; Yang, Z.; Deng, H.; Yu, H.; Ma, M.; Xiang, Z. Dynamic Visual Measurement of Driver Eye Movements. *Sensors* **2019**, *19*, 2217. [CrossRef] [PubMed]
4. Strobl, M.A.R.; Lipsmeier, F.; Demenescu, L.R.; Gossens, C.; Lindemann, M.; De Vos, M. Look me in the eye: Evaluating the accuracy of smartphone-based eye tracking for potential application in autism spectrum disorder research. *Biomed. Eng. Online* **2019**, *18*, 51. [CrossRef] [PubMed]
5. Shishido, E.; Ogawa, S.; Miyata, S.; Yamamoto, M.; Inada, T.; Ozaki, N. Application of eye trackers for understanding mental disorders: Cases for schizophrenia and autism spectrum disorder. *Neuropsychopharmacol. Rep.* **2019**, *39*, 72–77. [CrossRef]
6. Cruz, R.; Souza, V.; Filho, T.B.; Lucena, V. Electric Powered Wheelchair Command by Information Fusion from Eye Tracking and BCI. In Proceedings of the 2019 IEEE International Conference on Consumer Electronics (ICCE), Las Vegas, NV, USA, 11–13 January 2019.

7. Rupanagudi, S.R.; Koppisetti, M.; Satyananda, V.; Bhat, V.G.; Gurikar, S.K.; Koundinya, S.P.; Sumedh, S.K.M.; Shreyas, R.; Shilpa, S.; Suman, N.M.; et al. A Video Processing Based Eye Gaze Recognition Algorithm for Wheelchair Control. In Proceedings of the 2019 10th International Conference on Dependable Systems, Services and Technologies (DESSERT), Leeds, UK, 5–7 June 2019.
8. Kumar, A.; Netzel, R.; Burch, M.; Weiskopf, D.; Mueller, K. Visual Multi-Metric Grouping of Eye-Tracking Data. *J. Eye Mov. Res.* **2018**, *10*, 17.
9. Ahmed, H.M.; Abdullah, S.H. A Survey on Human Eye-Gaze Tracking (EGT) System "A Comparative Study". *Iraqi J. Inf. Technol.* **2019**, *9*, 177–190.
10. Vidal, M.; Turner, J.; Bulling, A.; Gellersen, H. Wearable eye tracking for mental health monitoring. *Comput. Commun.* **2012**, *35*, 1306–1311. [CrossRef]
11. Reddy, T.K.; Gupta, V.; Behera, L. Autoencoding Convolutional Representations for Real-Time Eye-Gaze Detection. In *Advances in Intelligent Systems and Computing*; Springer Science and Business Media LLC: Singapore, 2018; pp. 229–238.
12. Jafar, F.; Fatima, S.F.; Mushtaq, H.R.; Khan, S.; Rasheed, A.; Sadaf, M. Eye Controlled Wheelchair Using Transfer Learning. In Proceedings of the 2019 International Symposium on Recent Advances in Electrical Engineering (RAEE), Islamabad, Pakistan, 28–29 August 2019.
13. Deshpande, S.; Adhikary, S.D.; Arvindekar, S.; Jadhav, S.S.; Rathod, B. Eye Monitored Wheelchair Control for People Suffering from Quadriplegia. *Univers. Rev.* **2019**, *8*, 141–145.
14. Majaranta, P.; Bulling, A. Eye Tracking and Eye-Based Human–Computer Interaction. In *Human–Computer Interaction Series*; Springer Science and Business Media LLC: London, UK, 2014; pp. 39–65.
15. Hickson, S.; Dufour, N.; Sud, A.; Kwatra, V.; Essa, I. Eyemotion: Classifying Facial Expressions in VR Using Eye-Tracking Cameras. In Proceedings of the 2019 IEEE Winter Conference on Applications of Computer Vision (WACV), Waikoloa Village, HI, USA, 7–11 January 2019.
16. Harezlak, K.; Kasprowski, P. Application of eye tracking in medicine: A survey, research issues and challenges. *Comput. Med. Imaging Graph.* **2018**, *65*, 176–190. [CrossRef]
17. Fuhl, W.; Tonsen, M.; Bulling, A.; Kasneci, E. Pupil detection for head-mounted eye tracking in the wild: An evaluation of the state of the art. *Mach. Vis. Appl.* **2016**, *27*, 1275–1288. [CrossRef]
18. Liu, T.-L.; Fan, C.-P. Visible-light wearable eye gaze tracking by gradients-based eye center location and head movement compensation with IMU. In Proceedings of the 2018 IEEE International Conference on Consumer Electronics (ICCE), Las Vegas, NV, USA, 12–14 January 2018.
19. Robbins, S.; McEldowney, S.; Lou, X.; Nister, D.; Steedly, D.; Miller, Q.S.C.; Bohn, D.D.; Terrell, J.P.; Goris, A.C.; Ackerman, N. Eye-Tracking System Using a Freeform Prism and Gaze-Detection Light. U.S. Patent 10,228,561, 12 March 2019.
20. Sasaki, M.; Nagamatsu, T.; Takemura, K. Screen corner detection using polarization camera for cross-ratio based gaze estimation. In Proceedings of the 11th ACM Symposium on Eye Tracking Research & Applications, Denver, CO, USA, 25–28 June 2019.
21. Holland, J. Eye Tracking: Biometric Evaluations of Instructional Materials for Improved Learning. *Int. J Educ. Pedag. Sci.* **2019**, *13*, 1001–1008.
22. Chen, B.-C.; Wu, P.-C.; Chien, S.-Y. Real-time eye localization, blink detection, and gaze estimation system without infrared illumination. In Proceedings of the 2015 IEEE International Conference on Image Processing (ICIP), Quebec, QC, Canada, 27–30 September 2015.
23. Fuhl, W.; Santini, T.C.; Kübler, T.; Kasneci, E. Else: Ellipse selection for robust pupil detection in real-world environments. In Proceedings of the Ninth Biennial ACM Symposium on Eye Tracking Research & Applications, Charleston, SC, USA, 14–17 March 2016.
24. Fuhl, W.; Kübler, T.; Sippel, K.; Rosenstiel, W.; Kasneci, E. ExCuSe: Robust Pupil Detection in Real-World Scenarios. In *International Conference on Computer Analysis of Images and Patterns*; Springer: Cham, Germany, 2015; pp. 39–51.
25. Kassner, M.; Patera, W.; Bulling, A. Pupil: An Open Source Platform for Pervasive Eye Tracking and Mobile Gaze-based Interaction. In Proceedings of the 2014 ACM International Joint Conference on Pervasive and Ubiquitous Computing: Adjunct Publication, Seattle, WA, USA, 13–17 September 2014.
26. Javadi, A.-H.; Hakimi, Z.; Barati, M.; Walsh, V.; Tcheang, L. SET: A pupil detection method using sinusoidal approximation. *Front. Neuroeng.* **2015**, *8*, 4. [CrossRef] [PubMed]

27. Li, D.; Winfield, D.; Parkhurst, D.J. Starburst: A hybrid algorithm for video-based eye tracking combining feature-based and model-based approaches. In Proceedings of the 2005 IEEE Computer Society Conference on Computer Vision and Pattern Recognition (CVPR'05)-Workshops, San Diego, CA, USA, 21–23 September 2005.
28. Świrski, L.; Bulling, A.; Dodgson, N. Robust real-time pupil tracking in highly off-axis images. In Proceedings of the Symposium on Eye Tracking Research and Applications, Santa Barbara, CA, USA, 28–30 March 2012.
29. Mompeán, J.; Aragon, J.L.; Prieto, P.M.; Artal, P. Design of an accurate and high-speed binocular pupil tracking system based on GPGPUs. *J. Supercomput.* **2017**, *74*, 1836–1862. [CrossRef]
30. Naeem, A.; Qadar, A.; Safdar, W. Voice controlled intelligent wheelchair using raspberry pi. *Int. J. Technol. Res.* **2014**, *2*, 65.
31. Rabhi, Y.; Mrabet, M.; Fnaiech, F. A facial expression controlled wheelchair for people with disabilities. *Comput. Methods Programs Biomed.* **2018**, *165*, 89–105. [CrossRef]
32. Arai, K.; Mardiyanto, R.; Nopember, K.I.T.S. A Prototype of ElectricWheelchair Controlled by Eye-Only for Paralyzed User. *J. Robot. Mechatron.* **2011**, *23*, 66–74. [CrossRef]
33. Mani, N.; Sebastian, A.; Paul, A.M.; Chacko, A.; Raghunath, A. Eye controlled electric wheel chair. *Int. J. Adv. Res. Electr. Electron. Instrum. Eng.* **2015**, *4*. [CrossRef]
34. Gautam, G.; Sumanth, G.; Karthikeyan, K.; Sundar, S.; Venkataraman, D. Eye movement based electronic wheel chair for physically challenged persons. *Int. J. Sci. Technol. Res.* **2014**, *3*, 206–212.
35. Patel, S.N.; Prakash, V.; Narayan, P.S. Autonomous camera based eye controlled wheelchair system using raspberry-pi. In Proceedings of the 2015 International Conference on Innovations in Information, Embedded and Communication Systems (ICIIECS), Coimbatore, India, 19–20 March 2015.
36. Chacko, J.K.; Oommen, D.; Mathew, K.K.; Sunny, N.; Babu, N. Microcontroller based EOG guided wheelchair. *Int. J. Med. Health Pharm. Biomed. Eng.* **2013**, *7*, 409–412.
37. Al-Haddad, A.; Sudirman, R.; Omar, C. Guiding Wheelchair Motion Based on EOG Signals Using Tangent Bug Algorithm. In Proceedings of the 2011 Third International Conference on Computational Intelligence, Modelling & Simulation, Langkawi, Malaysia, 20–22 September 2011.
38. Elliott, M.A.; Malvar, H.; Maassel, L.L.; Campbell, J.; Kulkarni, H.; Spiridonova, I.; Sophy, N.; Beavers, J.; Paradiso, A.; Needham, C.; et al. Eye-controlled, power wheelchair performs well for ALS patients. *Muscle Nerve* **2019**, *60*, 513–519. [CrossRef]
39. Krafka, K.; Khosla, A.; Kellnhofer, P.; Kannan, H.; Bhandarkar, S.; Matusik, W.; Torralba, A. Eye Tracking for Everyone. In Proceedings of the 2016 IEEE Conference on Computer Vision and Pattern Recognition, Las Vegas, NV, USA, 26 June–1 July 2016.
40. George, A.; Routray, A. Real-time eye gaze direction classification using convolutional neural network. In Proceedings of the 2016 International Conference on Signal Processing and Communications (SPCOM), Bangalore, India, 12–15 June 2016.
41. Spinel IR Camera, Spinel 2MP full HD USB Camera Module Infrared OV2710 with Non-distortion Lens FOV 100 degree, Support 1920x1080@30fps, UVC Compliant, Support most OS, Focus Adjustable UC20MPD_ND. Available online: https://www.amazon.com/Spinel-Non-distortion-1920x1080-Adjustable-UC20MPD_ND/dp/B0711JVGTN (accessed on 10 March 2017).
42. He, D.-C.; Wang, L. Texture Unit, Texture Spectrum, and Texture Analysis. *IEEE Trans. Geosci. Remote. Sens.* **1990**, *28*, 509–512.
43. Lian, Z.; Er, M.J.; Li, J. A Novel Face Recognition Approach under Illumination Variations Based on Local Binary Pattern. In Proceedings of the International Conference on Computer Analysis of Images and Patterns, Seville, Spain, 29–31 August 2011.
44. Cs.columbia.edu. CAVE/Database: Columbia Gaze Data Set. 2017. Available online: http://www.cs.columbia.edu/CAVE/databases/columbia_gaze/ (accessed on 10 March 2017).
45. GI4E/Gi4E Database. Available online: Ttp://gi4e.unavarra.es/databases/gi4e/ (accessed on 10 March 2017).

46. UBIRIS Database. Available online: http://iris.di.ubi.pt/ (accessed on 10 March 2017).
47. Lakovic, N.; Brkic, M.; Batinic, B.; Bajic, J.; Rajs, V.; Kulundzic, N. Application of low-cost VL53L0X ToF sensor for robot environment detection. In Proceedings of the 2019 18th International Symposium Infoteh-Jahorina (INFOTEH), East Sarajevo, Srpska, 20–22 March 2019.

Article

Laryngeal Lesion Classification Based on Vascular Patterns in Contact Endoscopy and Narrow Band Imaging: Manual Versus Automatic Approach

Nazila Esmaeili [1,*], Alfredo Illanes [1], Axel Boese [1], Nikolaos Davaris [2], Christoph Arens [2], Nassir Navab [3] and Michael Friebe [1,4]

[1] INKA-Application Driven Research, Otto-von-Guericke University Magdeburg, 39120 Magdeburg, Germany; alfredo.illanes@med.ovgu.de (A.I.); axel.boese@ovgu.de (A.B.); michael.friebe@ovgu.de (M.F.)
[2] Department of Otorhinolaryngology, Head and Neck Surgery, Magdeburg University Hospital, 39120 Magdeburg, Germany; nikolaos.davaris@med.ovgu.de (N.D.); christoph.arens@med.ovgu.de (C.A.)
[3] Chair for Computer Aided Medical Procedures and Augmented Reality, Technical University Munich, 85748 Munich, Germany; navab@cs.tum.edu
[4] IDTM GmbH, 45657 Recklinghausen, Germany
* Correspondence: nazila.esmaeili@med.ovgu.de

Received: 29 May 2020; Accepted: 18 July 2020; Published: 19 July 2020

Abstract: Longitudinal and perpendicular changes in the vocal fold's blood vessels are associated with the development of benign and malignant laryngeal lesions. The combination of Contact Endoscopy (CE) and Narrow Band Imaging (NBI) can provide intraoperative real-time visualization of the vascular changes in the laryngeal mucosa. However, the visual evaluation of vascular patterns in CE-NBI images is challenging and highly depends on the clinicians' experience. The current study aims to evaluate and compare the performance of a manual and an automatic approach for laryngeal lesion's classification based on vascular patterns in CE-NBI images. In the manual approach, six observers visually evaluated a series of CE+NBI images that belong to a patient and then classified the patient as benign or malignant. For the automatic classification, an algorithm based on characterizing the level of the vessel's disorder in combination with four supervised classifiers was used to classify CE-NBI images. The results showed that the manual approach's subjective evaluation could be reduced by using a computer-based approach. Moreover, the automatic approach showed the potential to work as an assistant system in case of disagreements among clinicians and to reduce the manual approach's misclassification issue.

Keywords: laryngeal cancer; contact endoscopy; narrow band imaging; automatic classification; feature extraction; machine learning

---

1. Introduction

Laryngeal cancer is the second most frequent malignant tumor of the head and neck region [1]. The vast majority of primary laryngeal cancers are Squamous Cell Carcinomas (SCC) arising from the epithelial lining of the larynx, mostly as a result of tobacco and alcohol consumption. A total of 40% of these cancers are diagnosed at an advanced stage, which is associated with a poorer prognosis and quality of life [2]. The early diagnosis of laryngeal cancer is crucial to reduce patient mortality and preserve vocal fold function.

Specific changes in the morphology and three-dimensional orientation of the vocal fold's sub-epithelial blood vessels have proved to be associated with the development of benign and malignant laryngeal lesions. Several approaches have been proposed to describe and classify these vascular changes. Among the complex classification systems proposed by [3] and [4], the European

Laryngological Society (ELS) introduced a simplified classification that divides vascular changes into longitudinal and perpendicular classes [5,6]. Longitudinal Vascular Changes (LVC) spread along the length and width of the vocal fold and can be observed in all kinds of benign or malignant lesions. On the contrary, Perpendicular Vascular Changes (PVC) develop perpendicularly towards the mucosa, as a result of neoangiogenesis in laryngeal Papillomatosis, pre-malignant and malignant histopathologies.

The endoscopic detection and evaluation of vascular changes can provide complementary diagnostic information for clinicians to detect and differentiate between benign and malignant laryngeal lesions [7]. As a minimally-invasive endoscopic technique, Contact Endoscopy (CE) can provide real-time visualization of cellular and vascular structures of the laryngeal mucosa [8,9]. For the purpose of detecting and evaluating superficial vascular changes, several enhanced endoscopic techniques such as Narrow Band Imaging (NBI) have been combined with CE to ease the detection of vascular changes [10]. The use of enhanced CE showed promising results in the assessment of vascular patterns followed by indicative of various laryngeal pathologies [4,11,12].

Clinicians can receive useful information about the type and suspected histopathology of laryngeal lesions by evaluating LVC and PVC in enhanced CE images; however, it is a challenging task for them. There are similarities between vascular patterns of benign and malignant laryngeal lesions. The PVC with wide-angled turning points, as observed in laryngeal Papillomatosis can be difficult to distinguish from PVC with narrow-angled turning points, as observed in pre-malignant and malignant histopathologies [5,12–14]. Hence, the interpretation of vascular patterns in enhanced CE images requires an extensive learning curve from the clinicians to reduce the risk of subjective evaluation that can cause potential problems in differentiation between benign and malignant laryngeal lesions [4,10,12,15,16].

In this study, we first aimed to present the results of manual versus automatic classification of benign and malignant laryngeal lesions based on the vascular patterns in CE-NBI images. We then evaluated the issues of manual classification and subsequently showed how a computer-based approach can assist the clinicians to overcome these problems. A manual and an automatic classification approach were defined to conduct this evaluation. In the manual approach, six experienced and less experienced otolaryngologists individually evaluated PVC and LVC in CE-NBI images of patients and classified them into benign and malignant groups. An updated version of the algorithm proposed in [17,18] with 24 features and four supervised classifiers has been used to classify CE-NBI images into benign and malignant groups. The results of the two approaches were compared in terms of classification sensitivity and specificity. The potential of an automatic approach to assist the clinicians is presented through two evaluation strategies.

2. Material and Methods

2.1. Data Acquisition

CE-NBI images were extracted from video scenes of adult patients who received a microlaryngoscopy for benign, pre-malignant or malignant lesions of the vocal folds. Video scenes were captured using an Evis Exera III Video System with integrated NBI-filter (Olympus Medical Systems, Hamburg, Germany) and a rigid 30-degree contact endoscope (Karl Storz, Tuttlingen, Germany) with a fixed magnification of 60 to have a fixed camera–tissue distance. For each video scene, we selected the time intervals where the video quality was good enough to visualize the vessels. Then, one in every ten frames was extracted from the selected intervals in JPEG format images (1008 × 1280 pixels) to have unique and non-redundant CE-NBI images.

2.2. Dataset Generation

The CE-NBI dataset included 1632 extracted images of 68 patients. The patients' data were pseudonymized. Based on the WHO classification [19], histological diagnoses were used to label images as belonging to a benign or a malignant class. Table 1 shows the histopathologies with the number of patients and images used for the generation of the dataset.

Two image subsets were created from the CE-NBI dataset. The *Subset I* included a series of two to five randomly selected CE-NBI images of each patient—total of 336 images, ≈ 20% of the dataset. The *Subset II* included the rest of the CE-NBI images—a total of 1296 images, ≈ 80% of the dataset, and was used as the training set of the automatic approach. The *Subset I* was evaluated by the otolaryngologists in the manual approach and then used as the testing set for the automatic approach. Figure 1 presents some examples of CE-NBI images with LVC and PVC belonging to the generated dataset.

Table 1. Histopathologies used for the generation of the dataset.

Type of Lesion	Histopathology	Number of Patients	Number of Images
Benign	Cyst	3	90
	Polyp	5	71
	Reinke's edema	12	329
	Hyperkeratosis	4	82
	Squamous Hyperplasia	3	75
	Papillomatosis	11	286
	Amyloidosis	2	32
	Nodule	1	26
	Granuloma	1	28
	Fibroma	1	2
(Pre)Malignant	Mild Dysplasia	3	77
	Moderate Dysplasia	2	49
	Severe Dysplasia	3	68
	Carcinoma In Situ	9	249
	SCC	8	168
Total		**68**	**1632**

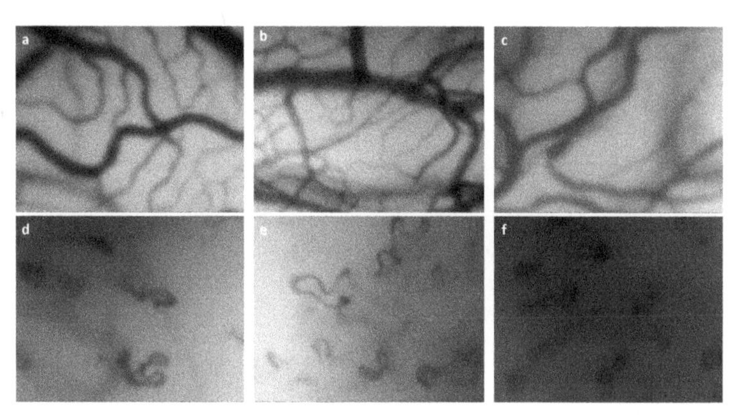

Figure 1. Examples of Longitudinal Vascular Changes (LVC) and Perpendicular Vascular Changes (PVC) in Contact Endoscopy (CE)-Narrow Band Imaging (NBI) images with different histopathologies: (**a**) Reinke's edema, LVC; (**b**) polyp, LVC; (**c**) amyloidosis, LVC; (**d**) severe dysplasia, PVC, (**e**) carcinoma in situ, PVC; (**f**) Squamous Cell Carcinomas (SCC), PVC.

2.3. Manual Approach

Three specialist and three resident otolaryngologists evaluated the images and classified the patients into benign and malignant groups. The residents had less than two years of experience in operating with CE-NBI images and the specialists worked for more than five years with such images. The otolaryngologists were blinded to the histologic diagnosis. They used the ELS guideline to independently visually evaluate the CE-NBI images of *Subset I* based on PVC appearance in the CE-NBI images, as explained in [12].

2.4. Automatic Approach

We used the algorithm presented in [17,18] to perform the automatic approach. The algorithm consists of a pre-processing step involving vessel enhancement and segmentation [20]. A feature extraction step was then applied to extract 24 geometrical features based on the consistency of gradient direction and the curvature level. Supervised classification step was conducted using the features and four classifiers to classify CE-NBI images into benign and malignant groups.

In this study, we made two main changes to the algorithm proposed in [17,18]. First, the Jerman filter [21] was used as pre-processing for the vessel enhancement step instead of the Frangi filter to overcome the problems related to the established enhancement function, not well adapted to natural variations of the vascular morphology. Second, the values of the tuning parameters of four classifiers including Support Vector Machine (SVM) with Polykernel and Radial Basis Function (RBF) [22], k-Nearest Neighbor (kNN) [23] and Random Forest Classifier (RFC) [24] were updated to have the optimum classification results with the current dataset.

In order to cover all the possible vascular structures, the vesselness parameter σ of the Jerman filter was set in the range of 0.5 mm to 2.5 mm with a step size of 0.5 mm. The parameter τ controlling the response uniformity was empirically set as 1.

The hyperparameter tuning process of all classifiers was updated using a grid search combined with 10-fold cross validation.

The performance of SVM is maily affected by the regulation parameter (C) and kernel parameter (γ). The regulation parameter together with Polykernel and RBF controls the trade-off between achieving a low error in training data. γ determines how quickly class boundaries dissipate when they get far from the support vectors in SVM with RBF. The range of C and γ values were set within the range of 0.001 to 1000 with a ten-fold increment. The SVM with RBF completed the high overall performance with $C = 1$ and $\gamma = 0.01$ and SVM with Polykernel indicated the best results with $C = 1$.

Euclidean Distance was applied to calculate the distance of a sample in the case of kNN. To select the optimum k, a range from 1 to 20 with the step size equal to one were used. kNN confirmed the best performance at $k = 5$.

The optimization for RFC was done by adjusting the depth of trees and the number of estimators. The range of depth of the trees was set from 1 to 20 with step size equal to one. For the number of estimators, values from 10 to 100 with an increase of five was defined. The classifier gave the best performance at a depth of 8 with 55 trees.

In all classification scenarios, Subset I and Subset II were used as the testing and training sets, respectively. CE-NBI images were labeled as 0 for benign and as 1 for malignant groups. Each classifier was trained using the images' labels and feature vectors that were computed form the CE-NBI images of the training set. For the testing, the features vectors computed from the CE-NBI images of the testing set were fed into the predictive model of each classifier and then the expected labels were collected.

3. Evaluation Strategy

3.1. Classification Performances of Manual and Automatic Approaches

The global performances of the manual and automatic classification were evaluated using two classification measurements: sensitivity and specificity.

In the manual classification, the otolaryngologists assessed the set of CE-NBI images in the *Subset I* and classified each patient's image set as benign or malignant. Following [12], the PVC-positive patients with the malignant histological diagnosis were considered as true positive cases. With this assumption, a confusion matrix was created and the average value of sensitivity and specificity of all otolaryngologists, specialists and residents, was calculated using the following parameters:

- True Positive: PVC-positive patients with malignant lesions.
- True Negative: PVC-negative patients with benign lesions.
- False Negative: PVC-negative patients with malignant lesions.
- False Positive: PVC-positive patients with benign lesions.

In the automatic classification, the classifiers classified each CE-NBI image of *Subset I* as benign or malignant. A confusion matrix was calculated for each classifier using the predicted and actual labels of the images. Then, sensitivity and specificity were calculated using the following parameters:

- True Positive: actual image label is malignant, predicted image label is malignant.
- True Negative: actual image label is benign, predicted image label is benign.
- False Negative: actual image label is malignant, predicted image label is benign.
- False Positive: actual image label is benign, predicted image label is malignant.

Based on the descriptions above, the sensitivity and specificity values can show the performances of classifiers/otolaryngologists to correctly classify malignant and benign images/patients.

3.2. *Comparison Procedure Between Manual and Automatic Classification*

In a routine clinical procedure, the otolaryngologist evaluates a set of CE-NBI images of a patient and then identifies a patient's lesion as benign or malignant. For the manual classification in this work, the clinicians performed a similar routine, making a decision based on a set of images belonging to a patient. Since the automatic classification does not classify a patient but an image, in order to compare automatic to manual classification we made the following assumption: if a given classifier correctly classifies more than half of the images of a patient, then the patient is considered as a correct classification performed by this classifier. Following the assumption, two procedures for comparing between manual and automatic classification were proposed.

The first comparison procedure consists of comparing both approaches based on the level of agreement/disagreement between clinicians for classifying a patient as benign or malignant. In this aim, patients were divided into three categories:

- Category I includes 29 patients. All otolaryngologists correctly classified these patients.
- Category II includes 26 patients. One to five otolaryngologists correctly classified these patients.
- Category III includes 13 patients. All otolaryngologists misclassified these patients.

The second comparison procedure aims to compare manual and automatic classifications in terms of their misclassification levels depending on the histopathologies. This evaluation was performed to analyze the histopathologies in benign and malignant groups that caused significant difficulties for otolaryngologists and then to see how the automatic approach behaves with these cases.

We divided the patients into the 15 groups presented in Table 1. For each histopathology, a misclassification percentage was computed per patient for the automatic and manual classification as follows:

- Misclassification percentage of all otolaryngologists per patient in each histopathology group:

$$\left(\frac{\text{Number of doctor(s) who misclassified the patients}}{\text{Total number of doctors} \times \text{Total number of patients}}\right) \times 100 \tag{1}$$

  where the total number of patients was the number of patients for the corresponding histopathology.
- Misclassification percentage of every classifier per patient in each histopathology group:

$$\left(\frac{\text{Number of misclassified patient(s)}}{\text{Total number of patients}}\right) \times 100 \tag{2}$$

- Misclassification percentage of all classifiers per patient in each histopathology group:

$$\left(\frac{\text{Number of misclassified patient(s) by all classifiers}}{\text{Total number of classifiers} \times \text{Total number of patients}}\right) \times 100 \quad (3)$$

4. Results and Discussion

Table 2 shows the global performances of the manual and automatic classification. In the manual approach, otolaryngology specialists showed a better performance than the otolaryngology residents. These results prove that the interpretation of CE-NBI images based on vascular patterns is subjective and highly depends on otolaryngologists' experience.

For the automatic approach, RFC with a sensitivity of 0.846 and SVM with RBF kernel with a specificity of 0.981 showed better results in comparison to the other classifiers.

The overall specificity values of otolaryngologists are low. This means that both groups had difficulties in distinguishing patients with benign histopathologies from malignant ones visually. This fact can be explained by the similarity between vascular patterns of benign and malignant histopathologies that can not be distinguished easily. For instance, Papillomatosis is a benign histopathology with similar vascular patterns than malignant histopathologies. This similarity leads to visually misclassify Papillomatosis as malignant. However, all four classifiers showed higher specificity than otolaryngologists proving the ability of automatic approach to overcome such a problem.

Table 2. General performance of manual and automatic approaches

Classification Measurements		Sensitivity	Specificity
Manual Classification (per patient)	Otolaryngology specialists	0.955	0.727
	Otolaryngology residents	0.630	0.609
	All otolaryngologists	0.818	0.630
Automatic Classification (per image)	SVM with polykernel	0.830	0.882
	SVM with RBF	0.806	0.981
	kNN	0.814	0.863
	RFC	0.846	0.895

Figure 2 shows the detailed results of the first comparison procedure consisting of comparing both approaches based on the level of agreement/disagreement between clinicians for classifying a patient as benign or malignant. A first visual inspection shows that the classifiers individually misclassified 1 to 2 images in some patients at the Category I, where all otolaryngologists correctly classified these patients. Nevertheless, based on the assumption made in Section 3.2, the automatic approach did not misclassify any patient of this category.

For the patients belonging to Category II, both manual and automatic increased their misclassification levels compared to Category I. In the automatic approach, it is possible to observe that several images belonging to a patient can be misclassified. However, if we consider the automatic classification per patient, only for one patient, two classifiers (SVM with polykernel and RFC) perform a misclassification. On the other hand, otolaryngologists showed a significant misclassification in some cases. For example, in the case of patients p26, p34 and p 72, five clinicians misclassified the patients, while the classifiers classified the patients correctly. These patients were diagnosed as Papillomatosis and Hyperkeratosis cases and belong to benign histopathologies. Figure 3a–c, displays the PVC vascular patterns in the CE-NBI images of these patients. As pointed out in the introduction, the difference between PVC in benign and in malignant histopathologies is not visually evident for the otolaryngologist. This causes a significant difficulty for the clinicians to distinguish benign from malignant cases based on the vascular patterns. Based on the results, the automatic approach showed the ability to identify this difference and then classify the patients correctly because it is capable of quantifying and differentiating these tiny differences. SVM with RBF did not show any misclassification per patient in this category.

For the Category III, where all otolaryngologists misclassified the patients, SVM with RBF misclassified fewer images compared to the other three classifiers. Concerning the classification per patient performed by the classifiers, it is possible to see that misclassifications were made for only two patients. Particularly, for patient p10 three classifiers failed in their classification. According to the histopathology, it corresponds to a patient presenting Hyperkeratosis. A set of CE-NBI images of this case is presented in Figure 3d. The type of vascular patterns of Hyperkeratosis can notably vary from one patient to another one. The CE-NBI dataset included 4 patients for this histopathology, presenting LVC and PVC vascular patterns. Due to this variation, the classifier's learning process using the proposed features [17,18] can be complicated. SVM with RBF showed no misclassification per patient in this Category.

These results show that the complexity of a manual analysis of a laryngeal lesion can be related to the type of histopathology and therefore we decided to perform a separated analysis based on the histopathology of the lesion. Table 3 presents the results of this second comparison procedure.

For the benign histopathologies, otolaryngologists showed high misclassification percentage of 83%, 77%, 46%, 33% and 27% for Fibroma, Papillomatosis, Hyperkeratosis, Squamous Hyperplasia and Polyp, respectively. Except for Fibroma, the misclassification level of each classifier is lower than the manual classification. Notably, in the case of Papillomatosis, the misclassification is significantly reduced in each classifier. If all classifiers are considered, the misclassification decreases from 77% to 7% in this histopathology. Papillomatosis causes classification difficulties to the otolaryngologists due to their vascular patterns that has similar characteristics to the malignant histopathologies. SVM with RBF and kNN seems to have the ability to solve this issue with 0% misclassification.

In the case of Fibroma, the misclassification percentage varied significantly among the four classifiers. This can be explained by the reduced number of images that the dataset contains for this type of histopathology (only one patient and two images).

In the malignant group, the otolaryngologists had the highest misclassification percentage of 61% for mild dysplasia. This histopathology can have PVC as well as LVC vascular patterns that usually appear in benign histopathologies. Hence, it is challenging for the otolaryngologists to classify patients with this condition as malignant visually. For this histopathology, the four classifiers performed well by classifying every patient correctly.

In general, SVM with RBF showed no patient misclassification for all histopathologies.

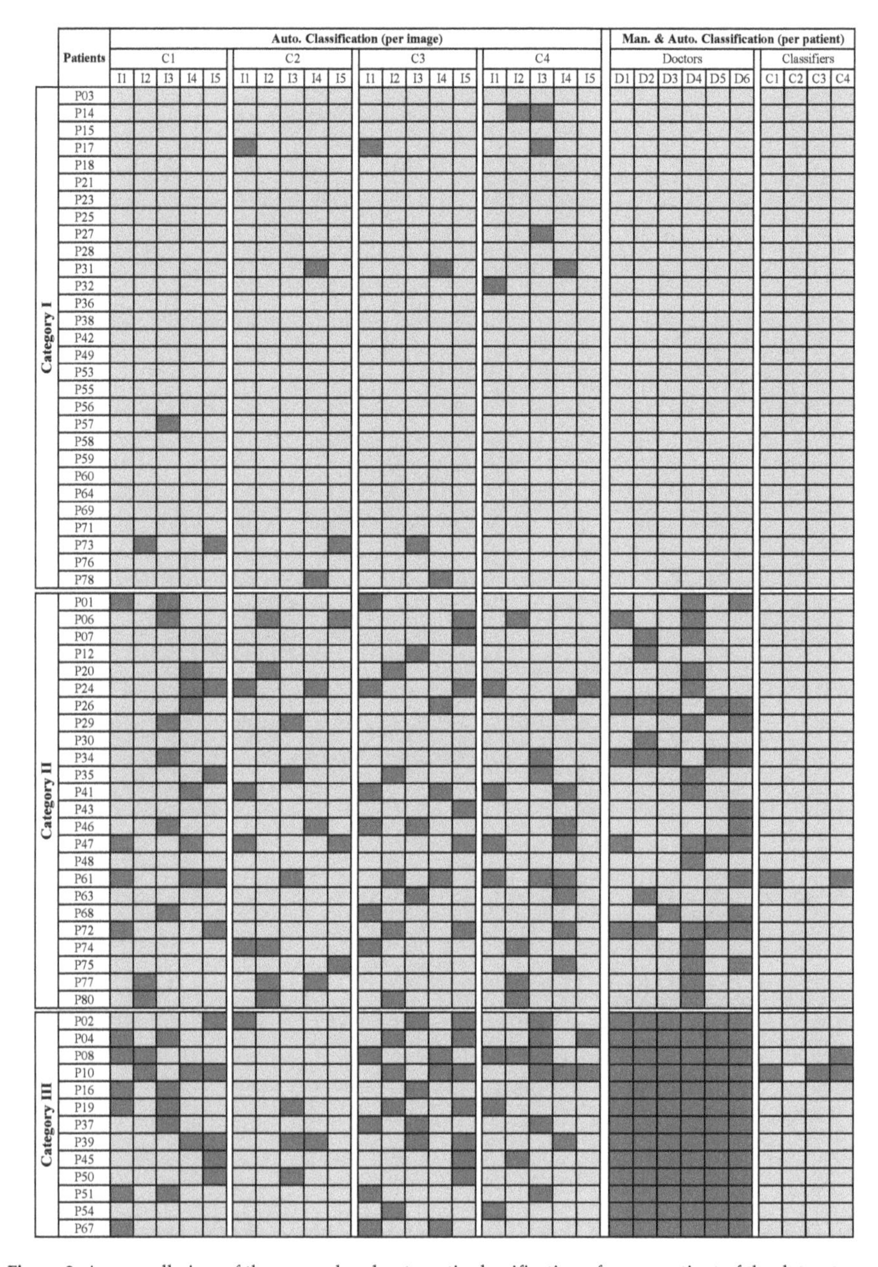

Figure 2. An overall view of the manual and automatic classification of every patient of the dataset; Green color: correct classification; Red color: misclassification. C1 to C4 represents the four classifiers; C1: Support Vector Machine (SVM) with polykernel, C2: SVM with Radial Basis Function (RBF), C3: k-Nearest Neighbor (kNN) and C4: Random Forest Classifier (RFC). I1 to I5 represent five testing images for each patient. D1 to D6 represent the six otolaryngologists.

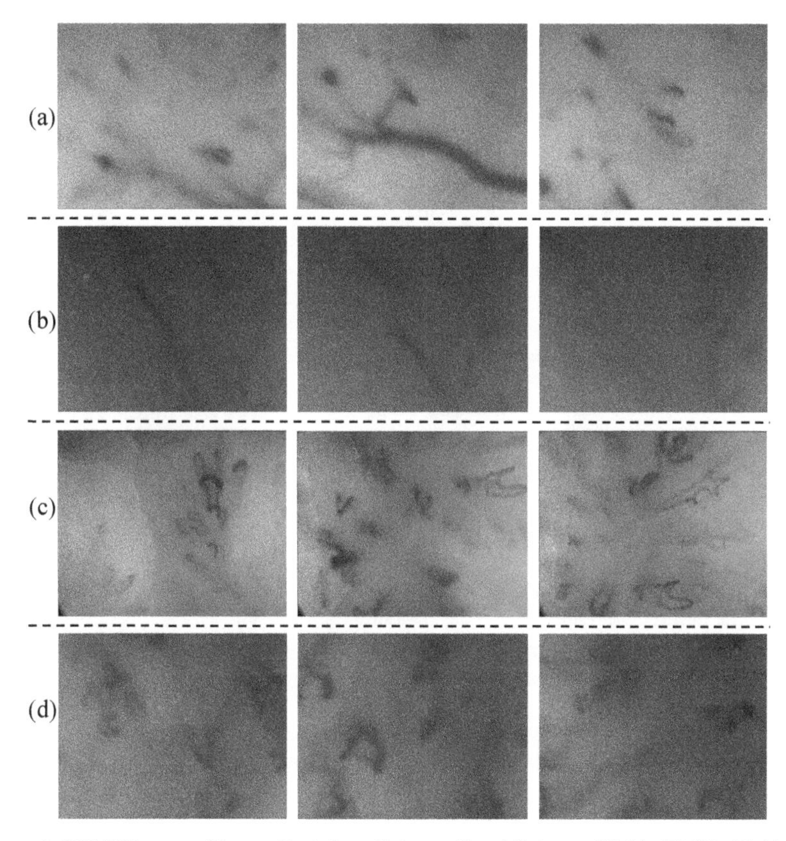

Figure 3. CE-NBI images of four patients from Category II and Category III: (**a**) p26, (**b**) p34, (**c**) p72 and (**d**) p10.

Table 3. Misclassification percentage of every histopathology category based on patient. C1 to C4 represent the four classifiers; C1: SVM with polykernel, C2: SVM with RBF, C3: kNN and C4: RFC.

Type of Lesions	Histopathology	Man. and Auto. Classification (per Patient)					
		Doctors	C1	C2	C3	C4	All Classifiers
Benign	Cyst	0%	0%	0%	0%	0%	0%
	Polyp	27%	0%	0%	0%	0%	0%
	Reinke's edema	7%	0%	0%	0%	0%	0%
	Hyperkeratosis	46%	25%	0%	25%	25%	19%
	Squamous Hyperplasia	33%	0%	0%	0%	0%	0%
	Papillomatosis	77%	9%	0%	0%	18%	7%
	Nodule	0%	0%	0%	0%	0%	0%
	Granuloma	0%	0%	0%	0%	0%	0%
	Amyloidosis	8%	0%	0%	0%	0%	0%
	Fibroma	83%	0%	0%	100%	100%	50%
(Pre)Malignant	Mild Dysplasia	61%	0%	0%	0%	0%	0%
	Moderate Dysplasia	17%	0%	0%	0%	0%	0%
	Severe Dysplasia	17%	0%	0%	0%	0%	0%
	Carcinoma In Situ	9%	0%	0%	0%	0%	0%
	SCC	25%	0%	0%	13%	0%	3%

5. Conclusions

Assessment of vascular patterns in CE-NBI images of vocal folds can provide valuable information for the clinicians to make the correct diagnostic decision before treatment. In this study, we showed how the evaluation of vascular patterns can be challenging for the otolaryngologists and how a computer-based approach can help clinicians ease this process.

In general, the otolaryngology specialists showed better classification performance than the residents in the manual approach. This proves that the interpretation of vascular patterns is subjective and depends on the clinicians' experience, as pointed out by several publications [4,10,12,15,16]. Both groups of otolaryngologists showed relatively low specificity on classifying a case as benign or malignant. This explains the difficulties in the visual classification of benign histopathologies. In the case of the benign group, otolaryngologists had the highest misclassification percentage for Papillomatosis and Hyperkeratosis. In the automatic approach, all four classifiers showed a higher specificity than both groups of otolaryngologists and showed significantly less misclassification percentage for Papillomatosis and Hyperkeratosis. The otolaryngology specialists showed significantly higher sensitivity than the residents. This means that specialists with more experience can easily detect PVC in CE-NBI images, while it is more challenging for the residents. In the malignant group, most of the misclassifications of otolaryngologists happened in the case of Mild Dysplasia and SCC. Although all classifiers showed lower sensitivity than otolaryngology specialists, they significantly reduced the misclassification percentage for Mild Dysplasia and SCC, compared to the otolaryngologists.

Two facts can explain the lower sensitivity and higher misclassification percentage that the classifiers show in the malignant group than the benign group. First, the CE-NBI dataset included more images in the benign group than in the malignant group (less training images were available for the malignant group). A significant part of CE-NBI images of the benign group belonged to the Papillomatosis with PVC patterns similar to those of malignant histopathologies. Second, the 24 features take only into account geometrical characteristics of the vascular patterns and no other characteristics that can also be important for the classification procedure. Due to these two points, it is possible that the algorithm shows some errors and classifies the CE-NBI images of malignant cases as benign. Hence, it is important to balance the number of CE-NBI images in the dataset for future works and develop new methods to improve the differentiation between wide and narrow angled points of PVCs.

The automatic approach showed its capacity to perform as an assistant system when there are disagreements among otolaryngologists or when they all misclassified the patients. SVM with RBF had the best performance and did not show any misclassification per patient in all the categories. This means that the combination of the proposed 24 features and SVM with RBF classifier, can provide valuable feedback for the clinicians to make decisions regarding the treatment planning. In general, the automatic approach has the potential to overcome the current issues in the field of enhanced CE and can operate as an assisting system to provide a more confident way for clinicians to learn as well as to make intraoperative decisions about the method and extent of surgical resection in patients with laryngeal cancer or benign vocal fold lesions in the routine surgical procedures.

Author Contributions: Conceptualization: N.E., A.I., A.B., N.D., C.A., N.N., M.F.; methodology: N.E., A.I.; software: N.E.; validation: N.E., A.I., N.D.; formal analysis: N.E., A.I.; investigation: N.E., N.D.; writing—original draft preparation: N.E.; writing—review and editing: A.I., A.B., N.D., C.A., N.N., M.F.; supervision, M.F., N.N.; project administration, N.E. All authors have read and agreed to the published version of the manuscript.

Funding: This research received no external funding.

Conflicts of Interest: The authors declare no conflicts of interest.

References

1. Siegel, R.L.; Miller, K.D.; Jemal, A. Cancer statistics, 2020. *CA Cancer J. Clin.* **2020**, *70*, 7–30. [CrossRef]
2. Tamaki, A.; Miles, B.A.; Lango, M.; Kowalski, L.; Zender, C.A. AHNS Series: Do you know your guidelines? Review of current knowledge on laryngeal cancer. *Head Neck* **2018**, *40*, 170–181. [CrossRef] [PubMed]
3. Ni, X.; He, S.; Xu, Z.; Gao, L.; Lu, N.; Yuan, Z.; Lai, S.; Zhang, Y.; Yi, J.; Wang, X.; et al. Endoscopic diagnosis of laryngeal cancer and precancerous lesions by narrow band imaging. *J. Laryngol. Otol.* **2011**, *125*, 288–296. [CrossRef] [PubMed]
4. Puxeddu, R.; Sionis, S.; Gerosa, C.; Carta, F. Enhanced contact endoscopy for the detection of neoangiogenesis in tumors of the larynx and hypopharynx. *Laryngoscope* **2015**, *125*, 1600–1606. [CrossRef] [PubMed]
5. Arens, C.; Piazza, C.; Andrea, M.; Dikkers, F.G.; Gi, R.E.T.P.; Voigt-Zimmermann, S.; Peretti, G. Proposal for a descriptive guideline of vascular changes in lesions of the vocal folds by the committee on endoscopic laryngeal imaging of the European Laryngological Society. *Eur. Arch. Otorhinolaryngol.* **2016**, *273*, 1207–1214. [CrossRef]
6. Mehlum, C.S.; Døssing, H.; Davaris, N.; Giers, A.; Grøntved, Å.M.; Kjaergaard, T.; Möller, S.; Godballe, C.; Arens, C. Interrater variation of vascular classifications used in enhanced laryngeal contact endoscopy. *Eur. Arch. Otorhinolaryngol.* **2020**, 1–8.
7. Sun, C.; Han, X.; Li, X.; Zhang, Y.; Du, X. Diagnostic performance of narrow band imaging for laryngeal cancer: A systematic review and meta-analysis. *Otolaryngol. Head Neck Surg.* **2017**, *156*, 589–597. [CrossRef]
8. Andrea, M.; Dias, O.; Santos, A. Contact endoscopy during microlaryngeal surgery: A new technique for endoscopic examination of the larynx. *Ann. Otol. Rhinol. Laryngol.* **1995**, *104*, 333–339. [CrossRef]
9. Arens, C.; Dreyer, T.; Glanz, H.; Malzahn, K. Compact endoscopy of the larynx. *Ann. Otol. Rhinol. Laryngol.* **2003**, *112*, 113–119. [CrossRef]
10. Piazza, C.; Cocco, D.; Del Bon, F.; Mangili, S.; Nicolai, P.; Peretti, G. Narrow band imaging and high definition television in the endoscopic evaluation of upper aero-digestive tract cancer. *Acta Otorhinolaryngol. Ital.* **2011**, *31*, 70.
11. Arens, C.; Voigt-Zimmermann, S. Contact endoscopy of the vocal folds in combination with narrow band imaging (compact endoscopy). *Laryngo-Rhino-Otologie* **2015**, *94*, 150. [PubMed]
12. Davaris, N.; Lux, A.; Esmaeili, N.; Illanes, A.; Boese, A.; Friebe, M.; Arens, C. Evaluation of vascular patterns using contact endoscopy and narrow-band imaging (CE-NBI) for the diagnosis of vocal fold malignancy. *Cancers* **2020**, *12*, 248. [CrossRef] [PubMed]
13. Šifrer, R.; Rijken, J.A.; Leemans, C.R.; Eerenstein, S.E.; van Weert, S.; Hendrickx, J.J.; Bloemena, E.; Heuveling, D.A.; Rinkel, R.N. Evaluation of vascular features of vocal cords proposed by the European Laryngological Society. *Eur. Arch. Otorhinolaryngol.* **2018**, *275*, 147–151. [CrossRef]
14. Šifrer, R.; Šereg-Bahar, M.; Gale, N.; Hočevar-Boltežar, I. The diagnostic value of perpendicular vascular patterns of vocal cords defined by narrow-band imaging. *Eur. Arch. Otorhinolaryngol.* **2020**, *277*, 1–9. [CrossRef]
15. Mannelli, G.; Cecconi, L.; Gallo, O. Laryngeal preneoplastic lesions and cancer: Challenging diagnosis. Qualitative literature review and meta-analysis. *Crit. Rev. Oncol.* **2016**, *106*, 64–90. [CrossRef] [PubMed]
16. Puxeddu, R.; Carta, F.; Ferreli, C.; Chuchueva, N.; Gerosa, C. *Enhanced Contact Endoscopy (ECE) in Head and Neck Surgery*; Endo-Press: Tuttlingen, Germany, 2018.
17. Esmaeili, N.; Illanes, A.; Boese, A.; Davaris, N.; Arens, C.; Friebe, M. A Preliminary Study on Automatic Characterization and Classification of Vascular Patterns of Contact Endoscopy Images. In Proceedings of the 2019 41st Annual International Conference of the IEEE Engineering in Medicine and Biology Society (EMBC), Berlin, Germany, 23–27 July 2019; pp. 2703–2706.
18. Esmaeili, N.; Illanes, A.; Boese, A.; Davaris, N.; Arens, C.; Friebe, M. Novel automated vessel pattern characterization of larynx contact endoscopic video images. *Int. J. Comput. Assist. Radiol. Surg.* **2019**, *14*, 1751–1761. [CrossRef]
19. Gale, N.; Hille, J.; Jordan, R.C.; Nadal, A.; Williams, M.D. Regarding Laryngeal precursor lesions: Interrater and intrarater reliability of histopathological assessment. *Laryngoscope* **2019**, *129*, E91–E92. [CrossRef]
20. Boese, A.; Illanes, A.; Balakrishnan, S.; Davaris, N.; Arens, C.; Friebe, M. Vascular pattern detection and recognition in endoscopic imaging of the vocal folds. *Curr. Dir. Biomed. Eng.* **2018**, *4*, 75–78. [CrossRef]

21. Jerman, T.; Pernuš, F.; Likar, B.; Špiclin, Ž. Enhancement of vascular structures in 3D and 2D angiographic images. *IEEE Trans. Med. Imaging* **2016**, *35*, 2107–2118. [CrossRef]
22. Hsu, C.W.; Chang, C.C.; Lin, C.J. *A Practical Guide to Support Vector Classification*; National Taiwan University: Taipei, Taiwan, 2003.
23. Peterson, L.E. K-nearest neighbor. *Scholarpedia* **2009**, *4*, 1883. [CrossRef]
24. Breiman, L. Random forests. *Mach. Learn.* **2001**, *45*, 5–32. [CrossRef]

Letter

Improving Temporal Stability and Accuracy for Endoscopic Video Tissue Classification Using Recurrent Neural Networks

Tim Boers [1,*], Joost van der Putten [1], Maarten Struyvenberg [2], Kiki Fockens [2], Jelmer Jukema [2], Erik Schoon [3], Fons van der Sommen [1], Jacques Bergman [2] and Peter de With [1]

1. Department of Electrical Engineering, Eindhoven University of Technology, Groene Loper 3, 5612 AE Eindhoven, The Netherlands; J.A.v.d.Putten@tue.nl (J.v.d.P.); fvdsommen@tue.nl (F.v.d.S.); P.H.N.de.With@tue.nl (P.d.W.)
2. Amsterdam University Medical Center, Meibergdreef 9, 1105 AZ Amsterdam, The Netherlands; m.r.struyvenberg@amsterdamumc.nl (M.S.); k.n.fockens@amsterdamumc.nl (K.F.); j.b.jukema@amsterdamumc.nl (J.J.); j.j.bergman@amsterdamumc.nl (J.B.)
3. Catharina Hospital, Michelangelolaan 2, 5623 EJ Eindhoven, The Netherlands; erik.schoon@catharinaziekenhuis.nl

* Correspondence: t.boers@tue.nl

Received: 31 May 2020; Accepted: 20 July 2020; Published: 24 July 2020

Abstract: Early Barrett's neoplasia are often missed due to subtle visual features and inexperience of the non-expert endoscopist with such lesions. While promising results have been reported on the automated detection of this type of early cancer in still endoscopic images, video-based detection using the temporal domain is still open. The temporally stable nature of video data in endoscopic examinations enables to develop a framework that can diagnose the imaged tissue class over time, thereby yielding a more robust and improved model for spatial predictions. We show that the introduction of Recurrent Neural Network nodes offers a more stable and accurate model for tissue classification, compared to classification on individual images. We have developed a customized Resnet18 feature extractor with four types of classifiers: Fully Connected (FC), Fully Connected with an averaging filter (FC Avg(n = 5)), Long Short Term Memory (LSTM) and a Gated Recurrent Unit (GRU). Experimental results are based on 82 pullback videos of the esophagus with 46 high-grade dysplasia patients. Our results demonstrate that the LSTM classifier outperforms the FC, FC Avg(n = 5) and GRU classifier with an average accuracy of 85.9% compared to 82.2%, 83.0% and 85.6%, respectively. The benefit of our novel implementation for endoscopic tissue classification is the inclusion of spatio-temporal information for improved and robust decision making, and it is the first step towards full temporal learning of esophageal cancer detection in endoscopic video.

Keywords: Barrett neoplasia; tissue detection; recurrent neural networks; upper GI tract

---

1. Introduction

Artificial intelligence (AI) systems exceeding expert performance have shortcomings when they are applied on data outside their training domains. At present, such AI systems lack a form of context awareness, which allows the model to reject data outside its learned feature space. Since medical examinations often include an extensive range of anatomical checks, there is a risk that AI-based automated lesion detectors will be applied outside the target domain. Potentially, when inexperienced clinicians are relying on the algorithm, this might lead to higher false positives and false negatives and thereby to malignancies in the diagnosis, which is to the detriment of patients. Assistive tools for automatic lesion detection should therefore be designed for robustness and accuracy with the standard clinical practice in mind.

In the field on gastroenterology, a Computer-Aided Detection (CAD) system has been developed for Barrett's neoplasia detection in white light endoscopic still images [1], achieving expert performance. Yet, this algorithm is restricted and validated on the visual features of a Barrett's esophagus (BE). In clinical practice, it is common to fully assess the esophagus from stomach to the healthy squamous region. Therefore, the current model should only be restricted to the analysis of the Barrett's region of the esophagus.

In order to facilitate the continuous analysis of the video signal during the full clinical protocol, a vast pool of new relevant and irrelevant features needs to be taken into account. For example, optical tissue deformation, which can be estimated through consecutive frames, is an inherent cell marker for testing malignant morphological changes, according to Guck et al. [2]. In contrast, when ambiguous frames are introduced, the model could become unstable according to Van der Putten et al. [3]. In order to deal with such ambiguity, the model should consider the context of prior frames for robust and reliable decision making. The consecutive frames in an endoscopy procedure do not differ substantially, and therefore information prior to an ambiguous frame can be exploited to make an accurate prediction. Such sequential models could be used to improve position tracking in during an endoscopy procedure. Accordingly, since Esophagus Adenocarcinoma (EAC) only occurs in a particular segment of the esophagus (i.e., in BE), frames that are captured outside this segment could be disregarded by an EAC detection algorithm, leading to a reduction in false alarms and an increased user confidence in the CAD system.

Practically, different approaches and algorithms have been applied on time-series data, including independent frame analysis, averaging over the temporal domain and hidden Markov models [4–8]. However, the absence of long-term memory in these models hampers the exploitation of long-distance interactions and correlations, which make the corresponding algorithms not suitable for learning long-distance dependencies typically found in clinical data. Since the employed, existing image-based classification networks are trained on still images in overview, the response on unseen non-informative frames is unknown. This implies that algorithms trained only on still images do not perform well on video signals without algorithm modifications. [9]

Recurrent Neural Networks (RNNs) can be used to provide a temporal flow of information. These networks have been widely used to learn the processing of sequential video data and are capable of dealing with long-term dependencies. In this type of artificial neural network, connections are formed between units and a directed cycle. This cycle creates an internal state of the network which allows it to exhibit and model dynamic temporal behavior without computation-intensive 3D convolutional layers. Recently, Yao et al. [10] demonstrated a state-of-the-art method for action recognition, which imposes Gated Recurrent Units (GRUs) on the deep spatiotemporal information extracted by a convolutional network. Furthermore, Yue et al. [11] and Donahue et al. [12] have successfully demonstrated the ability of RNNs to recognize activity, based on a stack of input frames. A similar approach could be followed for the classification of tissue in videos, thereby potentially leading to a more temporally stable algorithm, since it is able to exploit information existing in the temporal domain.

The literature describes a variety of methods to analyze video for classification tasks in endoscopy. The most basic form for video analysis describes a single frame based analysis for classification [13–15]. Other recent work on video analysis in endoscopy focuses on a frame-based analysis approach with additional post processing to yield some form of temporal cohesion. Byrne et al. [16] describe a frame-based feature extractor, which interpolates a confidence score between consecutive frames, in order to make a more confident prediction for colorectal polyp detection. De Groof et al. [9] implement a voting system for multiple frames on multiple levels. Yu et al. [17] describe a 3D convolutional model, in order to capture inter-frame correlations. Yet, 3D convolutions fail to capture long-term information. Harada et al. [18] propose an unsupervised learning method, which clusters frame-based predictions, in order to improve temporal stability in tissue classification. Yet, a clustering approach is not able to capture the consecutive or inter-frame correlation between frames. Frameworks that do actively learn spatiotemporal information with the implementation of RNNs are described by, Owais et al. [19] and Ghatwary et al. [20]. They demonstrate that the implementation of RNNs yield superior classification accuracies in endoscopic videos, but no quantitative results are reported on the stability of the employed models.

In this paper, we address the ambiguity in the classification of tissue in the upper gastrointestinal tract by introducing RNNs, as a first exploratory study to obtain a more robust system for endoscopic lesion detection. Our system is generally applicable for CAD systems in the gastrointestinal tract and can potentially serve as a pre-processing step that reduces the amount of false alarms for a wide range of endoscopic CAD systems. We hypothesize that by extending Resnet18 with RNNs, or more specifically, by employing Long Short Term Memory (LSTM) or Gated Recurrent Unit (GRU) as concepts, the model is able to actively learn and memorize information seen in earlier frames to make a more accurate prediction about the tissue class compared to networks without temporal processing.

Our contributions are therefore as follows. First, our work demonstrates that including temporal information in endoscopic video analysis leads to an improved classification performance. Second, We show that exploiting the concepts of LSTM and GRU outperform the conventional Fully Connected (FC) networks. Third, the proposed approach offers a higher stability and robustness in classification performance, so that it paves the way for applying automated detection during the complete clinical endoscopic procedure.

2. Materials and Methods

2.1. Ethics and Information Governance

This work and the involved local collection of data on implied consent, received national Research Ethics (IRB) Committee approval from the Amsterdam UMC (No. NTR7072). De-identifcation was performed in line with the General Data Protection Regulation (EU) 2016/679.

2.2. Datasets and Clinical Taxonomy

To train and evaluate the classification performance of our networks, we collected a dataset consisting of 82 endoscopic pullback videos from 75 patients, which were recorded prospectively in the Amsterdam UMC. Written informed consent was obtained from all participants. In total, 46 out of the 82 videos were derived from patients who were diagnosed with high-grade dysplasia. These videos were captured using White Light Endoscopy in full high-definition format (1280 $\times$ 1024 pixels) with the ELUXEO 7000 endosco py system (FUJIFILM, Tokyo, Japan). During the recording of a pullback video, the endoscope is slowly pulled from the stomach up to the healthy squamous esophagus tissue in one smooth sequential movement.

In our processing, we have sampled with 5 frames per second at a resolution of 320 $\times$ 256 pixels. Each resulting frame is manually labeled by one out of three experienced clinicians with respect to tissue class and informativeness. The five tissue classes are 'stomach', 'transition-zone Z-line', 'Barrett', 'transition-zone squamous' and 'squamous', see Figure 1. Frames are labeled as non-informative if they have: out-of-focus degradation, visible esophagus contractions, video motion blur, broad visibility of bubbles, or excessive contrast from lighting. For the training, we have only selected sequences in which the last frame is labeled informative. From the total dataset consisting of 20,663 frames, 19,931 are labeled as informative by a team of three clinical research fellows.

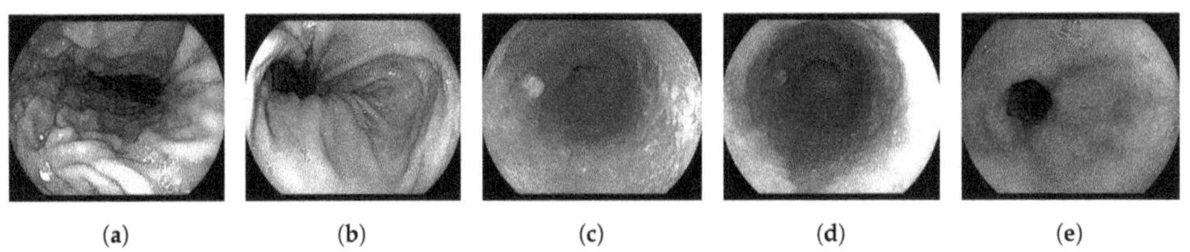

Figure 1. Visual examples for each tissue label to be classified by the model sourced from pullback videos. These pullback videos start recording from the (**a**) Stomach and stops, while pulling the endoscope with a constant speed, at the (**e**) Squamous area. (**a**) Stomach; (**b**) Transition-zone Z-line; (**c**) Barrett; (**d**) Transition-zone Squamous; (**e**) Squamous.

2.3. Network Architectures and Training Protocol

Our network architecture is split in two principal stages: a feature extractor followed by a classification network. The feature extraction is based on a modified ResNet18 network, containing four fully convolutional blocks. This model is applied in many fields for image classification, and should therefore allow for an easier comparison to other literature. Different classifiers are evaluated in our experiments. Since we only want to measure the effect of applying RNNs on the model robustness, we keep the classifiers as simplistic as possible. The following classifier configurations are tested: (1) two Fully Connected layers (FC) with a Rectified Linear Unit function in between, (2) a two-layer LSTM directly followed by a fully connected layer, and (3) a two-layer GRU directly followed by a fully connected layer. Both LSTM and GRU classifiers are assembled with two recurrent layers, since according to Graves et al. [21] depth is needed to increase the receptive field of an RNN. Both LSTM and GRU are trained with a hidden state size of 128 parameters. Since relevant literature also describes methods to exploit temporal information based on single image features, we also apply a simplistic averaging filter over 5 frames to the FC classifier in order to smooth the output over the temporal domain. This classifier is denoted as FC Avg(n = 5).

We introduce random under-sampling for training set stratification, by taking into account the asymmetry of the dataset. Doing so, each video frame is assigned a weight, so that the probability of sampling each class and case, is equal. Per iteration, we randomly sample 512 sequences consisting out of only one or 10 consecutive frames, to facilitate individual or temporal classification, respectively. The flow of data through our proposed model for temporal classification is illustrated in Figure 2.

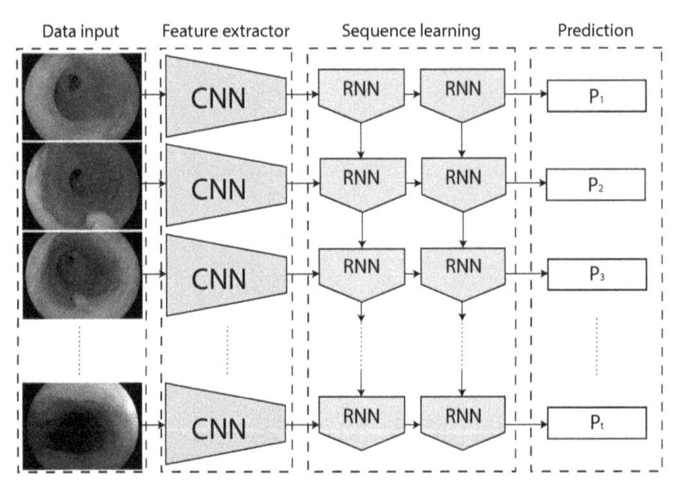

Figure 2. Data flow through the proposed model during training. Recurrent Neural Network (RNN) represents either Long Short-Term Memory (LSTM) or Gated Recurrent Unit (GRU) nodes.

The network is trained using Adam optimization with an initial learning rate of 10^{-4}. A cyclic cosine learning-rate scheduler is used to further control the learning rate. A cross-entropy loss is implemented to converge the neural network. For data augmentation, random-affine transformations are applied during training with rotations up to 5 degrees, translation and cropping of 2.5% of the image length and shearing up to 5 degrees. These parameters are kept constant per sequence.

3. Experiments

This section first describes the metrics for measuring the performance of our method, and then follows with the statistical analyses of the various classes and configurations to obtain a broad set of experimental results.

3.1. Metrics

The reported performance of the model is measured with two different metrics, i.e., stability and accuracy. The stability of the network is determined by the average amount of times the network switches from predicted label per video. Thus, if at time t_n, the model predicts label a, and at t_{n+1} the model predicts label b, the predicted labels switch from a to b. This change is counted as one domain switch. Since our dataset contains 5 labels, and each label should be passed only once, a perfect score would be 4 label switches. The accuracy is measured as the accuracy score averaged per patient, as shown in Equation (1). The label accuracy is averaged per patient, in order to normalize for a variable video length, and specified by

$$\text{Mean label Accuracy} = \frac{1}{N_p} \sum_{i=1}^{N_p} \text{Acc}(L_i), \tag{1}$$

where, L_i denotes the label of event i and Acc(.) is the accuracy function. In Equations (1) and (2), T_P is true positive, N_v is the number of frames, and N_p is the number of patient cases. True positives are defined as in Table 1 and specified by

$$\text{Label Acc(L)} = \frac{1}{N_v} \sum_{j=1}^{N_v} T_{Pj}. \tag{2}$$

The explanation of Table 1 is as follows. For all rows where double or triple labels are indicated, we mean that if the predicted label has a ground truth in one of the two/three labels, then the prediction label is considered correct. For example, if Barrett's tissue is predicted and deemed correct, this means that the ground-truth labels from the clinician annotation may be 'Transition Z-line', 'Barrett' or 'Transition Squamous'.

This mapping of the annotations labels from the ground truth is chosen to compensate for the ambiguity of the transition zones, since the transition zones adjacent to a distinct class are now mapped onto that class. Consequently, this metric will better separate non-metaplastic and metaplastic tissue. This is important because a metaplastic area is the region of interest to examine for Barrett's neoplasia.

Table 1. Correspondence between predicted labels and ground-truth annotations of the clinicians.

Predicted Label	True Positive If Label Is:
Stomach (St)	St
Transition Z-line (Tz)	Tz, B
Barrett (B)	Tz, B, Ts
Transition squamous (Ts)	B, Ts
Squamous (Sq)	Sq

3.2. Statistical Analysis

The network performance is evaluated using a fivefold cross-validation. The dataset is divided into five equally-sized folds, such that in each fold each individual patient is represented once. The performance reported in this paper is the average accuracy over all folds and patients. To substantiate the statistical significance, we will also perform a Wilcoxon signed-rank test [22]. This is a nonparametric test applied to matched-pair data, which tries to find a distribution centered around zero based on differences.

4. Results

The output stability is measured for four models, FC, FC Avg(n = 5), LSTM and GRU. We have found on average that the following networks switches from label: FC 43.27 (±23.83), FC Avg(n = 5) 18.48 (±9.76), LSTM 10.81 (±5.68) and GRU 11.91 (±6.33). These results demonstrate that the models implemented without RNNs, switch 2-4 more times from label within a single video. This is especially apparent in the video's in which the model has a bad performance, see Figure 3.

Table 2 displays the mean accuracy of the four different models for the five different tissue classes. The results show that by averaging over five consecutive frames, a performance improvement of 0.8% is obtained. The introduction of RNNs into the classification model results yield an increase of 3.7% in overall accuracy, as seen with the accuracies of 85.9% and 85.6% for LSTM and GRU, respectively, compared to 82.2% for FC. Detailed performances per model for each class are provided in confusion matrices in Figure 4. The Wilcoxon signed-rank test has found a $p < 0.001$ in all comparisons on accuracy of FC, LSTM and GRU classifiers, which confirms the statistical significance of the results.

Table 2. Mean accuracy per tissue class, and for various architecture configurations. Scores are averaged over all patient cases. The mean label accuracy is in correspondence with the label classification from Table 1. The classifiers reported are Fully Connected classifier, Fully Connected Averaged(n = 5), Long Short-Term Memory classifier, and Gated Recurrent Unit classifier.

Label	**N**	**Mean Accuracy (%)**			
		FC	**FC Avg(n = 5)**	**LSTM**	**GRU**
Stomach (St)	2593	59.6	62.2	60.7	**61.3**
Tran. Z-line (Tz)	2921	74.1	72.9	79.3	**79.7**
Barrett (B)	9444	95.7	96.1	98.0	**98.3**
Tran. squamous (Ts)	4215	79.5	81.3	**87.0**	83.9
Squamous (Sq)	755	58.3	59.9	63.8	**67.4**
Overall	19,931	82.2	83.0	**85.9**	85.6

An important observation is the accuracy of 98.3% on the Barrett's segment in the esophagus. A good performance on this label is crucial, since this model will be used in an a priori tissue classification to extend the robustness for lesion detection. High sensitivities are generally preferred in this field, since a false positive will only lead to an extra biopsy, while a false negative gives a severe detriment to the patient. The demonstrated accuracy score implies that roughly 1.7% of the images are rejected due to lesion detection caused by a false negative. We consider that this is acceptable because during inference, we are able to process up to 180 frames per second for real-time video analysis. This would mean that even if some frames are rejected, the time-gap between the analyzed frames would be small, so that the Barrett area will still be fully analyzed effectively.

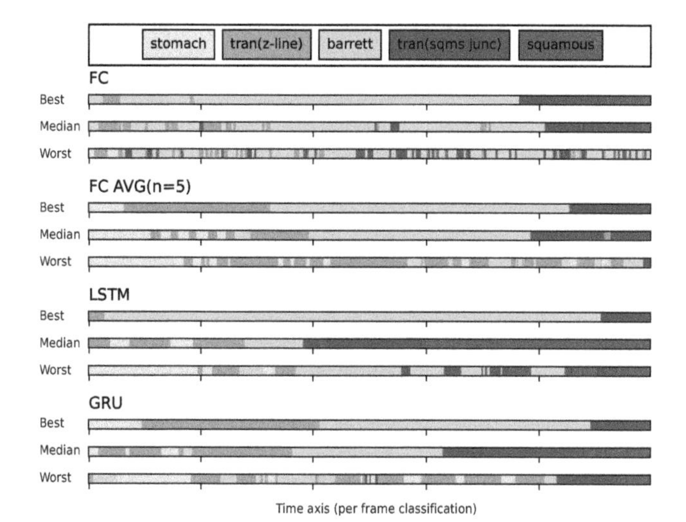

Figure 3. Each bar illustrates the predicted organ labels per frame over a time axis. The figure illustrates the instability of the compared networks architectures at three different performance levels, which are best, median and worst performance for each model respectively. The average domain switch for Fully Connected classifier, is 43.27, Fully Connected Averaged(n=5) 18.49, Long Short-Term Memory classifier 10.81 and Gated Recurrent Unit classifier 11.91

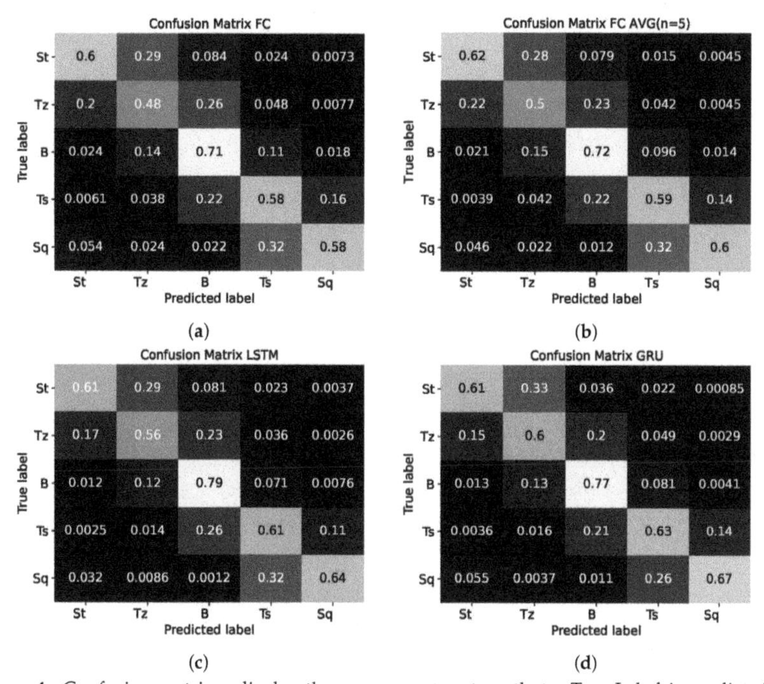

Figure 4. Confusion matrices display the average percentage that a True Label is predicted as a specific class. These values are normalized for the patient cases. (**a**) Fully Connected classifier, (**b**) Fully connected Averaged(n = 5), (**c**) Long Short-Term Memory classifier, and (**d**) Gated Recurrent Unit classifier.

5. Discussion and Conclusions

In this work, we have explored the use of Recurrent Neural Networks (RNNs) for true temporal analysis of endoscopic video. In particular, we have evaluated two popular RNN architectures (i.e., Long Short-Term Memory (LSTM) and Gated Recurrent Unit (GRU)) for tissue classification in endoscopic videos. This is a particularly interesting application, since current CAD systems show a relatively high number of false classifications for video frames captured outside the organ of interest. Reliably detecting the organ that is currently in view can therefore lead to an increased CAD performance. We demonstrate that by exploiting temporal information in latent space, much more stable classification behavior is observed than when simple frame averaging is used. Hence, the results confirm our hypothesis that by leveraging RNNs, we can stabilize the classification output from the model. Moreover, by learning the temporal flow we have also discovered an increase in the accuracy of all tissue classes. For the application of Barrett's cancer detection, the proposed system reliably detected the tissue of interest, i.e., the Barrett label, with an accuracy of 98.3%. These results are a proof of concept, and therefore the presented models do not yield the optimal results. In future work we will address this limitation by conducting an ablation study to find the optimal parameters.

The classification performance on the stomach and squamous tissue remains relatively poor. This discrepancy can be observed in Table 2 and is mostly caused by the definition of the label correspondence mapping in Table 1. Although the algorithm is able to approximate the tissue type, it often also guesses the neighboring tissue type. This error can be readily understood, as there is no hard defined transition on the visible border between tissue types, i.e., each view gradually transitions over time into the next one, resulting in the property that adjacent tissue areas (and labels) visually exhibit similar features (see Appendix A).

To address this transition ambiguity, a score based on the agreement between observers could be introduced. However, in our current training protocol, we only have one annotated label available per frame, originating from one of the three observers. By introducing multiple observers per frame, a score of agreement can be calculated (like simple majority voting), which can be used to train the future algorithm. Such an approach would take into account the ambiguity, and can then potentially also result into an additional score for ambiguity.

An other limitation is that the employed data is imbalanced at present. As can be seen in Table 2, the labels Stomach and Squamous are under-represented in the dataset. This imbalance is partly a reason for the poor performance on these classes. To overcome the limitation of available data, future efforts will focus on the collection of data, originating from other sources than videos alone.

In conclusion, our work has demonstrated that incorporating temporal information in endoscopic video analysis can lead to an improved classification performance. Exploiting the sequential bias present in endoscopic video (e.g., the order of the tissue types that are captured, in addition to a higher accuracy), also presents a more stable classification behavior over time. Although being directly applicable to EAC detection in BE patients, to likely enhance the CAD performance by reliably detecting the Barrett's tissue, our approach can be generalized and easily translated to similar endoscopic video analysis tasks. Future experiments should explore such novel applications and should focus on combining the proposed pre-processing system with several succeeding, and already established classification tasks.

Author Contributions: Conceptualization, T.B., J.v.d.P.; methodology, T.B.; validation, T.B.; resources, M.S. and K.F., J.J., E.S.; writing—original draft preparation, T.B.; writing—review and editing, P.d.W.; supervision, F.v.d.S.; funding acquisition, P.d.W. and J.B. All authors have read and agreed to the published version of the manuscript.

Funding: This research received no external funding.

Conflicts of Interest: The authors declare no conflict of interest.

Appendix A. Ambiguous Organ Classes

This section illustrates examples of ambiguous frames due to lacking a clearly defined transition between tissue classes.

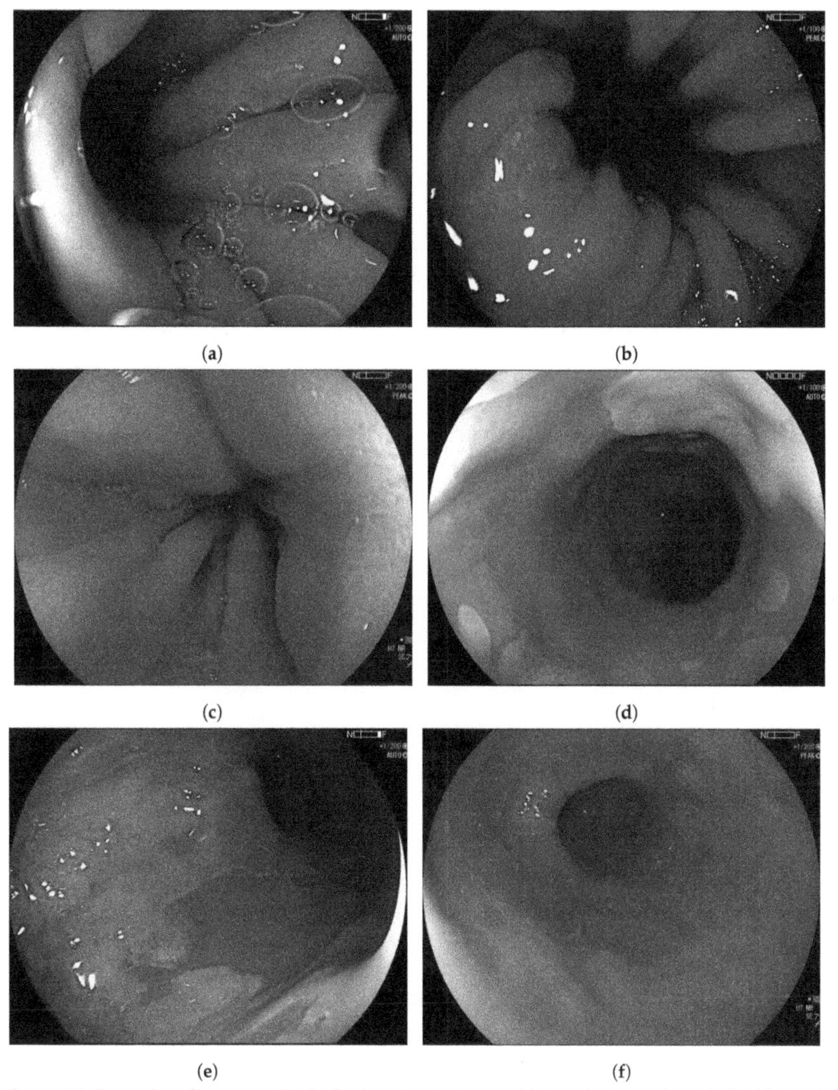

Figure A1. Examples of images with similar imaging features, which are lacking a hard definable border for a classification label. (**a**–**c**) contain images that represent the transition area from stomach to the Barrett's esophagus, which all contain features, like gastric folds, and (**a**,**b**) also include a view into the stomach. (**d**–**f**) contain representations seen around the transition from Barrett's esophagus to healthy squamous tissue, which all contain visual features from the Barrett's esophagus and squamous tissue. (**a**) Cardia of the stomach labeled as 'Stomach'. (**b**) Image captured after the esophageal sphincter labeled as 'transition Z-line'. (**c**) Contracted Barrett's esophagus labeled as 'Barrett'. (**d**) Barrett's esophagus with healthy squamous tissue falling into the field of view labeled as 'Barrett'. (**e**) Transition zone between Barrett's esophagus and healthy squamous tissue, labeled as 'transition squamous. (**f**) Mostly squamous tissue, with Barrett's esophagus in the background, labeled as 'squamous'.

References

1. Groof, J.D.; van der Sommen, F.; van der Putten, J.; Struyvenberg, M.R.; Zinger, S.; Curvers, W.L.; Pech, O.; Meining, A.; Neuhaus, H.; Bisschops, R.; et al. The Argos project: The development of a computer-aided detection system to improve detection of Barrett's neoplasia on white light endoscopy. *United Eur. Gastroenterol. J.* **2019**, *7*, 538–547. [CrossRef] [PubMed]
2. Guck, J.; Schinkinger, S.; Lincoln, B.; Wottawah, F.; Ebert, S.; Romeyke, M.; Lenz, D.; Erickson, H.M.; Ananthakrishnan, R.; Mitchell, D.; et al. Optical deformability as an inherent cell marker for testing malignant transformation and metastatic competence. *Biophys. J.* **2005**, *88*, 3689–3698. [CrossRef] [PubMed]
3. van der Putten, J.; de Groof, J.; van der Sommen, F.; Struyvenberg, M.; Zinger, S.; Curvers, W.; Schoon, E.; Bergman, J.; de With, P.H. Informative frame classification of endoscopic videos using convolutional neural networks and hidden Markov models. In Proceedings of the 2019 IEEE International Conference on Image Processing (ICIP), Taipei, Taiwan, 22–25 September 2019; pp. 380–384.
4. Wang, S.; Cong, Y.; Cao, J.; Yang, Y.; Tang, Y.; Zhao, H.; Yu, H. Scalable gastroscopic video summarization via similar-inhibition dictionary selection. *Artif. Intell. Med.* **2016**, *66*, 1–13. [CrossRef] [PubMed]
5. Giordano, D.; Murabito, F.; Palazzo, S.; Pino, C.; Spampinato, C. An AI-based Framework for Supporting Large Scale Automated Analysis of Video Capsule Endoscopy. In Proceedings of the 2019 IEEE EMBS International Conference on Biomedical & Health Informatics (BHI), Chicago, IL, USA, 19–22 May 2019; pp. 1–4.
6. Rezvy, S.; Zebin, T.; Braden, B.; Pang, W.; Taylor, S.; Gao, X. Transfer learning for Endoscopy disease detection and segmentation with mask-RCNN benchmark architecture. In Proceedings of the 2020 IEEE 17th International Symposium on Biomedical Imaging, Oxford, UK, 3 April 2020 ; p. 17.
7. Ali, H.; Sharif, M.; Yasmin, M.; Rehmani, M.H.; Riaz, F. A survey of feature extraction and fusion of deep learning for detection of abnormalities in video endoscopy of gastrointestinal-tract. *Artif. Intell. Rev.* **2019**, 1–73. [CrossRef]
8. Du, W.; Rao, N.; Liu, D.; Jiang, H.; Luo, C.; Li, Z.; Gan, T.; Zeng, B. Review on the Applications of Deep Learning in the Analysis of Gastrointestinal Endoscopy Images. *IEEE Access* **2019**, *7*, 142053–142069. [CrossRef]
9. van der Putten, J.; de Groof, J.; van der Sommen, F.; Struyvenberg, M.; Zinger, S.; Curvers, W.; Schoon, E.; Bergman, J.; de With, P.H.N. First steps into endoscopic video analysis for Barrett's cancer detection: Challenges and opportunities. In *Medical Imaging 2020: Computer-Aided Diagnosis*; International Society for Optics and Photonics: San Diego, CA, USA, 2020; Volume 11314, p. 1131431.
10. Yao, G.; Liu, X.; Lei, T. Action Recognition with 3D ConvNet-GRU Architecture. In Proceedings of the 3rd ACM International Conference on Robotics, Control and Automation, Chengdu, China, 11–13 August 2018; pp. 208–213.
11. Yue-Hei Ng, J.; Hausknecht, M.; Vijayanarasimhan, S.; Vinyals, O.; Monga, R.; Toderici, G. Beyond short snippets: Deep networks for video classification. In Proceedings of the IEEE Conference on Computer Vision and Pattern Recognition, Boston, MA, USA, 7–12 June 2015; pp. 4694–4702.
12. Donahue, J.; Anne Hendricks, L.; Guadarrama, S.; Rohrbach, M.; Venugopalan, S.; Saenko, K.; Darrell, T. Long-term recurrent convolutional networks for visual recognition and description. In Proceedings of the IEEE Conference on Computer Vision and Pattern Recognition, Boston, MA, USA, 7–12 June 2015; pp. 2625–2634.
13. Hashimoto, R.; Requa, J.; Tyler, D.; Ninh, A.; Tran, E.; Mai, D.; Lugo, M.; Chehade, N.E.H.; Chang, K.J.; Karnes, W.E.; et al. Artificial intelligence using convolutional neural networks for real-time detection of early esophageal neoplasia in Barrett's esophagus (with video). *Gastrointest. Endosc.* **2020**. [CrossRef] [PubMed]
14. Ali, S.; Zhou, F.; Bailey, A.; Braden, B.; East, J.; Lu, X.; Rittscher, J. A deep learning framework for quality assessment and restoration in video endoscopy. *arXiv* **2019**, arXiv:1904.07073.
15. van der Putten, J.; Struyvenberg, M.; de Groof, J.; Curvers, W.; Schoon, E.; Baldaque-Silva, F.; Bergman, J.; van der Sommen, F.; de With, P.H.N. Endoscopy-Driven Pretraining for Classification of Dysplasia in Barrett's Esophagus with Endoscopic Narrow-Band Imaging Zoom Videos. *Appl. Sci.* **2020**, *10*, 3407. [CrossRef]

16. Byrne, M.F.; Chapados, N.; Soudan, F.; Oertel, C.; Pérez, M.L.; Kelly, R.; Iqbal, N.; Chandelier, F.; Rex, D.K. Real-time differentiation of adenomatous and hyperplastic diminutive colorectal polyps during analysis of unaltered videos of standard colonoscopy using a deep learning model. *Gut* **2019**, *68*, 94–100. [CrossRef] [PubMed]
17. Yu, L.; Chen, H.; Dou, Q.; Qin, J.; Heng, P.A. Integrating online and offline three-dimensional deep learning for automated polyp detection in colonoscopy videos. *IEEE J. Biomed. Health Inform.* **2016**, *21*, 65–75. [CrossRef] [PubMed]
18. Harada, S.; Hayashi, H.; Bise, R.; Tanaka, K.; Meng, Q.; Uchida, S. Endoscopic image clustering with temporal ordering information based on dynamic programming. In Proceedings of the 2019 41st Annual International Conference of the IEEE Engineering in Medicine and Biology Society (EMBC), Berlin, Germany, 23–27 July 2019; pp. 3681–3684.
19. Owais, M.; Arsalan, M.; Choi, J.; Mahmood, T.; Park, K.R. Artificial intelligence-based classification of multiple gastrointestinal diseases using endoscopy videos for clinical diagnosis. *J. Clin. Med.* **2019**, *8*, 986. [CrossRef] [PubMed]
20. Ghatwary, N.; Zolgharni, M.; Janan, F.; Ye, X. Learning spatiotemporal features for esophageal abnormality detection from endoscopic videos. *IEEE J. Biomed. Health Inform.* **2020**. [CrossRef]
21. Graves, A.; Mohamed, A.R.; Hinton, G. Speech recognition with deep recurrent neural networks. In Proceedings of the 2013 IEEE international Conference on Acoustics, Speech and Signal Processing, Vancouver, BC, Canada, 26–31 May 2013; pp. 6645–6649.
22. Wilcoxon, F. Individual comparisons by ranking methods. *Biom. Bull.* **1945**, *1*, 80–83. [CrossRef]

MDPI
St. Alban-Anlage 66
4052 Basel
Switzerland
Tel. +41 61 683 77 34
Fax +41 61 302 89 18
www.mdpi.com

Sensors Editorial Office
E-mail: sensors@mdpi.com
www.mdpi.com/journal/sensors

CPSIA information can be obtained
at www.ICGtesting.com
Printed in the USA
LVHW020836171120
671904LV00011B/493

9 783039 434183